D1008341

Where to watch birds in

South America

THE *WHERE TO WATCH BIRDS* SERIES

Where to Watch Birds in Bedfordshire, Berkshire, Buckinghamshire, Hertfordshire and Oxfordshire
Brian Clews, Andrew Heryet and Paul Trodd

Where to Watch Birds in Cumbria, Lancashire and Cheshire
Jonathan Guest and Malcolm Hutcheson

Where to Watch Birds in Devon and Cornwall
David Norman and Vic Tucker

Where to Watch Birds in Dorset, Hampshire and the Isle of Wight
George Green and Martin Cade

Where to Watch Birds in East Anglia
Peter and Margaret Clarke

Where to Watch Birds in France
Ligue Française pour la Protection des Oiseaux

Where to Watch Birds in Ireland
Clive Hutchinson

Where to Watch Birds in Kent, Surrey and Sussex
Don Taylor, Jeffrey Wheatley and Tony Prater

Where to Watch Birds in Scotland
Mike Madders and Julia Welstead

Where to Watch Birds in Somerset, Avon, Gloucestershire and Wiltshire
Ken Hall and John Govett

Where to Watch Birds in South America
Nigel Wheatley

Where to Watch Birds in Southern Spain
Ernest Garcia and Andrew Paterson

Where to Watch Birds in Wales
David Saunders

Where to Watch Birds in the West Midlands
Graham Harrison and Jack Sankey

Where to Watch Birds in Yorkshire and North Humberside
John Mather

Where to watch birds in

South America

Nigel Wheatley

Christopher Helm

A & C Black · London

Christopher Helm (Publishers) Ltd, a subsidiary of
A & C Black (Publishers) Ltd, 35 Bedford Row, London WC1R 4JH

ISBN 0-7136-3909-1

A CIP catalogue record for this book is available
from the British Library

Printed and bound by Biddles Limited, Guildford,
Surrey in Great Britain

CONTENTS

ACKNOWLEDGEMENTS

Firstly, I would like to thank my friends, 'Chief' (Nick Cobb) and Alun Hatfield, for helping to persuade me, during a few long nights propped up against a bar, to visit South America in the first place, and two other friends who have accompanied me in the field there: Barry Stidolph and Nick Wall. Alun Hatfield has also been good enough to lend me many of his books, the price of which are beyond a writer's meagre means.

Once I had come up with the idea for the book I had to convince a publisher it would be popular. Fortunately Robert Kirk, the editor at Christopher Helm (Publishers) Ltd, has his finger on the pulse and deserves thanks for taking on the original idea and developing it. Once the contract had been sorted out I experienced my first headache. I needed to set up a word processor. That would not have been possible without Tony Morris, to whom I give my thanks for his patience, and my apologies for my computer illiteracy. Once I had started processing the words, well, to be more precise, the bird names, every species seemed to have at least three names. My second headache was solved by James Clements, to whom I am very grateful for giving me permission to use his Check List as the baseline for this book.

Most importantly of all though, it would have been impossible to write this book without the help of many birders who have been to South America and written up their experiences. It has to be said at this juncture that a tiny minority, some with a wealth of experience in South America, have not been willing to share their information, a sad reflection of the selfish competition which has crept into the birding world during recent years, and an aspect of birding which defies its very nature.

Fortunately such people are few and far between, and sharing comes naturally to the vast majority of birders. Those characters who have not just travelled to South America to see the wonderful birds, but been unselfish enough to record their experiences for the benefit of others in trip reports and books, which they have made generally available, are the real heroes of the birding world. These are the people who have made this book possible.

I would like to express my heartfelt thanks to the following birders who have kindly allowed me to use the information contained within their reports and books: 'Ando' (Paul Anderson), Mike Coverdale, Rolf A. de By, Dirk de Moes, Richard Evans, Mark Golley, Graeme Green, Richard Fairbank, Henk Hendriks, Clive Green, Phil Gregory, Wim Loode, Tim Marlow, Ian Mills, Erik Molgaard, Gerry Richards, Robert Ridgely, Paul Salaman, Richard Thomas, Olivier Tostain, Barry and Lyn Trevis, Frank van Groen, Richard Webb and, especially, Bruce Forrester, who supplied virtually all the information for Brazil, Nick and Daryl Gardner, and Jon Hornbuckle. Naturally, many thanks also go to the many birders who accompanied these people, helped to find the birds and contributed information.

I would also like to thank the birders and organisations who have helped in a smaller, but still useful, way: Chris Balchin, Birdquest,

Robin Brace, Cygnus, Dave Fisher (Sunbird), Rob Roberts, Richard Schofield, 'Spotter' (John Mason), Garry Stiles, the Tambopata Reserve Society, John Wall, Wings, Barry Wright, and especially Richard Ryan of Neotropic Bird Tours. (BirdLife) International officers have also assisted me in my task, particularly Martin Kelsey and David Wege.

Many birders have not only provided information but have been good enough to cover my 'country drafts' in honest red ink, helping to tighten up the book considerably. Special thanks go to: Thomas Brooks (Paraguay), Richard ffrench (Trinidad and Tobago), Bruce Forrester (Brazil), Tom Heijnen (Ecuador), Mark Johnston (Guyana), Guy Kirwan (Chile and Venezuela), Andrew Moon (Bolivia), and Jerry Warne (Argentina).

I have spoken to numerous other people on the telephone or in the field about minor, but important, matters, many of whom I have unwittingly omitted from the above list owing to my poor memory. I sincerely hope these people will accept my profuse apologies if I have failed to acknowledge their help in this edition, and hope they will let me know before the next edition!

Finally, writing a book of this nature is, for the most part, an obsessive labour of love, and the writer can all too easily forget the other aspects of their lives, except birds (naturally). With this in mind, special thanks must go to Georgie Malpass, since she has soothed the many headaches encountered *en route* to the final version of the book, and insisted on me taking the odd break.

INTRODUCTION

This book evolved from the questions I found myself asking when I began to plan my first trip to South America. I had arranged to go with three other birders so everyone's interests had to be taken into account. Where should we go to look for the birds we considered to be 'megas'; the birds that epitomise the continent: Torrent Duck, Agami Heron, Scarlet Ibis, Andean Condor, Harpy Eagle, Sungrebe, Sunbittern, Magellanic Plover, seedsnipes, Andean Avocet, Diademed Sandpiper-Plover, Inca Tern, macaws, Hoatzin, jacamars, toucans, White-plumed Antbird, cotingas, especially 'blue ones', fruiteaters and cock-of-the-rocks, and tanagers? Which countries held a good cross-section of South American families? Which countries were worth visiting in January? Where were the best sites in those countries? How far apart and accessible were they? How many could we visit in five weeks? All these questions and many more had to be considered carefully if the proposed trip was going to be a success.

Once we had decided, after great deliberation, to visit Venezuela and Chile, more questions had to be answered. Which route shall we take? How long should we spend at each site? Where would we stay? After three or four draft itineraries were circulated between the team, we booked the flights and the hire-cars, arranged accommodation and permits for the sites which required advanced bookings, and started swotting.

The time came, the birds, and most importantly the 'megas' flowed, virtually every one of them. An Andean Condor flew over a Diademed Sandpiper-Plover in the magnificent Lauca NP, Chile. Magellanic and Two-banded Plovers shared a pool on Tierra del Fuego. We were surrounded by White-plumed Antbirds and dazzled by male Spangled and Purple-breasted Cotingas on our first day along the road south from El Dorado in Venezuela. The next day we bumped into the lovely Red-banded Fruiteater and 'scoped a male Pompadour Cotinga for half an hour. No wonder this road has since been christened 'Paradise Road'. At other times, notably in the Venezuelan Andes, we were standing amongst the proverbial 'flock of ticks', when the simple act of looking through our bincoulars was enough to produce new bird after new bird, many of them stunningly beautiful. It was a fantastic birding experience and may never be matched.

The reason the trip was so successful was because we asked so many questions during the extensive planning stage. Without months of careful preparation it would not have been anywhere near as exciting.

This book's aim is to answer those questions birders ask themselves before venturing to South America for the first, or fortieth time. It is not meant to direct you to every site and bird in the minutest detail, but to be a guiding light. There can be no substitute for an up-to-date report. I urge readers to seek these reports out (see Useful Addresses, p. 409) once they have decided on their destination, and to write their own reports on their return.

Birders are notoriously hard to please so writing this book has been all-consuming. I began by compiling a list of sites and the species recorded at them, from every imaginable source. Birders reports were the major goldmine and without the generous permission of the writers (see Acknowledgements) this book would not exist. Whilst compiling the site lists I soon realised that every birder seemed to have a different name for every bird! Hence, I decided I needed a complete list of species names as a baseline for the species lists for each site. I chose *Birds of the World: A Check List*, by James Clements (Fourth Edition 1991 together with *Supplements)*, simply because the names and taxonomic order used seemed the most logical, and James was kind enough to allow me to use it.

The names of birds in this check list are spelt in 'American'. For example, 'coloured' is spelt 'colored' and 'grey' is spelt 'gray'. Although the 'English' names of New World birds should be spelt in the American way I have written them in English in this edition, simply because I think it is easier on the English eye and because my busy fingers refused *not* to press the 'u' in 'coloured' or to press the 'a' in 'grey' instead of the 'e', whilst whizzing, well, strolling, across the keyboard.

Using these databases virtually all, of the best birding sites, spectacular and endemic species, as well as those birds hard to see beyond the continent's boundaries, have been included, albeit in varying amounts of detail. Absolute coverage and precision would have resulted in the staggered publication of several thick volumes.

The result is detailed coverage of 206 sites, listed in the Contents, 105 site maps, endemic species lists (at the end of each country account) for those countries with endemics, and a Species Index (p. 412). Birders interested in a particular species can look it up in the Species Index, see the page number(s) and turn straight to the site(s) at which it occurs. Birders with an interest in a site or sites they have heard about can refer to the Contents, see the page number(s) and turn directly to the required site(s). Those readers interested in a particular country or region can refer to the list of sites given in the Contents and the country maps at the beginning of each country account.

After much consideration the book has taken the following shape. Countries are treated alphabetically (except for Antarctica which appears at the end of the country section). Details for each country are dealt with as follows:

The **Country Introduction** includes:

a brief **summary** of the features discussed in more detail below
size, in relation to England and Texas, and, in some cases, other South American countries
getting around, where the infrastructure of the country or archipelago is discussed
accommodation and food
health and safety, although general advice is given, it is best to check the latest vaccination requirements and personal safety levels, before planning your trip
climate and timing, where the best times to visit are given (these are summarised in the Calendar, (p. 408)
habitats
conservation, where the provision of national parks and reserves,

habitat destruction and numbers of threatened species are discussed
bird families, the number of regularly occurring families (out of 92 for the whole continent), Neotropical endemic families (out of 25) and South American endemic families (out of nine) are given (see Fig. 1, below)
bird species, where an approximate country species list is given, and a list of non-endemic specialities and spectacular species is intended to give a taste of what to expect on a well-planned, short trip (very rarely seen species are not usually included in this brief list)
endemic species,
expectations, where an idea of how many species to expect is given

---

FIG 1: BIRD FAMILIES IN SOUTH AMERICA

The 16 families listed below in bold, small type are endemic to the Neotropical region, which includes South America, Central America, and the Caribbean Islands. The nine families listed below in bold, capital type are endemic to South America. Together they make up 25 of the 27 families which are endemic to the Neotropical region (two other families occur in the Caribbean, but not South America: todies and Palmchat).

Tinamous
RHEAS
Grebes
Penguins
Albatrosses
Shearwaters and allies
Storm-Petrels
Diving-Petrels
Tropicbirds
Frigatebirds
Gannets and boobies
Cormorants
Anhingas
Pelicans
SCREAMERS
Waterfowl
Flamingos
Herons and egrets
Ibises and spoonbills
Storks
New World vultures
Osprey
Hawks, eagles and kites
Falcons
Cracids (Plain Chachalaca occurs in south Texas)
Wood-Partridges (Wood-Quails)
Rails and coots
Sungrebes
Sunbittern
Limpkin
TRUMPETERS
SERIEMAS
Jacanas
Painted-snipes
Sandpipers
Sheathbills
MAGELLANIC PLOVER
SEEDSNIPES
Thick-knees
Oystercatchers
Avocets and stilts
Plovers and lapwings
Gulls and terns
Skuas and jaegers
Skimmers
Pigeons and doves
Parrots and macaws
New World cuckoos
HOATZIN
Anis

Ground-Cuckoos (includes Greater Roadrunner of the USA)
Barn-Owls
Owls
OILBIRD (has occurred in Panama and Costa Rica)
Potoos
Nightjars
Swifts
Hummingbirds
Trogons
Kingfishers
Motmots
Jacamars
Puffbirds
American Barbets
Toucans
Woodpeckers
Woodcreepers
Furnariids
Antbirds
Antthrushes and Antpittas
GNATEATERS
Tapaculos
Cotingas
Sharpbill
Manakins
Tyrant flycatchers
Corvids (jays)
Vireos and allies (includes greenlets)
Waxwings and allies (Cedar Waxwing)
Dippers
Thrushes and allies
Mockingbirds and thrashers
Wrens
Gnatcatchers
Swallows
Larks (Horned Lark)
Old World sparrows (House Sparrow (introduced))
Wagtails and pipits
Fringillids (siskins)
New World warblers
Emberizids ('tanagers')
Icterids

Under each country the **Sites** are numbered, roughly along a more or less logical route through the country or in 'bunches'. Naturally, different birders will prefer their own routes, but I felt this was a better method than dealing with the sites alphabetically because those birders intending to visit just one region of a country will find all the sites in that region dealt with in the same section of the book.

Sites are dealt with as follows:

The **site name** usually refers to the actual site. However, if it is nameless, or it involves a number of birding spots which are in close proximity, the best city, town, village, lodge, or road name, from which to explore the site, however remote, is used.

The **site introduction** gives its general location within the relevant country before describing its size, main physical characteristics, habitat make-up and avifauna, including the number of species recorded (if known) and particular reference to endemic and speciality species best looked for at that site. In some cases the importance of the site is discussed in relation to similar sites.

The site's altitude (where known) is also given, in metres and feet. Both measurements are given since some birders prefer metres and some prefer feet, whereas only km readings are given for distances because most birders now have a grasp of kilometres. For those that do not, remember that one mile equals 1.6093, or roughly 1.6 kilometres, hence 10 miles equals 16 km and 100 miles equals 161 km.

Restrictions on access, if there are any, are also given in the site introduction. Such a negative start is designed to eliminate the extreme disappointment of discovering the site is inaccessible to all but the most well prepared and adventurous *after* reading about the wealth of avian riches present at the site. This is particularly true of some sites in Colombia and Peru, where drug-traffickers and terrorists are active and likely to shoot on sight someone with a pair of binoculars. In Brazil, a lot of sites require prior permission to visit. It is *imperative* to check the situation in these countries, and some others, with the relevant authorities before even considering a visit. I cannot stress the importance of this enough, having lost an adventurous friend in Peru who ventured into terrorist territory.

The **species lists** for the site follow the introduction and include:

Endemics: Species found only in the country in which the site is located. Some birds listed here may occur only at one or two sites within the country so it is important to make a special effort to see them. Such species are mentioned in the introduction. Others are more widespread but *still* endemic.

Specialities: Species which have (i) restricted ranges which cross country boundaries, or (ii) wider distributions throughout South America but are generally scarce, rare, or threatened, or (iii) have rarely been mentioned in the literature consulted in preparing this book.

Some of the species listed here may occur only at very few sites in South America so it is important to spend time looking for them. Such species are mentioned in the site introduction.

Others: Species that are well distributed but uncommon, spectacular or especially sought after for a variety of reasons.

Some species may appear under **Specialities** for one site and **Others** for another. This is because they are more likely to be seen at the site where they are listed under **Specialities**, and are rarely seen at sites where they are listed under **Others**. A good example is Harpy Eagle which, although rare, usually appears under **Others**, because it is so rarely seen.

Abundant widespread species such as Bananaquit are not listed (although some birders may argue that even this is a 'spectacular' species).

At the end of the bird lists there is also a list of **Other Wildlife**, where species of mammal, reptile, amphibian, fish, insect etc. are listed.

It is important to remember that no one, not even the most experienced observer, is likely to see all the species listed under each site in a single visit, or even over a period of a few days, or even, in some cases, during a prolonged stay of weeks or more! This is because many South American species, especially the forest birds, apart from being hyper-skulkers, are very thin on the ground.

The lists, particularly the **Others** section, is not comprehensive and many more species may have been recorded at the given site. Such species are common and likely to be seen at many sites in that region of the country or throughout a large part of South America. By restricting the numbers of species listed under **Others** I have hoped to avoid repetition.

Although you may not wish to take this book into the field once you have decided on your destination and itinerary, it may prove useful if you are prepared to scrawl all over it. For, by crossing out those species you have seen on a previous trip, or at a previously visited site, or already at the site you are at, you will be able to see what species you still need to look for. It is all too easy, in the haze of excitement generated by South America, to see a lot of good birds and be satisfied with your visit, only to discover later that you have missed a bird at that site which does not occur at any other, or that you have just left a site offering you the last chance to see a certain species (on your chosen route) and are unable to change your itinerary. Naturally, this use of the book may render it useless on your return home and you may have to buy another copy. This would please me, since it may help finance my next trip. However, someone has 'kindly' suggested using a pencil, and buying a rubber on your return home.

Within these lists those species which have been marked with an asterisk * have been listed as threatened or near-threatened by BirdLife International/IUCN in the Red Data Book, *Threatened Birds of the Americas* (Third Edition), N. Collar *et al.*, 1992. Please report any records of these species, and those described as rare, to BirdLife International, Wellbrook Court, Girton Road, Cambridge, CB3 0NA, England, UK. The book deals in amazing detail with all the threatened species so those birders interested in the rarest birds, many of which are some of South America's most spectacular species, should seriously consider scrutinising it as part of their pre-trip planning and research.

After the lists, directions to the site from the nearest large city, town or village, or previous site, are given in the **Access** section. Then the best trails, birding 'spots' and birds are dealt with. Directions and distances are usually given to the nearest kilometre (km) because speedometers vary so much; turnings are usually described as points of the compass rather than left or right so as not to cause confusion if travelling from a different direction from that dealt with. These directions are aimed at birders with cars. However, it is important to note that in most countries, buses and taxis go virtually anywhere there is habitation, and will drop the passenger off at birding sites on request.

I have decided not to repeat the vast amount of information regarding means of public transport given in reports, or compete with the mindboggling detail presented in the various guide books. These can be used by birders requiring them, thus allowing more room in this book to talk about birds.

Where access is limited, or permission required, this is stated and a contact address given (this may involve turning to the **Additional Information** section at the end of each country account). If the detail under this section seems scant this is usually because a severely endangered species is present at that site.

Originally I intended to include **Accommodation** in detail as well but, as mentioned under **Access**, it would be foolish to waste space on repeating all the information contained in general guide books. However, I have included the names of hotels etc. recommended by birders for their safety, economy, comfort, position and, especially, opportunities for birding in their grounds. These are graded as follows: (A) = over £10/$15 (usually a long way over); (B) = £5 to £10/$7.5 to

\$15; (C) = under £5/\$7.5 (all prices are per person per night in 1993).

Other Sites worth visiting nearby, if time allows, are also mentioned. These usually offer another chance to see those species already mentioned under the main sites, but, in some cases, include sites, especially new ones, where information is scant, and a deal of pioneer spirit is required. South American veterans may wish to find out more about such sites (and send me the details for inclusion in the next edition!).

At the end of each country account there is an **Additional Information** section which includes lists of organisations and books specifically relevant to that country.

Finally, each country account ends, where relevant, with a list of **Endemic Species.** The best sites to look for these birds, or notes on their status if they have not been mentioned, are given. Birders with a particular interest in these species may wish to plan their trip around this section. The sites have been 'bunched' into regions. These regions are stated for each species so that birders visiting these regions may make their own list of all the endemic birds which occur there. By referring to this section of the country accounts to make such a list, you may find there are some species which have not been mentioned under any specific sites. Now you can leave this book in the car, put your exploring boots on, and set out to find some sites for these rare endemics. Once successful, please send the details to me for inclusion in the next edition (see **Request**, p. 411).

It is important to remember that this book may not be as helpful as you expected. This is because some sites may not be there when you get there. South America is changing fast, mostly for the worst, as populations explode and selfish exploiters move in. For example, a number of sites are alongside or at the end of tracks and, more rarely, roads, which have been built by logging and mining companies. Whilst they enable birders to get to the remaining habitat, they also allow the mad axeman in (see **Conservation**, p. 30).

Still, a little uncertainty is what makes birding so fascinating. It would be a poor pastime if every bird was lined up on an 'x' on the map, or at a 'km reading'. Exploration is exciting and if you arrive at a site to find the *Polylepis* wood on the map has vanished, set out to find another one. There is immense satisfaction to be gained from finding your own birds, although a little guidance is often appreciated.

I hope this book offers that guidance, and helps you to find and enjoy the fabulous birds of South America.

INTRODUCTION TO BIRDING IN SOUTH AMERICA

South America is *the* bird continent. More than 3,000 species, almost a third of the world's birds, have been recorded in the thirteen countries and four archipelagos that comprise the region. That is 1,100 more birds than in the Orient, 800 more than in Africa, and over 2,000 more than in the Palearctic, the Nearctic, and Australasia.

Based on *Birds of the World: A Check List* (Fourth Edition), 1991, and *Supplements 1 and 2*, by James Clements, a total of 3,083 species have been recorded in South America. Over 800 (819) of these species also occur outside the continent, mainly in Central America, leaving over 2,000 (2,264) species which are totally endemic to the South American continent. Clements lists 9,700 species for the world so South America supports nearly a third (32%) of the world's birds, and over a fifth (23%) of the world's birds occur only here.

FIGURE 2: SPECIES LISTS OF THE WORLD'S AVIFAUNAL REGIONS (all figures are approximate)

Region	**List**	**% of World Total**	**% of South American Total** (approx. figures)
SOUTH AMERICA	3,083	32	-
AFRICA	2,280	24	74
ORIENT	1,900	20	62
PALEARCTIC	950	10	31
AUSTRALASIA	900	9	29
NEARCTIC	800	8	26

Habitat Diversity

The reason for this immense range of species lies with the continent's great diversity of habitats (see map opposite).

Off the shores of Peru and Chile the cold Humboldt current supports masses of seabirds and causes rain to fall offshore, leading to the formation of the Atacama desert, the driest on earth. In stark contrast, in west Colombia and north Ecuador, the annual rainfall may exceed 10 m, helping to sustain lowland tropical forest and montane temperate forest, where flocks of multi-coloured tanagers roam. The 10,000-km long Andes, which run the entire length of the continent on the west side and rise to 6,960 m (22,835 ft), support different vegetation zones according to altitude on both the west and east slopes. The luxuriant east Andean slope temperate and subtropical forests, especially in Ecuador and Peru, are two of the richest habitats in the world for birds. Above the treeline, the short grasslands of the paramo in the north and the puna in the south support their own specialist avifaunas as do the high-altitude lakes that lie within these zones.

East of the Andes there are two main mountain massifs: the Guianan (Tepui) highlands which rise in the Guianas, southeast Venezuela and

SOUTH AMERICA: MAIN PHYSICAL FEATURES AND HABITATS

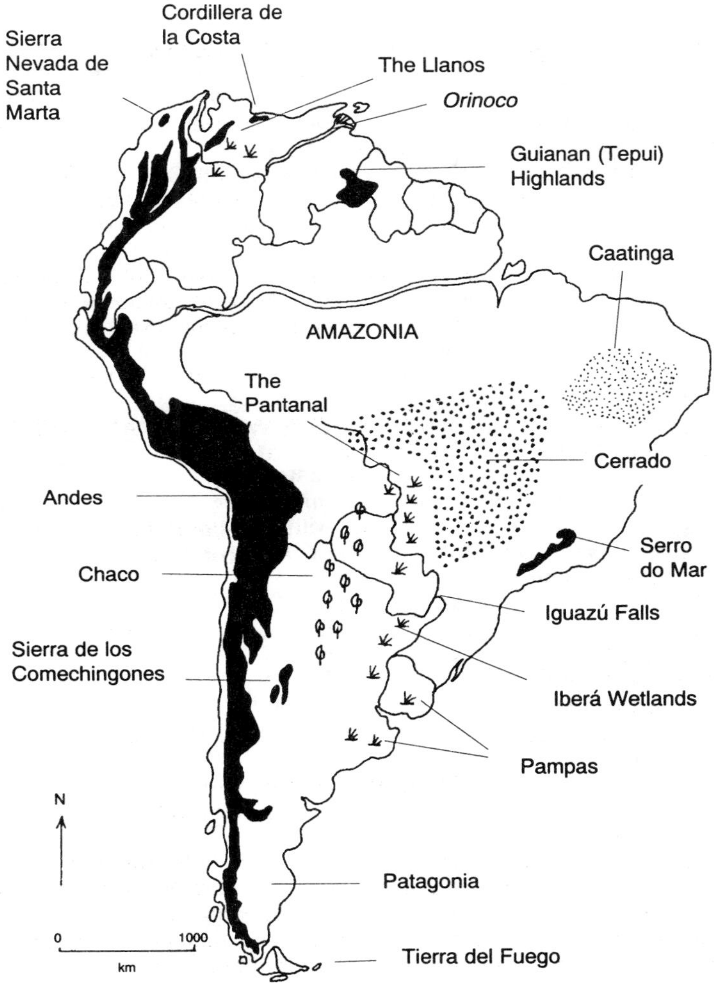

north Brazil, and the Serro Do Mar which runs along the southeast coast of Brazil. Both of these support their own range of endemic species, with the latter supporting one of the highest concentrations of unique birds, for its size, in the world.

The Amazonian lowland rainforest lies between these two mountain ranges and the northern Andes. This awesome stretch of forest once covered some 6 million km^2, an area the size of the USA, but it is disappearing fast. The variety of relief, rainfall and soils in Amazonia has resulted in a diversity of vegetation types, enhancing the already rich habitat. In east Peru and east Ecuador, or west Amazonia, just one square km may support an incredible 300 species. This is the richest place on earth for birds, the proverbial birder's paradise.

South America also possesses the world's biggest wetland, the Pantanal, which lies in west Brazil on the border with Bolivia. This seasonally-flooded grassland, and those to the north (the Llanos of Venezuela and Colombia), and south (the Pampas of Argentina and Uruguay), support millions of waterbirds. The grasslands of the central plateau in Brazil (cerrado), are drier, and the savanna landscape here resembles the great plains of Africa. At the southern tip of the continent *Nothofagus* (Southern Beech) forests cover the Andean foothills at the edge of the extensive cold, windswept grasslands (steppe) of Patagonia. All these habitats also support distinct avifaunas.

Country Lists

It is hardly surprising then that South America hosts the top six country lists in the world. Peru and Colombia (with a potential list of over 1,800) both support around 1,700 species, just 200 short of the whole of the Orient. Both these countries are smaller than nearby Mexico and yet support an extra 700 species.

FIGURE 3: COUNTRY SIZES

(**England** = 130,439 km^2, **Texas** = 688,681 km^2)

Country	**Size(km^2)**
BRAZIL	8,511,965
ARGENTINA	2,766,889
PERU	1,285,216
COLOMBIA	1,138,914
BOLIVIA	1,098,581
VENEZUELA	912,050
CHILE	756,945
PARAGUAY	406,752
ECUADOR	283,561
GUYANA	214,969
URUGUAY	176,215
SURINAME	163,265
GUYANE	91,000
THE FALKLANDS	12,173
THE GALÁPAGOS	7,845
TRINIDAD AND TOBAGO	5,130
NETHERLAND ANTILLES	920

FIGURE 4: COUNTRY SPECIES LISTS (All figures are approximate) **(see Fig. 5, p.24)**

Country	Species
COLOMBIA	1,700+
PERU	1,700
BRAZIL	1,661
ECUADOR	1,550
VENEZUELA	1,360
BOLIVIA	1,274
ARGENTINA	983
GUYANA	825
GUYANE	710
PARAGUAY	635
SURINAME	608+
CHILE	440
TRINIDAD AND TOBAGO	430
URUGUAY	404
NETHERLAND ANTILLES	236
THE FALKLANDS	185
THE GALÁPAGOS	136

Ecuador is a tiny country and yet it boasts the world's fourth biggest list (1,550). Kenya is often thought of as a small country with an incredible avian diversity (1,080), but Ecuador supports nearly 500 species more and is only half the size. Ecuador also has the greatest species diversity of any South American country. This is all thanks to its position, straddling the Andes on the equator. Worldwide, only Costa Rica and Nepal have a higher diversity.

Such high country lists are only approached outside South America by India's 1,300, a country which is nearly 12 times the size of Ecuador, and China's 1,195, a country 34 times larger than Ecuador.

Brazil, however, which is 2.6 times larger than India, supports only around 350 more species, and Argentina's list of 983 is similar to that of Thailand and Cameroon, despite being five times larger than both. Argentina, however, has no lowland rainforest.

Site Lists

Some of the sites described in this book support an astonishing diversity of birds. Many of the top sites (see Fig 6, p. 25) lie within Amazonian lowland rainforest, especially in Ecuador and Peru, and include the Tambopata Reserve in Peru which, arguably, supports the greatest concentration of species (587) in South America and the world. The richest site of all is Manu NP, also in Peru, where approximately 1,000 species have been recorded. However, this NP is far, far larger than Tambopata, and encompasses east Andean slope subtropical forest as well as Amazonian lowland rainforest.

Away from Amazonia and the east Andean slopes, the best site is Henri Pittier NP in Venezuela, where superb cloud forest and a variety of other habitats support over 500 species. On the west Andean slope in

Ecuador is the Nono–Mindo road and Mindo itself, where 430 species have been recorded over many years.

Birders in search of huge trip lists should read the Ecuador, Peru and Venezuela chapters first.

FIGURE 5: COUNTRY SPECIES TOTALS (see Fig 4, p. 23)

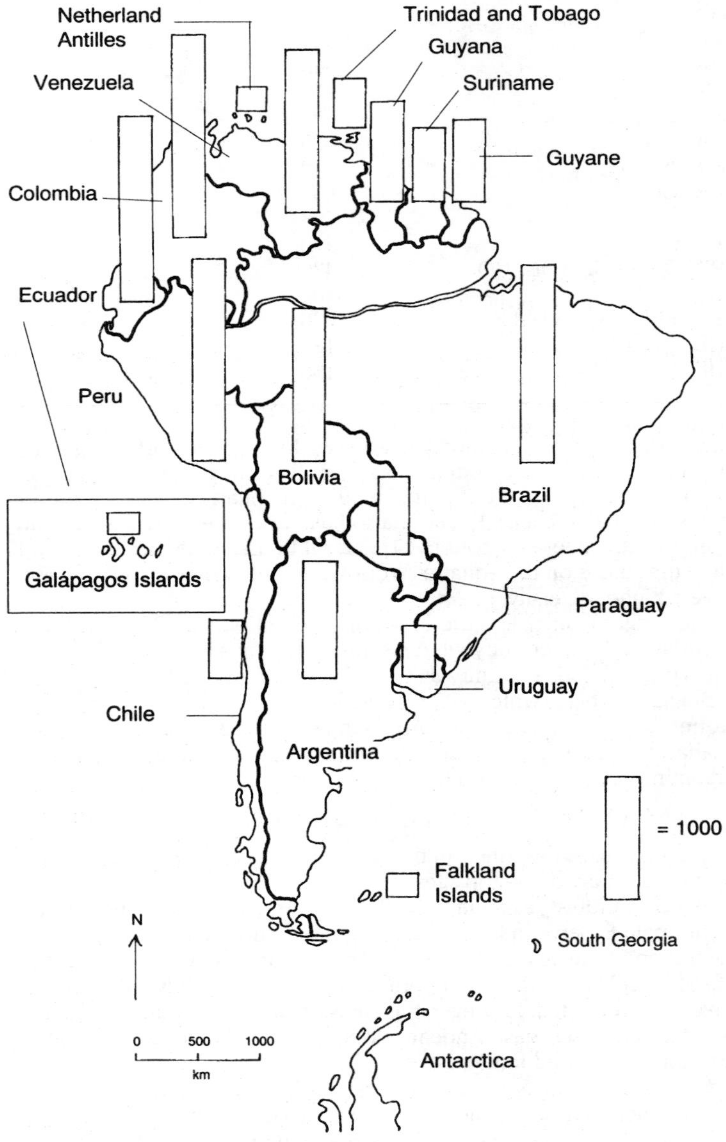

FIGURE 6: THE TOP TEN SITE LISTS IN SOUTH AMERICA

Site		List
1	Manu NP (Peru)	1,000 (approx.)
2	Noel Kempff Mercado NP (Bolivia)	700 (approx.)
3	Manaus (Brazil)	600+
4	Tambopata Reserve (Peru)	587
5	Podocarpus NP (Ecuador)	540 (potentially 800)
6	La Selva (Ecuador)	530
7	Henri Pittier NP (Venezuela)	500+
8	Iquitos (Peru)	500+
9	Amacayacu NP (Colombia)	500 (approx.)
10	Mindo (Ecuador)	430

Trip and Day Lists

Owing to habitat diversity and the ease with which birds may be seen, it is possible to record 650 species in 17 days in Kenya. In 21 days 730 is possible, and in November 1991 a 25-day tour led by Brian Finch recorded an amazing 797 species. In South America the richest places for birds tend to include lowland rainforest; birds are harder to find than in Kenya where such forest is limited in extent. However, Sunbird have recorded 585 species in 16 days in Venezuela, Birdquest have recorded 707 species in 21 days in Ecuador, and the unofficial 'World Bird Tour Record' belongs to the Danish Ornithological Society trip to Ecuador in August–September 1992. Led by the experienced Niels Krabbe and Erik Molgaard, this tour recorded a mindblowing 844 species in just 27 days! The fear of being overwhelmed in South America is a real one.

Such bamboozling totals are usually obtainable only on organised tours led by leaders equipped with masses of experience of sites, birds, birding, as well as tape-recorders which they use to lull out the many skulkers. However, thorough preparation by individuals can result in similar totals for small 'teams' of birders. I was fortunate to be a member of a team which recorded 598 species in Venezuela in just 22 days. All these birds were actually seen (we did not know any calls at the time) and I set eyes on 594 of them. The same period with another team in Kenya produced 532 species. In Peru, Jon Hornbuckle recorded 581 species in 21 days, and Barry and Lyn Trevis managed 653 in 33 days. In Brazil, Bruce Forrester has recorded between 493 and 624 on four trips, each lasting around 40 days.

Further south, it is still possible to record over 500 species in 25 days in Argentina, but in Chile 21 days may only produce only a maximum of 250.

South America once held the world day list record. In September 1986 experienced tropical birders, the late Ted Parker and Scott Robinson recorded 331 species in 24 hours at Cocha Cashu in Manu NP, southeast Peru. This was beaten shortly afterwards, in November 1986, by a team led by Terry Stevenson in Kenya which notched up 342 species with the aid of a small aeroplane. The record would no doubt return to Peru with the use of such transport, or perhaps it could be beaten with just a car in Ecuador. Incidentally, for the interest of budding record beaters, Kenya also boasts the two day record: Don Turner and David Pearson recorded 494 species, also in November 1986.

Family Diversity

Although South America hosts the most number of species, it supports only 92 regularly occurring families, 16 less than Africa (108), and the same as Australasia (92). Of the 27 Neotropical endemic families, 25 are represented in South America (the todies and Palm Chat occur only in the Caribbean) but nine families only are totally endemic to the continent: rheas, screamers, trumpeters, seriemas, Magellanic Plover, seedsnipes, Hoatzin, Oilbird and gnateaters (see Fig 1, p. 15).

FIGURE 7: COUNTRY FAMILY LISTS (out of 92)

Country	Families
PERU	86
BRAZIL	84
ECUADOR	82
ARGENTINA	80
COLOMBIA	79
VENEZUELA	79
GUYANA	77
BOLIVIA	74
GUYANE	74
SURINAME	74
PARAGUAY	69
URUGUAY	68
TRINIDAD AND TOBAGO	66
CHILE	56
NETHERLAND ANTILLES	40
THE GALÁPAGOS	38
THE FALKLANDS	28

FIGURE 8: COUNTRY NEOTROPICAL ENDEMIC FAMILY LISTS (out of 25)

Country	Families
BOLIVIA	23
BRAZIL	23
PERU	23
ECUADOR	21
ARGENTINA	20
COLOMBIA	20
GUYANA	20
VENEZUELA	20
GUYANE	19
SURINAME	19
PARAGUAY	18
TRINIDAD AND TOBAGO	12
URUGUAY	10
CHILE	7
THE FALKLANDS	2
NETHERLAND ANTILLES	0
THE GALÁPAGOS	0

FIGURE 9: COUNTRY SOUTH AMERICAN ENDEMIC FAMILY LISTS (out of 9)

Country	Families
BOLIVIA	8
BRAZIL	7
PERU	7
ARGENTINA	6
ECUADOR	6
COLOMBIA	5
GUYANA	5
GUYANE	4
PARAGUAY	4

SURINAME	4
URUGUAY	4
VENEZUELA	4
CHILE	3
TRINIDAD AND TOBAGO	1
NETHERLAND ANTILLES	0
THE FALKLANDS	0
THE GALÁPAGOS	0

Peru's 86 families include 23 of the 25 Neotropical endemic families and seven of the nine South American endemic families (only the seriemas and Magellanic Plover are absent) so it is the most family-diverse country in South America (86,23,7), closely followed by Brazil (84,23,7), where only Magellanic Plover and seedsnipes are absent. However, those birders seeking to see the most Neotropical and South American endemic families in one short trip may be somewhat surprised to learn that Bolivia (74,23,8) is the best destination. It is the only country that boasts eight of the nine South American endemic families (no Magellanic Plover), and only one (Sharpbill) of the 24 Neotropical endemic families is absent.

Owing to the dangers confronting visitors to Peru and Colombia (79,20,5), both fairly large countries, a newcomer to South America with approximately three weeks to spare, and in search of the best selection of its unique families would do well to choose Bolivia or Brazil. However, some pioneer spirit is necessary to see all eight of Bolivia's South American endemic families, and a certain amount of travelling to see Brazil's seven. Argentina (80,20,6) is also worth consideration but it too is a big country and difficult to cover in one short trip. It lacks only three South American endemic families: trumpeters, Hoatzin and Oilbird.

Tiny Ecuador (82,21,6) which could be covered in one short trip, is nearly as good. It has a high family list (82) and only three South American endemic families are missing: rheas, Magellanic Plover and seriemas. It is not surprising then that little Ecuador and Venezuela (79,20,4), where the same three families plus seedsnipes and gnateaters are absent, are the most popular destinations for South American newcomers. Both are safe, relatively small countries with good infrastructures and masses of birds. Guyana (77,20,5), however, is also a small country currently developing ecotourism, and almost on a par with Venezuela.

Birders who rightly fear being overwhelmed with species rather than families may consider Paraguay (69,18,4), Trinidad and Tobago (66,12,1) or Chile (56,7,3) as their first South American destination. To get the best out of Paraguay you will need to prepare extensively. The tiny archipelago of Trinidad and Tobago offers the chance to see 12 of the Neotropical endemic families. However, Oilbird is the only South American endemic family present. Chile supports rheas, Magellanic Plover and seedsnipes.

Endemic Species

Regions of the world with concentrations of restricted-range birds have been identified as Endemic Bird Areas (EBAs) in Birdlife International's

Putting Biodiversity on the Map, 1992. No less than a quarter (55) of the world's 221 EBAs are in South America, by far the most for any of the world's continents. The next highest total is 37, in Africa. Brazil alone supports 13, closely followed by Peru with ten, and Argentina, Bolivia, Colombia and Venezuela, all with three. Most of the other EBAs cross country boundaries but are concentrated between Colombia, Ecuador and Peru.

FIGURE 10: COUNTRY ENDEMIC LISTS **(see Fig. 11 p. 29)**

BRAZIL	179
PERU	104
COLOMBIA	59
VENEZUELA	41
GALÁPAGOS	25
BOLIVIA	18
ARGENTINA	16
ECUADOR	12
CHILE	8
GUYANE	1
THE FALKLANDS	1
TRINIDAD AND TOBAGO	1
GUYANA	0
NETHERLAND ANTILLES	0
PARAGUAY	0
SURINAME	0
URUGUAY	0

EBA distribution is reflected in country endemics lists. Brazil has by far the most endemic species, a total of 179. Only New Guinea (332) and Australia (306) boast a higher degree of endemism, whilst the Philippines (172) are close behind. Peru's also has over 100 endemics (104), whilst adjacent Ecuador has only 12, some way behind Colombia's 59. Whilst it is possible to see a lot of Brazilian endemics on a short trip, it is difficult to repeat such a performance in Bolivia, Peru and Colombia, even over a long period of time, owing to the remoteness and rarity of so many species. However, a higher success rate is likely in Argentina, Chile, the Galápagos Islands, and Venezuela.

Exploration

A lot of birders have seen over 1,500 species in South America, some over 2,000 and a select few over 2,500 (an incredible 80%). No doubt these 'big-listers' will now be seeking out those endemics with very restricted ranges; the birds hiding in the nooks and crannies of the continent. Furthermore, new areas of South America are opening up almost daily (which has made writing this book turn into something resembling a nightmare), especially in Bolivia, Brazil and Colombia, and many virtually unknown, rare and spectacular birds remain to be rediscovered, or . . . even discovered.

FIGURE 11: COUNTRY ENDEMIC TOTALS (see Fig. 10, p. 28)

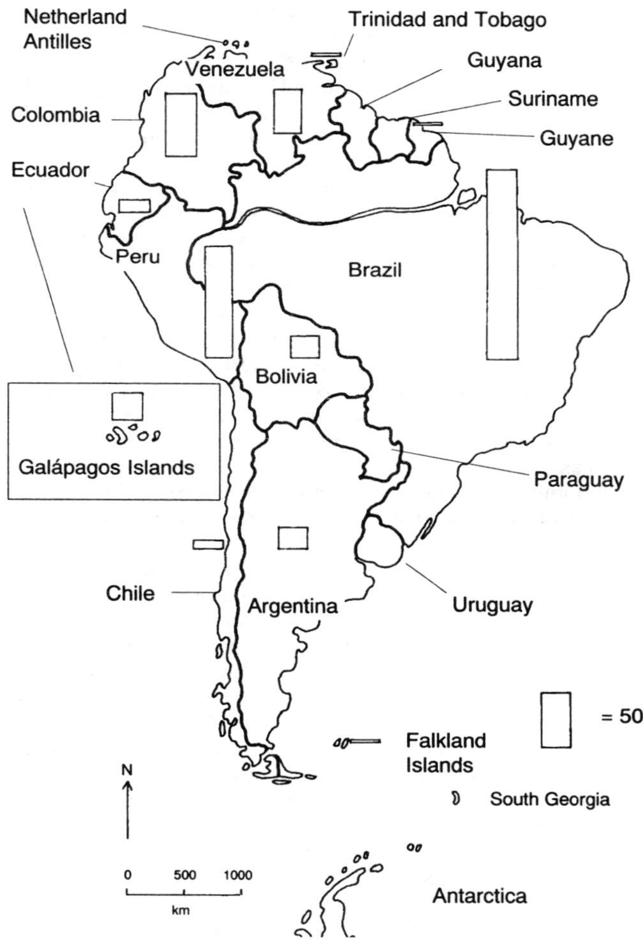

New species are still being described by the keenest researchers and birders determined to get to the less well known localities, some of which are actually only just off the beaten track. Birders in search of virtually unknown or new birds, with experience and time to spare, need not look far for places to explore. Some painstaking research into old records, habitat preferences, present-day distribution of those habitats, and a finger on the pulse to find out which areas are opening up, may lead the pioneering birder to yet more birds and who knows, some obscure antbird named after them.

CONSERVATION

Over 95% of Brazil's original Atlantic forest has been destroyed. In 1988 alone nearly 50,000 km^2 of Amazonian lowland rainforest was destroyed by deliberate fires. Over 50% of Ecuador's remaining forest may be lost by the year 2000. Nearly half of Colombia's endemic birds are threatened with extinction. These sad statistics merely underline what most of us already know: many South American birds and their habitats are disappearing fast.

A ceaselessly increasing human population is putting severe pressure on the planet Earth. Currently we are spoiling our home and the birds' home, destroying it even, and for what? Improved quality of life? No. When the birds stop singing because their trees have been cut down and there is nowhere else for them to go, except a cage, our quality of lives will be very poor indeed.

In South America, forest is felled mainly to create space for small farms and cattle ranches, and to open up areas for colonisation. Many 'mad axemen' are also employed by logging, mining and oil companies, and to make way for coffee, coca and fruit plantations. All these reasons to fell forest are a result of human population pressures, in South America as elsewhere.

Burgeoning populations lead people, quite understandably in many cases, to seek short-term, rather than long-term survival methods. This way of life has a particularly disastrous effect in tropical forests where tree removal and inappropriate land use, including the use of chemicals, lead to soil degradation, erosion and poor water quality. Some bird species have adapted and will adapt, especially open-country birds which naturally do better when trees are felled, but some of these species suffer, in turn, from hunting and the loss of specific grass species to preferred agricultural crops.

Where the population problem is critical, local people are forced to look for new land away from the currently sustainable areas. In South America this 'new' land almost always means forest which, once cleared to make way for crops or cattle, soon degenerates into a virtual desert, exacerbated by the use of chemicals and overgrazing. Just a few years after a patch of forest has been felled the soil is almost dead, and regeneration is either not possible or painfully slow. The once sedentary farmers, who have now become nomads, move on to the next patch of forest and the same process happens all over again. This is not sustainable land use. The only way to inactivate this 'slash-and-burn' self-destruct button is to curb population growth so the farmers, who are currently forced deeper and deeper into the virgin forest, can return to the sustainable areas. A shrinking world population, together with less greed, would ultimately also lead to less logging and mining, and decreasing demand for coffee, coca and fruits.

Meantime, as the forest remnants shrink and become ever more isolated, they become too small and remote to sustain viable populations of birds. Many of the species that do survive are hunted for food and the cagebird trade, especially the larger birds such as cracids and parrots. In many areas reforestation is taking place, but with inappropriate non-native tree species. In South America for example, fast-growing *Eucalyptus* species are being used. These trees support few birds, which is not surprising considering they are totally different from the original habitat.

Away from the forests the problems are equally severe. Rivers, streams, lakes, bogs and marshes not only contain poor quality water, draining down from chemically-treated highland farms, mining operations etc., they are also subject to siltation and damming. Lowland lakes, marshes and grasslands, are already disappearing thanks to various drainage schemes. More people need more food so many lowland wetlands are potential agricultural land in the eyes of government officials. In South America, even remote, seasonally-flooded grasslands such as the Llanos and the Pantanal are not sacred, despite their immense importance for birds.

Since stemming population growth, the only realistic route to saving the earth and its birds, seems impossible and, indeed, undesirable to some governments, smaller initiatives such as integrating local needs with sustainable use of the natural habitat, through such schemes as ecotourism, need to be implemented. I hope this book, in its own small way, will help to convince governments, via encouraging ecotourism to the sites described (and many more), that their sustainable natural resources, especially birds and their habitats, are the logical solutions to their long-term economic and social problems.

It is important that governments also become aware of the fact that whole ecosystems, not individual sites or species, need to be preserved. Large areas with wide altitudinal ranges and the highest possible biodiversity, where the local people can enjoy a rich quality of life, must be set aside now.

Birders look forward to the day when the 'remote tracks' lead not to logging camps, mines or coffee plantations, but to rustic lodges run by local people, surrounded by forests and marshes full of birds.

GENERAL TIPS

A basic understanding of Spanish would be an asset in virtually all South American countries, although it is possible to struggle through most countries by picking up the language as you go along. In Brazil, however, a form of Portuguese is the most widely spoken language and it is especially difficult to get by there without some understanding. An 'understanding' unfortunately means being able to comprehend the answer as well as stating the question correctly.

Early morning is usually the best time for birding anywhere in the world and South America is no exception, although this continent is rather special and virtually any time of the day, especially in montane sites such as temperate forest, where it never gets really hot, can be productive. However, high noon in hot lowlands, such as Amazonia and the big wetlands of the Llanos and the Pantanal, can be quiet, and it is wise to take a siesta before a late afternoon bash (the evening owl and potoo search, and the appropriate celebration of another great day in the field). Cotingas are good examples of birds which seem to perch on exposed branches to rest and preen towards the end of the day.

Cotingas and a host of other birds are also attracted to fruiting trees. Whilst such trees may seem devoid of life as the birds feed quietly, they can slowly come to life, and standing next to one for a few hours may be more productive than frantic wandering.

Rain is prevalent in most South American forests. If anything, many birds, especially the colourful tanagers of temperate forests, prefer damp conditions so it is worth staying in the field during the rain. An umbrella is awkward to hold but red ones attract hummers (as do red T-shirts). Better still perhaps, wear a red waterproof with a good hood. Airtight binoculars are essential in these damp, misty and often downright wet forests, as well as in lowland rainforests where humidity is very high. The pure nitrogen type are better than the air-water-vapour models.

Most people can travel up to 3,000 m (9,843 ft) in a short time. Most can manage 4,500 m (14,764 ft) after a night at 3,000 m (9,843 ft). Going straight to 4,500 m (14,764 ft), however, can be very dangerous. One in three people who attempt this usually end up with a severe headache or, even worse, feel dizzy and begin to vomit. If that person then turns blue or coughs up pink mucus they should descend immediately to below 3,000 m (9,843 ft). These symptoms may be alleviated by eating lightly, or by drinking coca tea, although the latter is probably illegal, and possibly addictive! When birding at high altitudes take it very easy, use maximum sun-block, beware of snow blindness, and remember that the thermometer plummets as soon as the sun goes down.

It is worth taking an altimeter if you are unsure of the terrain you are visiting, or if the only information you have on a certain species is the altitude at which it has been recorded. Vehicles may be a problem at high altitudes. If you have difficulty starting your car engine in the early mornings try pouring petrol directly into the carburetter.

Using tapes of bird calls and songs helps to confirm identification, but this devious birding method is not appropriate for use with threatened species, if any species at all. It distracts birds from feeding and causes them undue excitement.

GLOSSARY

***Araucaria* forest**: Araucaria is the scientific name for the Monkey-puzzle tree. Araucaria forests contain large stands of this species. They are found in Argentina, far south Brazil and Chile.
Caatinga: Extremely arid scrub and woodland in northeast Brazil.
Campo: Low grassland with some scrub but no trees.
Cerrado: Open, tall grassland savanna with scattered low trees. This habitat is present on the central plateau of Brazil and in extreme northeast Bolivia.
Chaco: Dense, dry scrub and woodland present from extreme south Bolivia, through west Paraguay to north Argentina.
Elfin forest/woodland: A stunted, exposed, high-elevation form of temperate forest.
Gallery forest/woodland: Waterside forest and woodland where forested areas merge into more open areas. For example, at the edge of Amazonia.
Igapo forest: Amazonian lowland rainforest which is nearly always flooded.
Llanos: Seasonally-flooded grasslands in central Venezuela and north-east Colombia.
Matorral: Scrub and woodland on the low coastal hills in central Chile.
***Nothofagus* forest**: *Nothofagus* is the scientific name for the Southern Beech tree, which dominates forests on the low Andean slopes in south Argentina and south Chile.
Pampas: Grassland (which is often wet) in south Brazil, Uruguay and east Argentina.
Pantanal: Seasonally-flooded grassland on the borders of Bolivia, Brazil and Paraguay.
Paramo: Wet, windswept grassland above the treeline in the north Andes, from southwest Venezuela, through Colombia and Ecuador, to north Peru.
Patagonian steppe: Windswept low scrub and grass in south Argentina and south Chile.
***Polylepis* woodland**: A low, gnarled woodland dominated by *Polylepis* trees (characterised by their flaky reddish bark), which occur in pockets above the treeline in the Andes, from Ecuador south to extreme north Chile.
Puna: Seasonally dry, windswept grassland above the treeline in the Andes, from central Peru to south Chile.
Savanna: Seasonally wet or completely dry grasslands with and with-out trees.
Secondary: This term usually refers to forested areas which have been cleared but have since partly regenerated, albeit usually in frag-ments alongside newly established land use.
Subtropical forest: Usually luxuriant forest between 1,000 (3,281 ft) and 2,500 m (8,202 ft).
Temperate forest: Forest above subtropical forest, usually between 2,300 and 2,500 m (7,546–8,202 ft) and 3,100 and 3,500 m (10,171-11,483 ft). Although not strictly true, in this book, temperate forest

includes 'cloud forest' (unless otherwise stated) which, in reality, is the term applied to those forests often enveloped in damp cloud but situated in an otherwise dry region.
Terra Firme forest: Amazonian lowland rainforest which is not usually flooded.
Várzea forest: Amazonian lowland rainforest which is occasionally flooded.
Yungas forest: Subtropical forests on the very steep eastern slopes of the Andes in Bolivia and north Argentina.

MAPS

Birders tend to be more interested in the birds rather than distances and directions when in the field, hence, although every effort has been made to make the maps in this book as precise as possible, they may not be entirely accurate, at least in terms of exact scale and compass points.

The main purpose of the maps is to facilitate birding at the sites so more often than not 'direction-pointers' such as rivers and buildings, and on-site detail such as trails, have been exaggerated and are not drawn to scale.

Each country account begins with a map of sites, and there are also some maps of regions within countries to show how 'bunches' of sites are distributed. These are intended to aid birders in their production of trip routes and itineraries during the planning stage.

It is important to remember that South America is a fast-changing continent so trails may have become tracks, tracks may have become roads, buildings may have been knocked down, buildings may have been put up, marshes, lakes and rivers may have dried up, rivers may have changed course, signs may have fallen down, habitats may have been totally destroyed and some sites may no longer exist by the time you come to use these maps.

Do not blame the birders who have been before you, and do not blame me; blame the ever increasing human population on the planet if you arrive at a site which no longer exists or looks totally different from what you have perceived as a result of reading this book. And remember, there is no substitute for up-to-the-minute information (see Trip Reports and Useful Addresses, p. 409).

The map symbols used are as follows:

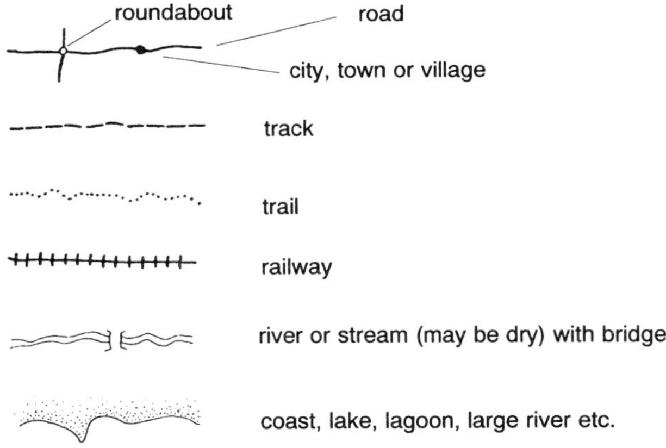

Key to maps (continued)

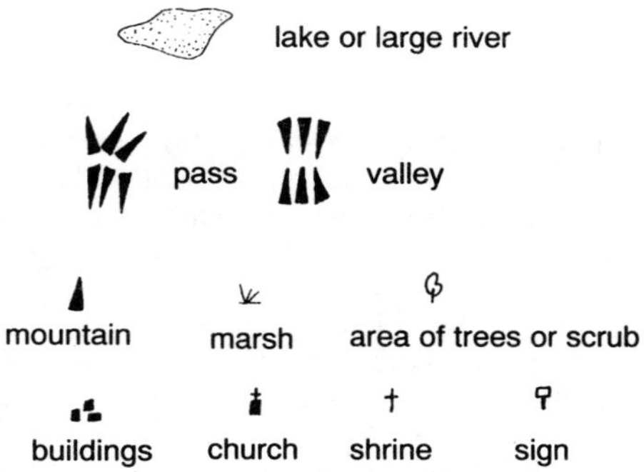

lake or large river

pass valley

mountain marsh area of trees or scrub

buildings church shrine sign

ARGENTINA

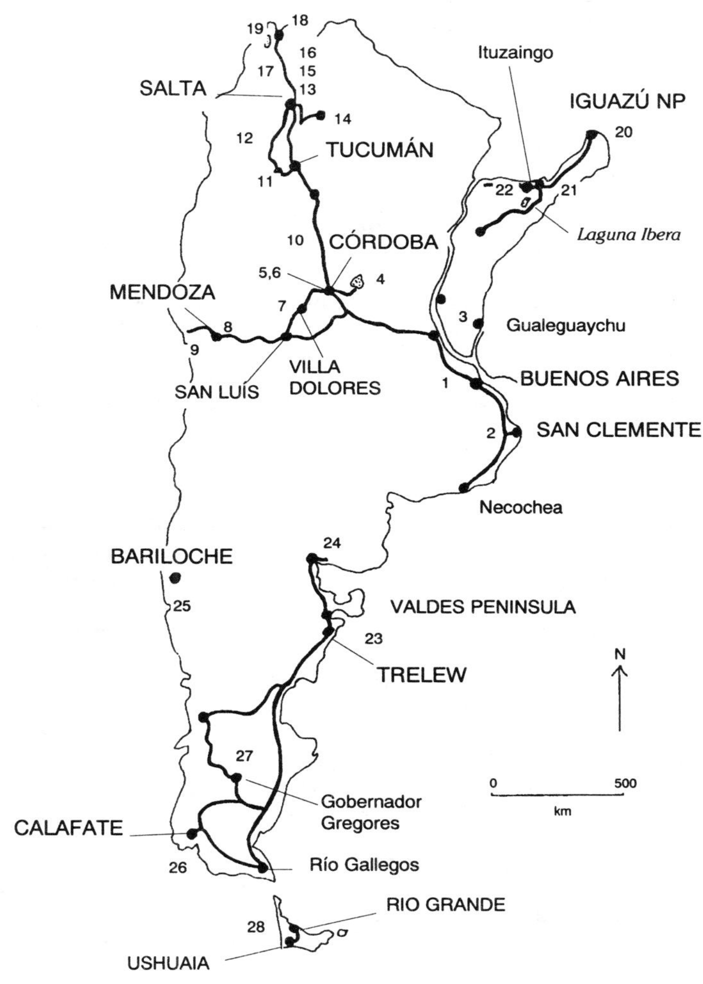

1	Costanera Sur ER	6	Icho Cruz
2	San Clemente	7	Chancani
3	Entre Ríos	8	Potrerillos
4	Laguna Mar Chiquita	9	Laguna Horcones
5	Pampa de Achala	10	Salinas Grandes

11 Taffi del Valle
12 Cafayate–Salta Road
13 Palomitas
14 Joaquín V González
15 El Rey NP
16 Calilegua NP
17 Salta–Humahuaca Road
18 La Quiaca
19 Lago Pozuelos
20 Iguazú NP
21 San Martín
22 Iberá Marshes
23 Puerto Madryn and Trelew
24 San Antonio Oeste
25 Bariloche
26 Calafate
27 Strobel Plateau
28 Tierra del Fuego

INTRODUCTION

Summary

Argentina is a friendly, modern country with a good infrastructure, a field guide and many spectacular settings in which to watch birds. Although it is basically a temperate country with no lowland rainforest, it is over 3,000 km long, and there is actually a good range of habitats, which support a superb variety of South American bird families. It would be difficult to do justice to all these habitats in one short trip, so the dilemma for most Argentine first-timers is whether to go north or south. The north supports a number of high-Andean specialities whilst many species restricted to Patagonia occur in the south.

Size

Argentina is big. It is the eighth largest country in the world, over twenty times the size of England, four times the size of Texas and nearly ten times larger than Ecuador, a total of 2,766,889 km^2. It is a mammoth 3,460 km from the high Andes on the border with Bolivia in the northwest to the windswept shores of Tierra del Fuego in the south.

Getting Around

Although the infrastructure is highly developed, only one thirdof the many roads are paved. There is an extensive, cheap and efficient bus service. Most of the cities and larger towns are accessible by air from Buenos Aires, making the various airpasses very useful in this big country. There is also a cheap but slow rail network which covers all but Patagonia.

Accommodation and Food

High quality, good value accommodation is fairly extensive. Camping is very popular and roadside service stations often provide facilities. Argentina is a carnivore's paradise, but vegetables are scarce.

Health and Safety

Argentina is a clean and hygienic country. The only precaution necessary against disease is for malaria if visiting the low-lying subtropical areas in the north. Very rarely, a Reduvid bug, which looks like a small cockroach and normally lives only in squalid houses in the northwest, bites someone and gives them Chaga's disease which is very nasty and difficult to treat. Beware of altitude sickness in the high Andes of the northwest. This is one of the safest countries in South America and violent crime against tourists is very rare.

Climate and Timing

The climate is temperate in the south, hence the austral winters (April to September) are very cold. In contrast the north is more subtropical and it can be very hot and humid from Buenos Aires north in the summer, especially December to February. At the same time, flooding may occur in the northern Andes. Hence the winter, June to August, is a fairly good time to visit the north, although the Ibera wetlands and Misiones are particularly wet in August; and the austral spring, October to November, is the best time to visit the whole country, especially the south.

Habitats

Thanks to its vast size Argentina contains a wide variety of habitats.

The Andes run along its western and northern boundaries with Chile and Bolivia. In the south they are low and glaciated with *Nothofagus* (Southern Beech) forest on the lower slopes. In the north they are high and dry, with high-altitude lakes surrounded by puna below the peaks, arid cacti-clad intermontane valleys below there, and temperate and subtropical 'yungas' forests on the wetter lower slopes. Due west of Buenos Aires and east of the Andes, near the city of Córdoba, lies an isolated rugged mountain range known as the Sierra de los Comechingones. Three birds are endemic to these mountains and the surrounding area.

Much of north Argentina is covered by chaco, which is wet in the east, with palm savanna and woodland, and dry in the west. The western portion of the chaco was once covered in scrubby woodland, dominated by *Mesquite* and *Quebracho*, but it has been seriously degraded over the years and in many places is now dominated by thick, nasty thorn scrub, known to many as 'El Impenetrable'. The eastern chaco ends at the huge Iberá wetlands in the east.

Remnant *Araucaria* woodland (Monkey-puzzle trees) is present in the northeast finger of the country, where there is also subtropical forest surrounding Iguazú falls, the world's largest waterfall.

The rolling grasslands and marshy flatlands which form the pampas cover much of central Argentina. As with the chaco, the east is wet and still relatively wild while the west is dry and mainly turned over to agriculture.

South of the pampas lies the arid, lake-dotted, windswept Patagonian steppe, covering over a third of the country. Spectacular glaciated mountains form the western boundary, the famous Valdés peninsula is on the east coast and the *Nothofagus* (Southern Beech) forests, lakes and wild coast of Tierra del Fuego form the southern extremity.

Conservation

37 threatened species occur in Argentina, none of which are endemic. Three (four) endemics are near-threatened.

Bird Families

Of the 92 families which regularly occur in South America 80 are represented, fewer than most of the countries to the north. They include 20 of the 25 Neotropical endemic families and six of the nine South American endemic families. There are no trumpeters, Hoatzin, Oilbird, Sunbittern or barbets. Both rheas, both seriemas, Magellanic Plover and seedsnipes occur.

Well-represented families include tinamous, grebes, flamingos,, and furnariids, of which there are over 70 species.

Bird Species

983 species have been recorded, well below the countries to the north and over 700 fewer than Peru and Colombia. However, this is merely a reflection of the country's complete lack of lowland rainforest, rather than the lack of birds.

Non-endemic specialities and spectacular species include Hooded Grebe*, which was, until recently, an endemic, King and Magellanic Penguins, Magellanic Diving-Petrel, Ruddy-headed Goose*, Spectacled Duck*, Andean* and Puna* Flamingos, Crowned Eagle*, Spot-winged Falconet, Red-faced Guan*, Black-fronted Piping-Guan*, Dot-winged Crake*, Horned Coot*, American Painted-snipe, Eskimo Curlew*, Snowy Sheathbill, Magellanic Plover, White-bellied Seedsnipe, Andean Avocet, Two-banded Plover, Rufous-chested and Tawny-throated Dotterels, Magellanic Woodpecker, Red-tailed Comet, Toco Toucan, Lark-like Brushrunner, Crested Gallito, Black-throated Huet-huet, Olive-crowned Crescent-chest, Spotted Bamboowren*, Strange-tailed Tyrant*, seven out of eight monjitas, Yellow Cardinal*, and Canary-winged* and Yellow-bridled Finches.

Endemics

16 species occur only in Argentina, a slightly higher total than Ecuador, but well below the giddy heights set by Brazil and Peru. These include three canasteros, Sandy Gallito, Rusty-backed Monjita, Cinnamon Warbling-Finch and Tucuman Mountain-Finch*. Three endemics are restricted to the Sierra de los Comechingones area near Córdoba.

Near endemics include Rufous-throated Dipper*, which is hard to see in Bolivia, the only other country in which this species occurs.

Expectations

Well-prepared birders, who cover only the north, can expect to see approximately 300 species in three weeks. An extra 100 species are possible if another week is spent in the south. Exceptionally, over 550 species may be possible on an extensive five-week trip, including 15 of the 16 endemics, and over 50 furnariids.

BUENOS AIRES

Buenos Aires, the capital of Argentina, lies on the huge estuary of the Río Paraná and Río Uruguay, known as the Río de la Plata, at the northern edge of the open grasslands, lakes and marshes of the pampas. There are a couple of small sites well worth visiting if you are passing through the capital, including the Costanera Sur Ecological Reserve, which is just a short walk from the city centre.

COSTANERA SUR ECOLOGICAL RESERVE (BUENOS AIRES WILDLIFE REFUGE)

Map p. 42

The reed-fringed pools, pampas grass and scrubby woodland of this excellent reserve, situated in the old part of the capital, just 30 minutes walk from the city centre, support a fine selection of birds which form an excellent introduction to South American birds for newcomers to the continent. Over 200 species have been recorded including the scarce Black-headed Duck*, the rare Dot-winged Crake*, and Curve-billed Reedhaunter, although these are all rather rare visitors.

Specialities

Black-headed Duck*, Rufous-sided Crake, Giant Wood-Rail, Dot-winged Crake*, Snowy-crowned Tern, Sulphur-bearded Spinetail, Freckle-breasted Thornbird, Curve-billed Reedhaunter, Rufous-capped Antshrike, Red-crested Cardinal, Unicoloured Blackbird, Brown-and-yellow Marshbird.

Others

White-tufted and Great Grebes, Southern Screamer, Lake Duck, Black-necked Swan, White-cheeked Pintail, Rosy-billed Pochard, Whistling Heron, Stripe-backed Bittern, Bare-faced and Plumbeous Ibises, Snail Kite, Chimango Caracara, Plumbeous Rail, Spot-flanked Gallinule, White-winged, Red-gartered and Red-fronted Coots, Limpkin, Wattled Jacana, Southern Lapwing, Grey and Brown-hooded Gulls, Picazuro Pigeon, Picui Ground-Dove, Monk Parakeet, Guira Cuckoo, Scissor-tailed Nightjar, Glittering-bellied Emerald, Campo Flicker, Sooty-fronted and Yellow-chinned Spinetails, Wren-like Rushbird, Small-billed Elaenia, Many-coloured Rush-Tyrant, Spectacled and Yellow-browed Tyrants, Rufous-bellied Thrush, Masked Gnatcatcher, White-rumped Swallow, Masked Yellowthroat, Blue-and-yellow Tanager, Long-tailed Reed-finch, Black-and-rufous and Black-capped Warbling-Finches, Great Pampa-Finch, Rusty-collared and Double-collared Seedeaters, Epaulet Oriole, Yellow-winged and Chestnut-capped Blackbirds.

Other Wildlife

Coypu.

Access

The reserve entrance is opposite Estados Unisos on the east side of Avenida Antart Argentina, southeast of the city centre. The Black-headed Duck* usually occurs on the larger pools and Rufous-sided Crake near the warden's house.

Accommodation: Crillon (A), Hotel Novel (B).

The **Paraná delta**, 40 km northwest of Buenos Aires, is one of the few places in the world where Dot-winged Crake* occurs, as well as Sulphur-throated Spinetail, Curve-billed Reedhaunter, Many-coloured Rush-Tyrant and Crested Doradito. Curve-billed and Straight-billed* Reedhaunters have been recorded at nearby **Ing. R. Otamendi**. Bird the track which runs alongside an overgrown ditch for 1 km behind the railway station.

COSTANERA SUR ECOLOGICAL RESERVE

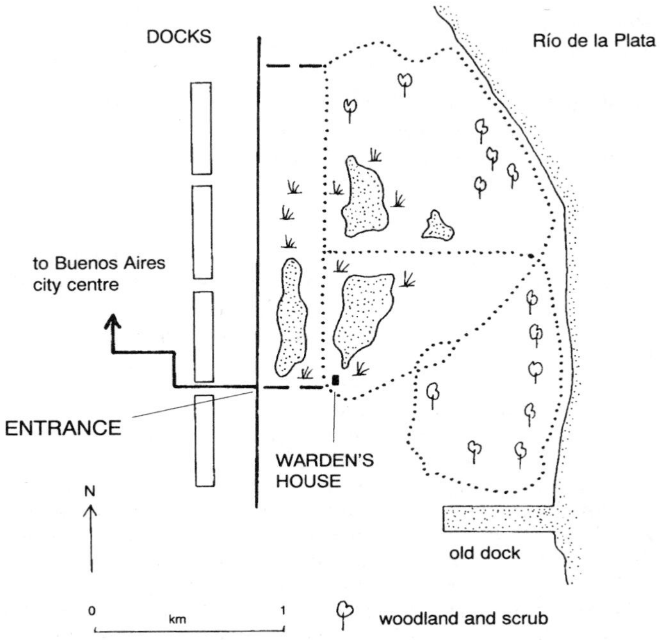

The thickets, overgrown marshes and remnant woodlands on the shores of Samborombón Bay, southeast of Buenos Aires, *en route* to San Clemente, support Giant Wood-Rail, Checkered Woodpecker, Short-billed Canastero and Curve-billed Reedhaunter. The area around **Attalaya**, near Magdalena, 110 km southeast of Buenos Aires, is best. Bird the disused railway track.

SAN CLEMENTE

Map opposite

This small seaside resort, 330 km southeast of Buenos Aires, is an ideal base from which to explore the pampas grasslands, marshes, coastal dunes and mudflats nearby. The area is a popular birding site for Argentine birders, and there is an observatory on the Punta Rasa peninsula just to the north, a migrant hotspot. Even a short trip into the pampas alone is likely to produce over 100 species in a day.

Specialities

Spotted Nothura, Dot-winged Crake*, American Painted-snipe, Snowy Sheathbill, Olrog's Gull*, Snowy-crowned Tern, Golden-breasted Woodpecker, Hudson's Canastero, Bay-capped Wren-Spinetail*, Warbling Doradito, White-banded Mockingbird, Red-crested Cardinal, Scarlet-headed Blackbird.

San Clemente is a good site for Snowy-crowned Tern

Others

Red-winged Tinamou, Greater Rhea*, Great Grebe, Black-browed Albatross, Southern Fulmar, White-chinned Petrel, Southern Screamer, Coscoroba Swan, Chiloe Wigeon, White-cheeked Pintail, Chilean Flamingo, Stripe-backed Bittern, Maguari Stork, Long-winged and Cinereous Harriers, Two-banded Plover, Tawny-throated Dotterel, Guira Cuckoo, White-throated Hummingbird, Campo Flicker, Tufted Tit-Spinetail, Wren-like Rushbird, Firewood-gatherer, White-tipped Plantcutter, Southern Beardless-Tyrannulet, Small-billed Elaenia, Sooty Tyrannulet, Many-coloured Rush-Tyrant, Spectacled and Yellow-

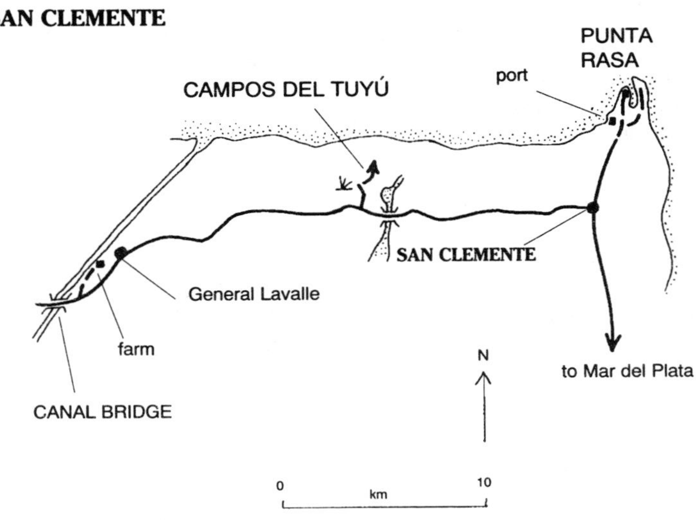

browed Tyrants, Southern Martin, Blue-and-yellow Tanager, Long-tailed Reed-Finch, Double-collared Seedeater, Chestnut-capped Blackbird.

Other Wildlife

Six-banded Armadillo.

Access

Concentrate on birding the peninsula just north of San Clemente, known as **Punta Rasa**. 2 km after entering the reserve take the left fork for 3 km to the Biological Station, observatory and lighthouse. White-tipped Plantcutter occurs in the scrub, and Hudson's Canastero and Bay-capped Wren-Spinetail* in the saltmarsh, usually to the left along-side this road, while White-throated Hummingbird may be found in the small wood, just before the station. Olrog's Gull* occurs on the mudflats near here; or take the right fork for 4 km to the point. A small marsh on the left near the point supports Dot-winged Crake* and Bay-capped Wren-Spinetail*. Punta Rasa can be good for migrants in the austral spring (October–November) and autumn (March–April), and seabirds pass offshore when strong onshore winds are blowing.

San Clemente Port, 4 km from the town centre, next to the marine park (Mundo Marino) is a good site for Olrog's Gull*.

Stripe-backed Bittern and American Painted-snipe occur at the **Campos del Tuyú Reserve**. The entrance is on the north side of the road, 12 km west of San Clemente. The snipe occurs in the marsh on the left, 1 km along the entrance road. The reserve is another 9 km further on and full of birds.

Many-coloured Rush-Tyrant and Warbling Doradito occur at the canal bridge west of General Lavalle (28 km west of San Clemente).

Accommodation: Hotel Fontainbleau (A); Hotel Piedras (San Martín) (C).

Pelagic trips on local fishing boats can be arranged at **Necochea**, a port 500 km due south of Buenos Aires. Albatrosses and White-chinned Petrel have been recorded on such trips. White-throated Hummingbird occurs in Necochea and Rockhopper Penguins occasionally appear (usually in May) on the beach at Costa Bonita, 16 km to the east.

Accommodation: Hotel Flamingo (C).

ENTRE RÍOS

The rarely visited state of Entre Ríos, north of Buenos Aires, supports a number of rare and threatened birds, especially seedeaters. Recent DNA studies suggest that Narosky's Seedeater*, endemic to this tiny area, is a subspecies of Marsh Seedeater*.

Endemics

Narosky's (Marsh) Seedeater*.

Specialities

Spotted Nothura, Giant Wood-Rail, Sickle-winged Nightjar*, Golden-breasted Woodpecker, Straight-billed Reedhaunter*, Brown Cachalote, Black-and-white Monjita*, Red-crested Cardinal, Dark-throated*, Marsh*, Grey-and-chestnut* and Chestnut* Seedeaters, Glaucous-blue Grosbeak, Saffron-cowled* and Scarlet-headed Blackbirds.

Others

Red-winged Tinamou, Greater Rhea*, Southern Screamer, Ringed Teal, Bare-faced Ibis, Striped Owl, Little and Scissor-tailed Nightjars, Suiriri Flycatcher, Many-coloured Rush-Tyrant, Grey Monjita, White-naped Xenopsaris, Long-tailed Reed-Finch, Black-and-rufous and Black-capped Warbling-Finches, Solitary Cacique, Epaulet Oriole, Chestnut-capped and White-browed Blackbirds.

Access

Directions for these sites are scant owing to the sensitive species present. Good sites include **El Palmar NP**, situated on the Uruguay border and some 400 km north of Buenos Aires. Turn east 45 km north of Colón

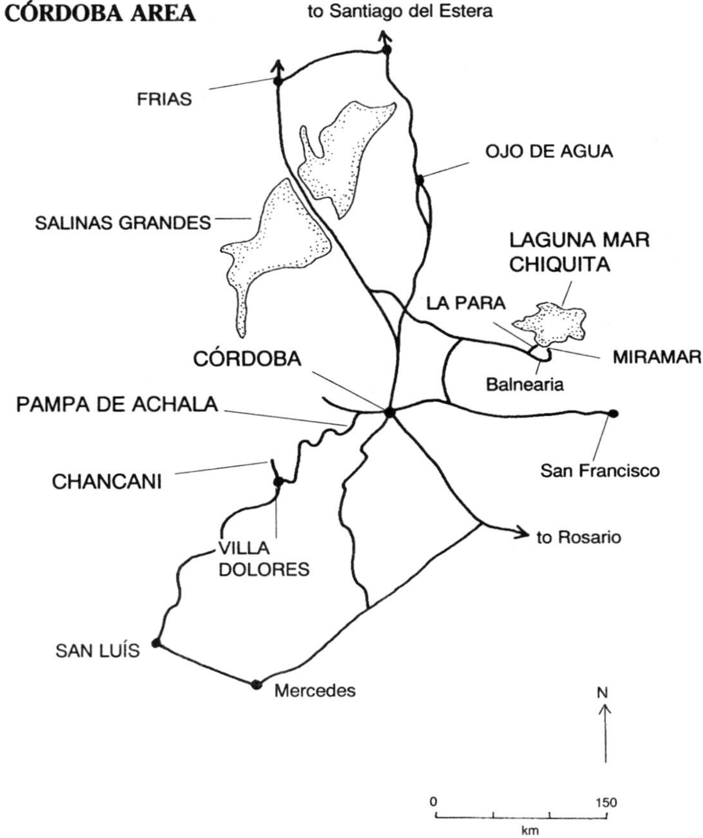

(route 14) opposite the 'Colón Hotel Palmar' sign. Spotted Nothura and Grey Monjita occur along the 10 km to a fork. Turn right here and bird the marsh 2 km further on, where Brown Cachalote and Chestnut Seedeater* occur. The track continues to a river where Scissor-tailed Nightjar occurs. The roadside marshes east of **Va Féderal** support Greater Rhea*, Southern Screamer and Giant Wood-Rail. Goodies including Sickle-winged Nightjar*, Black-and-white Monjita*, Marsh*, Grey-and-chestnut* and Narosky's* (Marsh*) Seedeaters, and Saffron-cowled Blackbird* occur in the **Puerto Boca** wetlands near Gualeguaychu.

CÓRDOBA AREA

Map p. 45

Argentina's 'second' city, some 700 km northwest of Buenos Aires, lies close to some superb birding sites including Laguna Mar Chiquita, where the almost extinct Eskimo Curlew* has been seen, and the Sierra de los Comechingones where three very localised endemics occur.

LAGUNA MAR CHIQUITA

Maps below and opposite

Although famous for the most recent sightings of the virtually extinct Eskimo Curlew*, and rightly so, this huge salt lake and its surrounds, some 200 km northeast of Córdoba, also supports thousands of shore-birds, as well as masses of waterfowl and some very localised species such as Olive-crowned Crescent-chest and two species of doradito.

LAGUNA MAR CHIQUITA

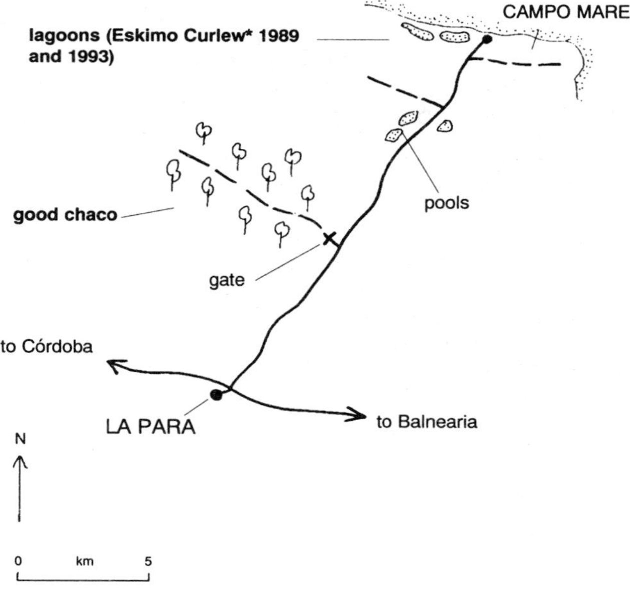

Specialities

Ringed Teal, Spotted Rail, Eskimo Curlew*, Little Nightjar, White-fronted and Checkered Woodpeckers, Chaco Earthcreeper, Crested Hornero, Chotoy Spinetail, Short-billed Canastero, Little Thornbird, Lark-like Brushrunner, Brown Cachalote, Crested Gallito, Olive-crowned Crescent-chest, White-tipped Plantcutter, Crested and Dinelli's* Doraditos, Stripe-capped Sparrow, Many-coloured Chaco-Finch.

LAGUNA MAR CHIQUITA

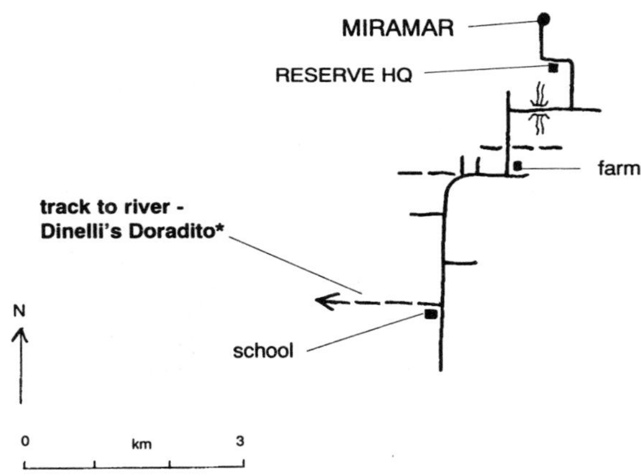

Others

Great Grebe, Southern Screamer, Coscoroba Swan, Stripe-backed Bittern, Plumbeous Rail, White-winged and Red-gartered Coots, Pale-breasted Spinetail, Wren-like Rushbird, Pearly-vented Tody-Tyrant, Spectacled Tyrant, White-naped Xenopsaris, Epaulet Oriole, Yellow-winged and Chestnut-capped Blackbirds.

Access

Four Eskimo Curlews* were seen at Campo Mare, 17.5 km northeast of La Para, in October 1989 and two in 1993. The chaco scrub west of the road northeast out of La Para to Campo Mare supports specialities such as Olive-crowned Crescent-chest.

Another excellent birding area lies near Miramar, a popular resort and the best base from which to bird the area. Turn west 8 km south of Miramar (alongside the school) and head towards the river mouth, looking for Short-billed Canastero *en route* and the two doraditos in the tamarisk at the river mouth.

Three Eskimo Curlews* were seen northeast of Miramar in November 1988.

Accommodation: Miramar: Hotel Las Vegas.

PAMPA DE ACHALA

Map below

West of Córdoba the road to San Luís crosses a grey granite plateau known as the Pampa de Achala, in the Sierra de los Comechingones, a range of rugged hills isolated from the Andes. The three endemics restricted to this sierra and its surroundings all occur here, as well as the scarce Black-crowned Monjita.

Endemics

Comechingones and Olrog's Cinclodes, Cordoba Canastero.

Specialities

Red-tailed Comet, Golden-breasted Woodpecker, Buff-breasted Earthcreeper, Black-crowned Monjita, Short-billed Pipit.

Others

Andean Tinamou, Andean Condor, Red-backed Hawk, Cordilleran Canastero, Rufous-naped Ground-Tyrant, Spectacled Tyrant, Plumbeous Sierra-Finch, Plain-coloured Seedeater.

Access

Olrog's Cinclodes occurs along the creek running parallel to the Córdoba–Villa Dolores road opposite the Parador El Condor café, 84 km southwest of Córdoba. From here, turn north on the old dirt road to Córdoba. Just past the radio station, 1 km along this road to the north, check the rocky area with tussock grass for Cordoba Canastero and Buff-breasted Earthcreeper.

PAMPA DE ACHALA

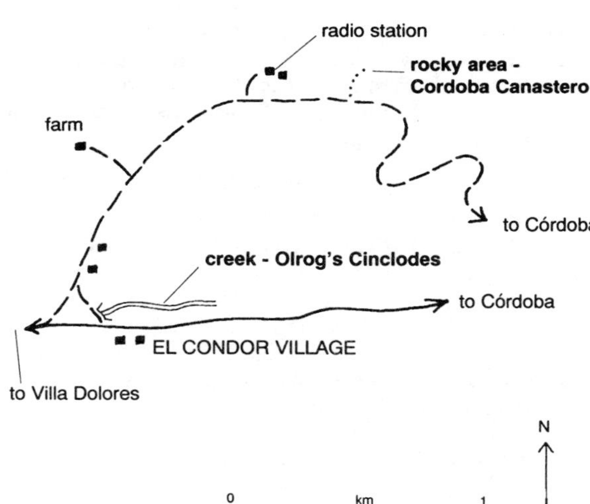

ICHO CRUZ

Map below

This popular resort, just one hour's drive southwest of Córdoba on the road to Villa Dolores, is a good site for the endemic Cinnamon Warbling-Finch and a number of chaco specialities hard to see elsewhere in South America.

Endemics

Olrog's Cinclodes, Cinnamon Warbling-Finch.

Specialities

Spot-winged Falconet, Grey-hooded Parakeet, Blue-tufted Starthroat, Spot-backed Puffbird, Checkered and Golden-breasted Woodpeckers, Scimitar-billed Woodcreeper, Chaco Earthcreeper, Crested Hornero, Short-billed Canastero, Lark-like Brushrunner, Brown Cachalote, Little Thornbird, Southern Scrub-Flycatcher, Greater Wagtail-Tyrant, Black-crowned Monjita, White-banded Mockingbird, Short-billed Pipit, Stripe-capped Sparrow, Many-coloured Chaco-Finch, Black-and-chestnut Warbling-Finch.

Others

Andean Tinamou, Spot-winged Pigeon, Guira Cuckoo, Scissor-tailed Nightjar, Glittering-bellied Emerald, White Woodpecker, Sooty-fronted and Pale-breasted Spinetails, White Monjita, Rufous-naped Ground-Tyrant, Crowned Slaty Flycatcher, White-tipped Plantcutter, Southern Martin, Hooded Siskin, Black-capped Warbling-Finch, Golden-billed Saltator, Long-tailed Meadowlark.

Access

Cinnamon Warbling-Finch occurs in the chaco south of the Córdoba–Villa Dolores road opposite the Icho Cruz police station. Walk north through Icho Cruz town and beyond the camp-site to the chaco alongside the Río San Antonio for 3 km. Spot-winged Falconet and Spot-backed Puffbird occur here.

ICHO CRUZ

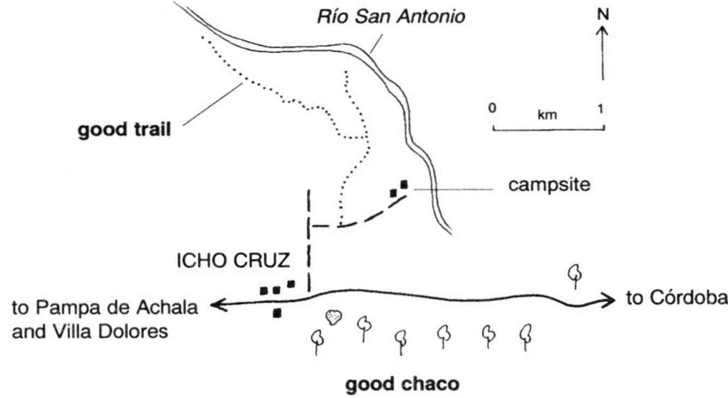

CHANCANI (PARQUE Y RESERVA NATURAL)

Map p. 45

This excellent reserve, some 200 km southwest of Córdoba near Villa Dolores, supports many of the speciality species which also occur at Icho Cruz, as well as Black-legged Seriema and Black-crested Finch.

Specialities

Brushland Tinamou, Spot-winged Falconet, Black-legged Seriema, Blue-tufted Starthroat, White-fronted, Golden-breasted and Cream-backed Woodpeckers, Scimitar-billed Woodcreeper, Chaco Earthcreeper, Crested Hornero, Little Thornbird, Lark-like Brushrunner, Brown Cachalote, Crested Gallito, Greater Wagtail-Tyrant, Black-crowned Monjita, Cinereous Tyrant, Many-coloured Chaco-Finch, Black-crested Finch.

Others

Blue-crowned Parakeet, Dark-billed Cuckoo, Tropical Screech-Owl, Narrow-billed Woodcreeper, Suiriri Flycatcher, Tawny-crowned Pygmy-Tyrant.

Access

Head north out of Villa Dolores (187 km southwest of Córdoba) to the town of Chancani, looking out for Spot-winged Falconet and Black-legged Seriema *en route*. Turn east 4 km north of Chancani to the reserve HQ. The woodland around here and the nearby warden's house supports Blue-tufted Starthroat and Cream-backed Woodpecker. Take the track east, then north, from the warden's house, for the best birding though. The track east from the sign 'Picada Central' along here is good for Black-legged Seriema.

MENDOZA

This pleasant city lies at the foot of the Andes, 1,060 km west of Buenos Aires. It is near two good sites, one of which supports the localised endemic Steinbach's Canastero, and another where the rare Creamy-rumped Miner occurs.

POTRERILLOS

This charming resort, 58 km north of Mendoza, on the road to Puente del Inca and Uspallata, is a good site for the endemic Steinbach's Canastero.

Endemics

Steinbach's Canastero.

Specialities

Red-tailed Comet, White-sided Hillstar, Brown-capped Tit-Spinetail, Subtropical Doradito, Rufous-sided Warbling-Finch.

Others

Andean Swift, Black-billed Shrike-Tyrant, Spot-billed Ground-Tyrant, White-winged Black-Tyrant, Double-collared and Band-tailed Seedeaters.

Access

En route from Mendoza turn left just after the river, just before the police station, towards Chacritas. Steinbach's Canastero occurs on the escarpment to the right of this road, 2.5 km from the Mendoza–Uspallata road.

LAGUNA HORCONES

Situated within the confines of Aconcagua NP, the surrounds of this lake, 166 km west of Mendoza, are a good site for Creamy-rumped Miner, although the area was threatened by hotel developments in 1990.

Specialities

Creamy-rumped Miner, Grey-flanked Cinclodes, Cinereous, White-browed and Black-fronted Ground-Tyrants, Greater Yellow-Finch.

Others

Crested Duck, Andean Condor, Mountain Caracara, Grey-breasted Seedsnipe, Rufous-banded Miner, Scale-throated Earthcreeper, Cordilleran Canastero, Grey-hooded, Mourning and Ash-breasted Sierra-Finches.

Access

The lake is 6 km northwest of Puente del Inca (2,720 m/8,924 ft). Take the track north from the main road a few km west of Puente del Inca, after crossing the bridge over the Río Horcones. This track leads to Cerro Aconcagua, the highest mountain in South America (6,960 m/22,835 ft). Creamy-rumped Miner occurs in the flat area to the west of the track, 1 km north of the lake, a few km from the main road.

SALINAS GRANDES **Map p. 45**

The roads north from Córdoba to Frias and Santiago del Estero pass through a large inland saltmarsh known as Salinas Grandes and the Salinas de Ambargasta, the type locality for the endemic Salinas Monjita*. There is also some good chaco around Frias where a number of specialities occur.

Endemics

Salinas Monjita*.

Specialities

Spotted Nothura, Elegant Crested-Tinamou, Spot-winged Falconet, Black-legged Seriema, Spot-backed Puffbird, Chaco Earthcreeper, Crested Gallito.

Others

Dark-billed and Striped Cuckoos, Little and Scissor-tailed Nightjars, Narrow-billed Woodcreeper, Black-capped Warbling-Finch.

Access

Salinas Monjita* occurs at Salinas de Ambargasta, 35 km northwest of Ojo de Agua. The area around Puente Saladillo is reportedly good for this species though it is rare and difficult to find. There is a maze of tracks worth exploring around Frias.

TUCUMÁN–SALTA AREA

Map below

There are many excellent birding sites between Tucumán, the largest city in north Argentina, and Salta, a fine colonial city in the Andean foothills (1,200 m/3,937 ft), as well as north and east of Salta. Both cities are connected by air to most cities in Argentina.

Accommodation: Salta: Hotel Petit (B)

TUCUMÁN–SALTA AREA

TAFFI DEL VALLE

Map below and p. 52

This is a good place from which to bird some excellent subtropical 'yungas' forest and the high slopes of Aconquija, 97 km west of Tucumán. Four endemics occur here including the rare and localised Tucuman Mountain-Finch*.

Endemics

Bare-eyed Ground-Dove, Cordoba Canastero, Yellow-striped Brush-Finch, Tucuman Mountain-Finch*.

Specialities

White-faced Dove, Grey-hooded Parakeet, Tucuman Parrot*, Blue-capped Puffleg, Red-tailed Comet, Dot-fronted Woodpecker, Buff-breasted Earthcreeper, Brown-capped Tit-Spinetail, Puna and Scribble-tailed Canasteros, White-browed Tapaculo, Subtropical Doradito, Grey-bellied and Lesser (April-Sept) Shrike-Tyrants, Plush-crested Jay, Rufous-throated Dipper*, Short-billed and Hellmayr's Pipits, Brown-capped Redstart, Rust-and-yellow Tanager, Rusty-browed and Black-and-chestnut Warbling-Finches.

Others

Andean Tinamou, Torrent Duck, Fasciated Tiger-Heron, Andean Condor, Long-winged Harrier, Mountain Caracara, Andean Lapwing, Black-winged Ground-Dove, Mitred and Mountain Parakeets, Andean Swift, Fork-tailed Woodnymph, White-bellied Hummingbird, Andean Hillstar, Ocellated Piculet, Andean Flicker, Slender-billed Miner, White-winged Cinclodes, Creamy-breasted Canastero, Buff-browed Foliage-gleaner, Mottle-cheeked Tyrannulet, D'Orbigny's and White-browed Chat-Tyrants, Black-billed Shrike-Tyrant, Andean Slaty-Thrush (Oct–Dec), Paramo Pipit, Fawn-breasted Tanager, Purple-throated Euphonia, Plumbeous and Band-tailed Sierra-Finches, Greenish Yellow-Finch, Black-backed Grosbeak.

TAFFI DEL VALLE (EL INFERNILLO)

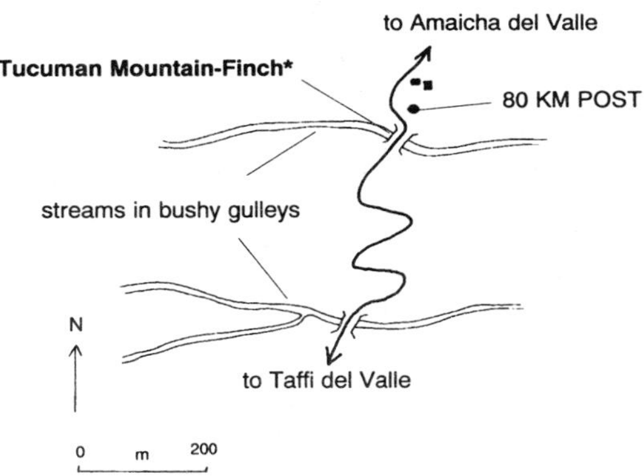

Access

Head south out of Tucumán on the road to Concepción. Turn west after 46 km on to Route 307. Rufous-throated Dipper* occurs on the **Río los Sosas**, 28 km along this road (23 km before and below Taffi). Here there is a layby and picnic area at a bridge over a tributary of the Río los Sosas. It is possible to reach the main river via a trail opposite the southern entrance to the layby. White-browed Tapaculo occurs along the tributary. White-bellied Hummingbird, Red-tailed Comet and Yellow-striped Brush-Finch also occur along this road.

Subtropical Doradito occurs in the overgrown fields to the west of the road 2 km below Taffi. Grey-hooded Parakeet occurs in the town plaza.

Bare-eyed Ground-Dove, Brown-capped Tit-Spinetail, Puna Canastero, White-browed Tapaculo and Tucuman Mountain-Finch* occur in the vegetated gullies at **El Infernillo** (Map p. 53) (2,850 m/9,350 ft), the highest part of the road between Taffi and Amaicha del Valle, 20 km northwest of Taffi. The gully at km post 80 is especially good. Bare-eyed Ground-Dove also occurs here.

Accommodation: Taffi: Hotel Colonial (C).

CAFAYATE–SALTA ROAD

Map p. 52

This road, which runs between Tucumán and Salta, passes through arid valleys, between Cafayate and Cachi, which support the endemic Sandy Gallito as well as the scarce Burrowing Parrot.

Endemics

Sandy Gallito.

Specialities

Dusky-legged Guan, Burrowing and Tucuman* Parrots, Red-tailed Comet, Slender-tailed Woodstar, White-fronted Woodpecker, Rock and Buff-breasted Earthcreepers, Brown-capped Tit-Spinetail, Short-billed Canastero, Spot-breasted Thornbird, Slaty Elaenia, White-tipped Plantcutter, Brown-capped Redstart, Sayaca Tanager, Rufous-sided Warbling-Finch, Rufous-bellied Saltator*.

Others

Ornate and Andean Tinamous, Andean Condor, Black-chested Buzzard-Eagle, Tawny-throated Dotterel, Mitred Parakeet, Scaly-headed Parrot, Andean Hillstar, Giant Hummingbird, Rufous-banded Miner, Straight-billed Earthcreeper, Plain-mantled Tit-Spinetail, Stripe-crowned Spinetail, Cordilleran Canastero, Yellow-billed Tit-Tyrant, White-winged Black-Tyrant, Spotted Nightingale-Thrush, Black Siskin, Common Diuca-Finch, Rusty Flower-piercer, Black-backed Grosbeak, Golden-billed Saltator.

Access

From Tucumán continue north from Taffi del Valle to Cafayate, Cachi and Salta. From Salta head south on route 68, then turn west on to route 33, 60 km south of Salta at El Carril, and head towards Cachi. Bird at various places along this road, which is not paved beyond Chicoana. The

Río Chicoana is a good place to see Dusky-legged Guan whilst, higher up, well-vegetated gulleys support Rufous-bellied Saltator* and flowering tobacco trees attract the brilliant Red-tailed Comet. Most of the furnariids occur on the cacti-clad slopes beyond and below the Cumbres del Obispo (Bishop's Heights) (3,500 m/11,483 ft). Sandy Gallito and Burrowing Parrot occur in the cacti-clad arid valleys between Cachi and Cafayate.

Red-tailed Comet, Slender-tailed Woodstar and Saffron-billed Sparrow occur at **Quebrada San Lorenzo**, a good site just 18 km northwest of Salta (1,450 m/4,757 ft). Bird the trail which starts opposite the Restaurant Quebrada and runs along the Río San Lorenzo.

PALOMITAS

Map p. 52

The nearest chaco to Salta lies east of Palomitas. The much degraded scrubby woodland, full of cacti, supports some chaco specialities including Black-legged Seriema, but this site is not as rich as Joaquín V Gonzáles, much further east (see next site).

Specialities

Brushland Tinamou, Chaco Chachalaca, Black-legged Seriema, Spot-backed Puffbird, White-fronted, Checkered, Golden-breasted and Cream-backed Woodpeckers, Great Rufous Woodcreeper, Little Thornbird, Greater Wagtail-Tyrant, Black-crowned Monjita (April-Sept), Stripe-capped Sparrow, Many-coloured Chaco-Finch.

Others

Red-legged Seriema, Sooty-fronted, Pale-breasted and Stripe-crowned Spinetails, Common Thornbird, White-tipped Plantcutter, Pearly-vented Tody-Tyrant, Suiriri Flycatcher, Tawny-crowned Pygmy-Tyrant, White-winged Black-Tyrant, Black-capped Warbling-Finch, Blue-black Grassquit, Golden-billed Saltator, Ultramarine Grosbeak.

Access

To reach Palomitas from Salta head east 43 km to Route 34, then turn south for some 25 km. Turn east here on to a 2-km track, just north of the 'Montenegro' sign, to the village of Palomitas. Continue east beyond Palomitas along the main track and explore this and other side tracks. The area 14 km east of Palomitas is particularly good. Black-legged Seriema occurs here, alongside Red-legged. Birding is best at dawn; the thermometer shoots up when the sun appears whilst avian activity shoots down.

JOAQUÍN V GONZÁLEZ

Map p. 52

This small town is the ideal base from which to look for chaco specialities in the hot, dusty, degraded scrubby woodland, otherwise known as 'El Impenetrable'. There are many goodies here, not least Crowned Eagle*.

Specialities

Tataupa and Brushland Tinamous, Spotted Nothura, Quebracho Crested-Tinamou, Crowned Eagle*, Spot-winged Falconet, Chaco Chachalaca, Blue-crowned Parakeet, Ash-coloured Cuckoo, Ashy-tailed Swift, Blue-tufted Starthroat, Spot-backed Puffbird, Checkered and Cream-backed Woodpeckers, Scimitar-billed and Great Rufous Woodcreepers, Crested Hornero, Short-billed Canastero, Little Thornbird, Lark-like Brushrunner, Brown Cachalote, Stripe-backed Antbird, Crested Gallito, Small-billed Elaenia, Greater Wagtail-Tyrant, Lesser Shrike-Tyrant, Cinereous Tyrant, Plush-crested Jay, White-banded Mockingbird, Many-coloured Chaco-Finch, Black-crested Finch, Ringed Warbling-Finch.

Others

Blue-fronted Parrot, Nacunda Nighthawk, Little Nightjar, Stripe-crowned Spinetail, Common Thornbird, Great and Variable Antshrikes, Pearly-vented Tody-Tyrant, Suiriri Flycatcher, Large Elaenia, Tawny-crowned Pygmy-Tyrant, White Monjita, Crowned Slaty Flycatcher, Creamy-bellied Thrush, Masked Gnatcatcher, Black-capped Warbling-Finch, Golden-billed Saltator.

Access

To reach **Joaquín** continue south past the Palomitas turning for approximately 50 km, then turn east on to Route 16. The town of Joaquín is approximately 100 km further on. Just south of town turn northeast on to Route 41 to Santo Domingo. Bird the tracks to the east after 14 km and 17 km. The cool dawn is the best birding time.

Beyond Joaquín, *en route* to Roque Saenz Peña is the best area for Quebracho Crested-Tinamou and Spot-winged Falconet.

Accommodation: Hotel Majorca (C).

EL REY NP

Map p. 52

Situated at the foot of the Andes on the edge of the chaco, 190 km east of Salta, El Rey NP supports a wide variety of habitats where the endemic Yellow-striped Brush-Finch occurs. Most of the species present here, and more, also occur in Calilegua NP (see next site), which is easier to get to.

Endemics

Yellow-striped Brush-Finch.

Specialities

Tataupa Tinamou, Spot-winged Falconet, Chaco Chachalaca, Dusky-legged Guan, Rufous-sided Crake, White-faced Dove, Green-cheeked Parakeet, Tucuman Parrot, Hoy's Screech-Owl, Ashy-tailed Swift, Red-tailed Comet, Dot-fronted and Cream-backed Woodpeckers, Giant and Rufous-capped Antshrikes, Slaty Elaenia, Andean Tyrant, Plush-crested Jay, Brown-capped Redstart, Rusty-browed Warbling-Finch, Lined Seedeater, Golden-winged Cacique.

Others

Red-winged Tinamou, Southern Screamer, Maguari Stork, Andean

Condor, King Vulture, Bicoloured Hawk, Red-legged Seriema, Mitred Parakeet, Striped Cuckoo, Blue-crowned Trogon, Toco Toucan, White-barred Piculet, Golden-olive Woodpecker, Sooty-fronted Spinetail, Streaked Xenops, Giant Antshrike, Pearly-vented Tody-Tyrant, Sooty and Mottle-cheeked Tyrannulets, Dusky-capped and Swainson's Flycatchers, Spotted Nightingale-Thrush, Mountain Wren, Two-banded and Pale-legged Warblers, Saffron-billed Sparrow, Stripe-headed Brush-Finch, Orange-headed Tanager, Black-backed Grosbeak.

Access

From Salta head for Palomitas (p. 55) and continue south along Route 34 to the junction with Route 5. Turn east on to Route 5 in the direction of Las Lajitas, then turn north just before Las Viboras. The last 45 km is a rough track which is susceptible to flooding (April in particular). Rufous-sided Crake occurs in the marsh opposite the old airstrip before reaching HQ. King Vulture, Chaco Chachalaca and Tucuman Parrot occur around the hotel and HQ clearing. Rufous-capped Antshrike, Andean Tyrant, Spotted Nightingale-Thrush and Yellow-striped Brush-Finch occur on the Sendero Pozo Verde trail which starts at HQ.

A new species, Hoy's Screech-Owl, has been recorded from El Rey NP.

CALILEGUA NP

Map p. 52

This large (76,320 ha) NP near Jujuy encompasses a wide range of habitats, from savanna through subtropical 'yungas' and temperate forests to Alder woodland. The avifauna is similar to that found in El Rey NP, except for the park's major speciality, the rare Red-faced Guan*.

Endemics

Yellow-striped Brush-Finch.

Specialities

Tataupa Tinamou, Red-faced* and Dusky-legged Guans, White-faced Dove, Golden-collared Macaw, Green-cheeked Parakeet, Tucuman Parrot*, Ashy-tailed Swift, Blue-capped Puffleg, Red-tailed Comet, Slender-tailed Woodstar, Dot-fronted and Cream-backed Woodpeckers, Ochre-cheeked Spinetail, Giant Antshrike, Black-capped Antwren, White-throated Antpitta, Slaty Elaenia, Buff-banded and White-bellied Tyrannulets, Plush-crested Jay, Brown-capped Redstart, Fulvous-headed Brush-Finch, Rusty-browed Warbling-Finch, Golden-winged Cacique.

Others

Brazilian Teal, Black-and-white Hawk-Eagle, Black-and-chestnut Eagle, Barred Forest-Falcon, White-throated Quail-Dove, Striped Cuckoo, Planalto Hermit, Speckled Hummingbird, Blue-crowned Trogon, Toco Toucan, White-barred Piculet, Golden-olive Woodpecker, Buff-browed Foliage-gleaner, Sepia-capped Flycatcher, Ochre-faced Tody-Flycatcher, White-throated, Sooty and Mottle-cheeked Tyrannulets, Cliff and Euler's Flycatchers, Rufous Casiornis, Swainson's Flycatcher, Green-backed and White-winged Becards, Spotted Nightingale-Thrush, Andean Slaty-Thrush, Mountain Wren, Southern Martin, Two-banded and Pale-legged Warblers, Saffron-billed Sparrow, Stripe-headed Brush-

Finch, Orange-headed Tanager, Black-backed Grosbeak, Golden-billed Saltator, Crested Oropendola.

Access

To reach the park from Salta head east out of the city to Route 34 (43 km), then turn north and remain on Route 34 for approximately 125 km to Libertador General San Martín. Just north of the town, beyond the river, turn west on the track to Valle Grande. The park entrance and first ranger station is 10 km along here. White-throated Antpitta occurs at the second ford (Río Tres Cruces). The best place for Red-faced Guan* is between the first obvious stream crossing, 7 km above the second ranger station, and the obelisk, 1 km above the 'Río Jordan' sign, just before the park boundary at 1,700 m (5,577 ft). This area is a solid three-hour trek from the second ranger station, should the car refuse to be coaxed across the Río Tres Cruces ford!

SALTA–HUMAHUACA ROAD

Map p. 52

North of Salta, the road to Bolivia passes through good subtropical 'yungas' forest between Salta and Jujuy, and near Jujuy, it crosses the Río Yala, one of the few accessible rivers on which the rare Rufous-throated Dipper* occurs.

Endemics

Bare-eyed Ground-Dove.

Specialities

Red-faced Guan*, Grey-hooded Parakeet, Rothschild's Swift*, Red-tailed Comet, Slender-tailed Woodstar, Cream-backed Woodpecker, Spot-breasted Thornbird, White-tipped Plantcutter, Cinereous and Andean Tyrants, Rufous-throated Dipper*, Brown-backed Mockingbird, Stripe-capped Sparrow, Fulvous-headed Brush-Finch, Rust-and-yellow Tanager, Black-hooded Sierra-Finch, Black-and-chestnut Warbling-Finch, Rufous-bellied Saltator.

Others

Red-legged Seriema, Guira Cuckoo, Andean Swift, White-bellied and Giant Hummingbirds, White-barred Piculet, Sooty-fronted Spinetail, Creamy-breasted Canastero, Mouse-coloured Tyrannulet, D'Orbigny's Chat-Tyrant, White Monjita, Dusky-capped Flycatcher, Crowned Slaty Flycatcher, Crested Becard, Black Siskin, Orange-headed Tanager, Mourning Sierra-Finch, Black-capped Warbling-Finch, Band-tailed Seedeater.

Access

Look out for Rothschild's Swift*, Mouse-coloured Tyrannulet and Rufous-bellied Saltator* along this road, especialy around Jujuy. Bird the roadside forest between Salta, La Caldera and Jujuy. There are a few good trails here. Red-legged Seriema occurs on hillsides to the east of this road. The reservoir just south of El Carmen may be worth a quick look.

Red-tailed Comet and Rufous-throated Dipper* occur in **Yala NP** (map opposite).

The rare Rufous-throated Dipper frequents the Río Yala near the Salta–Humuhuaca road*

Head north out of Jujuy on Route 9 for 13 km, then turn west on to theroad (which becomes a track after a few km) to Lagunas de Yala (signposted). Look for an obscure trail on the south side of the track 6 km along the Lagunas road, opposite the first white pumping station. This leads to the Río Yala where the dipper occurs.

Between Yala and Humahuaca the roadside fields just south of Tumbaya, and the cactus scrub to the west, 25 km north of Tumbaya, are worth a look. More cactus scrub, to the west 4 km north of

SALTA–HUMAHUACA ROAD (YALA NP)

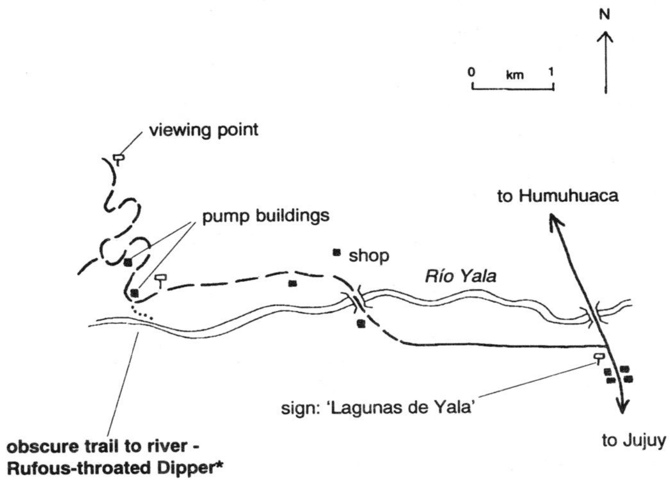

Humahuaca, supports Red-tailed Comet.

The small indian town of Humahuaca (2,939 m/9,642 ft) lies in the Andean rain shadow, where the thorn scrub holds a number of goodies including White-tipped Plantcutter and Brown-backed Mockingbird. It makes a good base for the night too.

Accommodation: Jujuy: Residencial San Carlos (C).

LA QUIACA

This small town on the Bolivian border, at 3,442 m (11,293 ft), is a good site for Citron-headed Yellow-Finch*, a species otherwise confined to Bolivia.

Specialities

Brown-capped Tit-Spinetail, Brown-backed Mockingbird, Black-hooded Sierra-Finch, Puna and Citron-headed* Yellow-Finches.

Others

Tawny-throated Dotterel, Scale-throated Earthcreeper, Plain-mantled Tit-Spinetail, White-tipped Plantcutter, Mourning and Ash-breasted Sierra-Finches, Common Diuca-Finch, Bright-rumped Yellow-Finch, Bay-winged Cowbird.

Access

Citron-headed Yellow-Finch* occurs at the compound behind the market and beyond the water pumping station.

Accommodation: Hotel Frontera (C).

LAGO POZUELOS

Map opposite

This large (10,000 ha), high-altitude (3,670 m/12,041 ft) Andean lake, and the nearby Laguna Larga, is one of the few sites in South America where the rare Horned Coot* occurs. August is the best time for this species, when at least 1,000 have been present on Laguna Larga. In the right conditions it is also possible to see the two rare Andean flamingos on Lago Pozuelos.

Specialities

Andean* and Puna* Flamingos, Giant and Horned* Coots, Andean Avocet, Puna Plover, Puna Miner, Rock Earthcreeper, Cinnamon-bellied Ground-Tyrant (April–Sept), Short-billed Pipit, Black-hooded and Red-backed Sierra-Finches.

Others

Ornate and Puna Tinamous, Lesser Rhea*, Silvery Grebe, Andean Goose, Crested Duck, Chilean Flamingo, Andean Condor, Long-winged Harrier, Mountain Caracara, Wilson's Phalarope, Grey-breasted Seedsnipe, Tawny-throated Dotterel, Andean Lapwing, Andean Gull, Golden-spotted Ground-Dove, Andean Swift, Andean Flicker, Common, Rufous-banded and Slender-billed Miners, Straight-billed Earthcreeper,

Plain-mantled Tit-Spinetail, Cordilleran and Creamy-breasted, Streak-fronted Thornbird, D'Orbigny's Chat-Tyrant, Black-billed Shrike-Tyrant, Plain-capped Ground-Tyrant, Andean Negrito, Black Siskin, Mourning and Ash-breasted Sierra-Finches, Puna, Bright-rumped and Greenish Yellow-Finches.

Other Wildlife

Vicuña.

LAGO POZUELOS

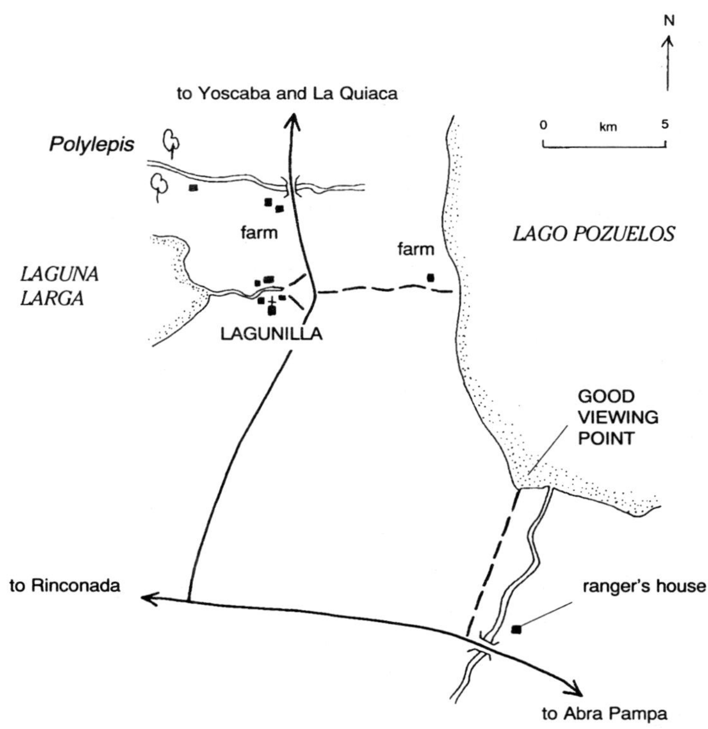

Access

Head west from Abra Pampa, approximately halfway between Humahuaca and La Quiaca, on the road to Rinconada. Bare-eyed Ground-Dove, Streak-fronted Thornbird, D'Orbigny's Chat-Tyrant and Red-backed Sierra-Finch (in ravines) occur alongside this road.

To view the southern end of Lago Pozuelos turn north on to a 6 km track, just past the ranger's house and bridge, to the shore. Lesser Rhea occurs here. Continuing west turn north to Lagunilla on the road to Yoscaba at the next junction. There is a 5-km track east from here to the western shore of Lago Pozuelos. Laguna Larga lies 3 km to the west of Lagunilla and is often a better site for Giant and Horned* Coots than Lago Pozuelos. Shorebirds sometimes occur in the area in vast numbers, and up to 10,000 Wilson's Phalaropes have been recorded.

Rock Earthcreeper and D'Orbigny's Chat-Tyrant occur in the *Polylepis* woodland a few km northwest of Lagunilla.

Take food, warm clothes and a good sleeping bag to Lagunilla, and beware of soft mud when approaching the lakes.

Accommodation: Abra Pampa: Residencial (C).

IGUAZÚ NP

Map opposite

Situated on the border with Brazil in extreme northeast Argentina, Iguazú falls, the biggest in the world, are surrounded by subtropical forest with large stands of bamboo, some of which lies in Iguazú NP (492 km^2). The Argentine side of the falls is better than the Brazilian side, although Helmeted Woodpecker* is recorded more regularly in Brazil. Four days here should produce over 100 species.

Specialities

Solitary Tinamou*, Grey-bellied Goshawk*, Rusty-margined and Dusky-legged Guans, Black-fronted Piping-Guan*, Rufous-sided Crake, Slaty-breasted Wood-Rail, Ash-throated Crake, Maroon-bellied Parakeet, Pileated Parrot*, Pearly-breasted Cuckoo, Rusty-barred and Buff-fronted* Owls, Sickle-winged Nightjar*, Great Dusky Swift, Rufous-capped Motmot, Rusty-breasted Nunlet, Saffron* and Spot-billed Toucanets, Red-breasted Toucan, Yellow-fronted, White-spotted, Helmeted* and Robust Woodpeckers, Black-capped Foliage-gleaner, Tufted and Large-tailed Antshrikes, Bertoni's Antbird, Streak-capped Antwren, White-shouldered Fire-eye, Spotted Bamboowren*, Blue Manakin, Grey-hooded Flycatcher, Drab-breasted Bamboo-Tyrant, Southern Antpipit, Southern Bristle-Tyrant*, Sao Paulo* and Bay-ringed* Tyrannulets, Eared Pygmy-Tyrant, Russet-winged Spadebill*, Greenish Schiffornis, Plush-crested Jay, Creamy-bellied Gnatcatcher*, Chestnut-headed, Ruby-crowned and Green-headed Tanagers, Blackish-blue Seedeater*, Green-winged Saltator.

Others

Brown Tinamou, Rufous-thighed Kite, Blackish Rail, Scaled and Pale-vented Pigeons, White-eyed Parakeet, Blue-winged Parrotlet, Canary-winged Parakeet, Grey Potoo, Grey-rumped Swift, Gilded Sapphire, Black-throated and Surucua Trogons, Chestnut-eared Aracari, Toco Toucan, Green-barred and Blond-crested Woodpeckers, Scaled and Lesser Woodcreepers, Ochre-breasted and White-eyed Foliage-gleaners, Plain Xenops, Spot-backed Antshrike, Rufous-winged Antwren, Short-tailed Antthrush, Red-ruffed Fruitcrow, Sharpbill, Wing-barred Manakin, White-throated Spadebill, Long-tailed Tyrant, Sirystes, Three-striped Flycatcher, Black-crowned Tityra, Rufous-crowned Greenlet, Rufous bellied and Creamy-bellied Thrushes, Neotropical River Warbler,

IGUAZÚ NP

Chestnut-vented Conebill, Magpie, Guira and Black-goggled Tanagers, Red-crowned Ant-Tanager, Purple-throated Euphonia, Blue-naped Chlorophonia, Swallow-Tanager, Lesser Seed-Finch.

Access

Bird the extensive network of boardwalks, trails and tracks around the falls and in Iguazú NP. Solitary Tinamou*, Rusty-breasted Nunlet, Short-tailed Antthrush, Drab-breasted Bamboo-Tyrant, Southern Antpipit and

Toco Toucan is one of many excellent birds occurring around the world's biggest waterfall at Iguazú on the Argentina/Brazil border

Southern Bristle-Tyrant* occur along the 4-km long Sendero Macuco trail, which heads north from the park office just west of the Hotel Internacional (Toco Toucan occurs in the grounds of this hotel). The junction of this trail with the Sendero Yacaratia trail is a good site for Robust Woodpecker, Bay-ringed Tyrannulet* and Blackish-blue Seedeater* (in the bamboo). The marsh south of the park office, with hide, holds rails and crakes. Black-fronted Piping-Guan* occurs along the 6-km trail leading east from the Puerto Canoas car park at the start of the waterfall boardwalk. Red-ruffed Fruitcrow, Spot-backed Antshrike and White-shouldered Fire-eye occur along the trail leading south from this car park to the El Timbo ranger station, which is a good site for Rusty-barred Owl.

A number of species, such as Helmeted Woodpecker*, Sharpbill and Southern Bristle-Tyrant* are more likely to be seen on the Brazilian side of the falls (p. 112). The best birding areas are the entrance road to Iguaçu NP and, especially, the Poco Preto trail, which starts behind the hospedaje just past km post 13, 2.5 km from the Iguaçu NP entrance. Permission is needed from the nearby Policia Florestal to walk this trail.

The very rare Brazilian Merganser* and Black-fronted Piping-Guan* have been recorded at **Arroyo Uruguai**, a Fundacion Vida Silvestre reserve, not far from the falls.

Accommodation: Puerto Iguazú (23 km northwest of falls): Residencial Lilian (B). Iguazú Falls: Hotel Internacional (A).

CORRIENTES/MISIONES

The vast wet grasslands of Corrientes state, known as the Iberá marshes, in north Argentina, support a number of endangered species, whilst the subtropical forests of Misiones state, near the Brazilian border, including those around Iguazú falls, hold a number of birds not found elsewhere in Argentina. Many of these species have ranges centred on south Brazil where they are somewhat easier to find.

SAN MARTÍN

Map below

The remnant subtropical forest at this site supports a number of specialities including the elegant hummingbird known as Plovercrest.

Specialities

Maroon-bellied Parakeet, Plovercrest, White-spotted, Yellow-browed* and Robust Woodpeckers, Planalto and Scaled Woodcreepers, Rufous-capped, Chicli, Grey-bellied and Olive Spinetails, Sharp-billed Treehunter, Large-tailed Antshrike, Rufous Gnateater, Blue Manakin, Planalto Tyrannulet, Southern Bristle-Tyrant*, Eared Pygmy-Tyrant, Azure Jay*, Chestnut-headed, Ruby-crowned, Diademed and Chestnut-backed Tanagers, Red-rumped Warbling-Finch, Glaucous-blue Grosbeak.

Others

Surucua Trogon, Yellow Tyrannulet, Small-billed Elaenia, Black-goggled Tanager, Ultramarine Grosbeak, Chopi Blackbird.

Access

This site is situated 10 km southwest of the village of San Martín on the Posadas–Oberá road (route 14), where the road crosses the Río Arroyo Martinez Grande. Bird the forest alongside this river either side of the road, especially to the north.

Remnant *Araucaria* forest and rivers around **Montecarlo**, further

SAN MARTÍN

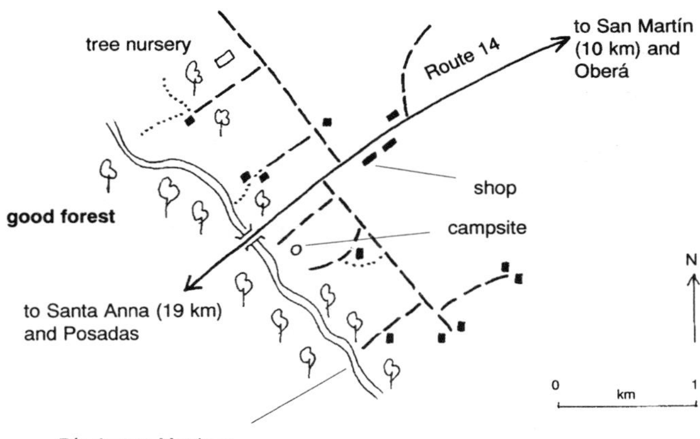

north on the Obera–Bernardo de Irigoyen road, support, Red-spectacled* and Vinaceous* Parrots, Plovercrest, Araucaria Tit-Spinetail and Azure Jay*.

IBERÁ MARSHES

Argentina's biggest wetland, covering 20,000 km^2, from the border with Paraguay south, contains rivers, reedbeds, lagoons, *Yatay* palm groves, dry scrub and gallery forest. Approximately 300 species occur here, including a superb assortment of rare and scarce tyrants and seedeaters, as well as Crowned Eagle*, Ochre-breasted Pipit*, Yellow Cardinal* and Saffron-cowled Blackbird*.

Specialities

Spotted Nothura, Crowned Eagle*, Giant Wood-Rail, Nanday Parakeet, Bay-capped Wren-Spinetail*, Rufous-capped Antshrike, Sooty Tyrannulet, Cock-tailed*, Strange-tailed* and Streamer-tailed Tyrants, Black-and-white Monjita*, Ochre-breasted Pipit*, White-rimmed Warbler, Yellow* and Yellow-billed Cardinals, Lesser Grass-Finch, Rusty-collared, Capped, Dark-throated*, Marsh*, Grey-and-chestnut* and Chestnut* Seedeaters, Saffron-cowled* and Unicoloured Blackbirds, Yellow-rumped Marshbird, Scarlet-headed Blackbird.

Others

Greater Rhea*, Southern Screamer, Plumbeous Ibis, Maguari Stork, Jabiru, Lesser Yellow-headed Vulture, Long-winged and Cinereous Harriers, Great Black-Hawk, Wattled Jacana, Collared Plover, Large-billed Tern, White Woodpecker, Narrow-billed Woodcreeper, Chotoy Spinetail, Greater Thornbird, Firewood-gatherer, Grey and White Monjitas, White-headed Marsh-Tyrant, Black-capped Donacobius, Yellowish Pipit, Long-tailed Reed-Finch, Double-collared and Ruddy-breasted Seedeaters, Epaulet Oriole, Chestnut-capped and White-browed Blackbirds, Bay-winged Cowbird.

Other Wildlife

Capybara, Howler Monkey, Swamp and Pampa Deer, Maned Wolf.

Access

Laguna Iberá, west of the rough road between Posadas and Mercedes, is a good place to see some the area's specialities. The 123-km stretch of road between Route 14 and the town of Colonia C. Pellegrini is susceptible to flooding but one of the few sites in South America where Strange-tailed Tyrant* occurs. Scour the roadside fences for this species. Saffron-cowled Blackbird* also occurs along this road, usually near the junction with route 14, and Black-and-white Monjita* has been seen 8 km east of Pellegrini.

Colonia C. Pellegrini is best reached from Mercedes to the south, on a much better road. At Colonia C. Pellegrini, Scarlet-headed Blackbird occurs in the marsh adjacent to the Centro Turistico on the north side of the road west of the Bayley bridge. Giant Wood-Rail also occurs around the Centro Turistico. Strange-tailed Tyrant*, Chotoy Spinetail and Lesser Grass-Finch all occur opposite here on the south side of the road. White Woodpecker occurs in the woodland at the end of a track to the north of the road, just to the east of the bridge.

Accommodation: Centro Turistico (basic, free).

Ituzaingo, west of Posadas, is another good base from which to explore the fringes of this largely inaccessible region. Black-and-white Monjita*, Strange-tailed Tyrant*, Lined, Capped, Marsh*, Grey-and-chestnut* and Chestnut* Seedeaters, and Saffron-cowled Blackbird* occur around here and alongside the road west to Resistencia.

PUERTO MADRYN AND TRELEW AREA **Map p. 68**

Either of the pleasant towns of Puerto Madryn and Trelew would make a good base from which to bird the Valdés peninsula to the north and Punta Tombo to the south. Six endemics occur in this area as well as the famous colonies of sea mammals and penguins.

Endemics

Chubut Steamerduck*, Band-tailed Earthcreeper, Patagonian Canastero, White-throated Cachalote, Rusty-backed Monjita, Carbonated Sierra-Finch.

Specialities

Elegant Crested-Tinamou, Magellanic Penguin, Snowy Sheathbill, Burrowing Parrot, Greater Wagtail-Tyrant, Black-crowned Monjita, Grey-bellied and Lesser Shrike-Tyrants, Patagonian Sierra-Finch, Patagonian Yellow-Finch.

Others

Darwin's Nothura, Lesser Rhea*, Guanay Cormorant, Imperial and Rock Shags, Coscoroba Swan, White-cheeked Pintail, Rosy-billed Pochard, Red-backed Hawk, Hudsonian Godwit*, White-rumped Sandpiper, Blackish Oystercatcher, Two-banded Plover, Tawny-throated Dotterel,

The Trelew area is a good site for Two-banded Plover

Dolphin Gull, South American Tern, Chilean Skua, Scale-throated Earthcreeper, Plain-mantled Tit-Spinetail, Lesser Canastero, Many-coloured Rush-Tyrant, Patagonian Mockingbird, Common Diuca-Finch, Grey-headed and Mourning Sierra-Finches, Long-tailed Meadowlark.

Other Wildlife

Southern Right and Killer Whales, Southern Fur and Elephant Seals, Southern Sea Lion, Guanaco, Mara, Hairy Armadillo.

Access

The **Valdés peninsula** is a 100-km long headland which supports colonies of breeding seabirds and sea mammals, as well as three Argentine avian endemics. It was here that the BBC filmed the Killer Whales skimmimg on to the beach to attack seals. Band-tailed Earthcreeper and Patagonian Canastero occur just north of the junction between the main peninsula road and the road to Puerto Pirámides. Snowy Sheathbill occurs at Puerto Pirámides Observatory, 5 km from town, and at Punta Norte, attending the mammal colonies. Punta Cantor, to the south, supports Southern Elephant Seal, Killer Whale and Chubut Steamerduck*. The bays at the base of the peninsula are good

PUERTO MADRYN AND TRELEW AREA

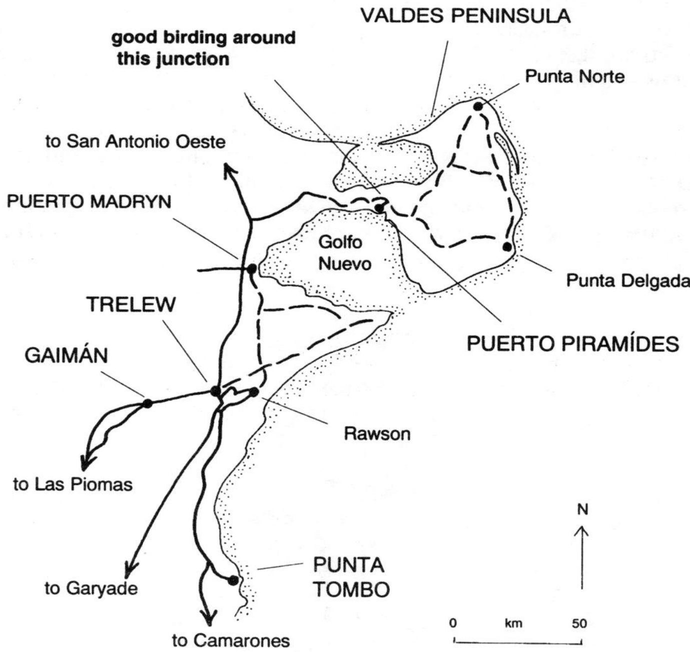

for waders including Hudsonian Godwit* and up to 10,000 White-rumped Sandpipers.

Punta Tombo, 130 km south of Trelew, is a 3-km long peninsula which supports a million Magellanic Penguins. This huge concentration occasionally attracts unusual penguins such as Macaroni. It is also the best site for the endemic Chubut Steamerduck*, and Lesser Rhea*, Guanay Cormorant, Imperial and Rock Shags, Dolphin Gull and South America Tern also occur here. Nearby Isla Escondida also holds breeding seabirds. Band-tailed Earthcreeper, Patagonian Canastero and White-throated Cachalote all occur around **Dos Pozos**, 95 km south of Trelew *en route* to Punta Tombo. The area around two derelict buildings is the best.

Lago de Trelew behind the main bus terminal in Trelew is good for waterfowl.

Burrowing Parrot occurs in the Chubut valley near **Gaimán**, 18 km west of Trelew on Route 25 to Esquel. Head south out of Gaimán, just south of Route 25 cross the river and turn west onto Route 10, checking the roadside cliffs up to 10 km from Gaimán. Rusty-backed Monjita also occurs in this area.

Accommodation: Puerto Madryn: Peninsula Valdes Hotel (A). Valdés peninsula: Punta Delgada Lighthouse Lodge (A); camping at Puerto Pirámides (C). Trelew: Residencial Argentino (C).

SAN ANTONIO OESTE

Map p. 70

Six of the 16 Argentine endemics, and the endangered Yellow Cardinal* occur near the town of San Antonio Oeste, some 250 km north of Puerto Madryn.

Endemics

Band-tailed Earthcreeper, Patagonian Canastero, White-throated Cachalote, Sandy Gallito, Rusty-backed Monjita, Carbonated Sierra-Finch.

Specialities

Burrowing Parrot, Checkered Woodpecker, White-crested Tyrannulet, Greater Wagtail-Tyrant, Black-crowned Monjita, Grey-bellied and Lesser Shrike-Tyrants, Hudson's Black-Tyrant*, White-banded Mockingbird, Yellow Cardinal*.

Others

Darwin's Nothura, Tawny-throated Dotterel, Spot-winged Pigeon, Lesser Canastero, White-tipped Plantcutter, Yellow-billed Tit-Tyrant, White Monjita, Patagonian Mockingbird, Southern Martin, Ringed Warbling-Finch, Golden-billed Saltator.

Access

White-throated Cachalote and Sandy Gallito occur around the lagoon 10 km northwest of San Antonio Oeste on the west side of the road to Larmarque and in the scrub between the roads to General Conesa and Viedma 10 km from San Antonio Oeste. Burrowing Parrots roost at the road junction of the roads to Puerto Madryn, Viedma amd General Conesa, 8 km northwest of San Antonio Oeste.

SAN ANTONIO OESTE

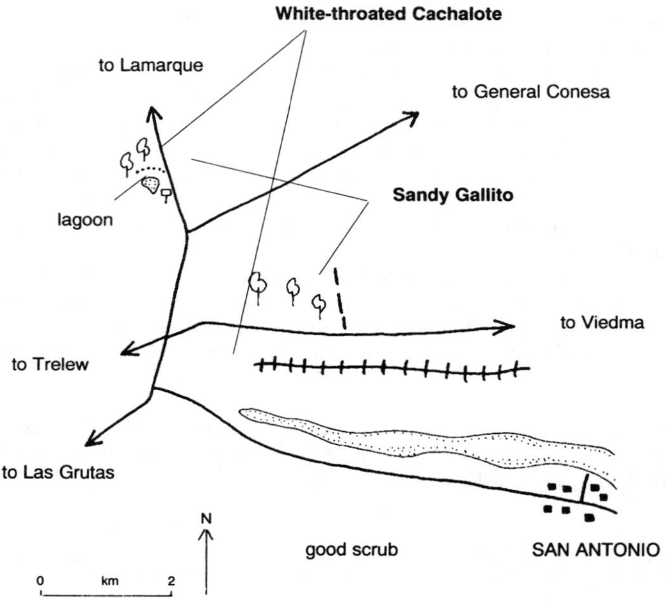

Burrowing Parrot, Patagonian Canastero, Sandy Gallito, White-banded Mockingbird, Carbonated Sierra-Finch and Cinnamon Warbling-Finch occur at **Balneario de las Grutas**, a beach resort 17 km southwest of San Antonio Oeste. Head to the end of the beach road and beyond, through the campsite, to the scrub, concentrating on birding the streambeds.

The scrub around **Laguna del Monte**, 45 km east of San Antonio Oeste, supports six endemics: Band-tailed Earthcreeper, White-throated Cachalote, Sandy Gallito, Rusty-backed Monjita, Carbonated Sierra-Finch, and Cinnamon Warbling-Finch, the rare Hudson's Black-Tyrant* and the very rare Yellow Cardinal*, as well as Checkered Woodpecker, Lesser and Short-billed Canasteros, Greater Wagtail-Tyrant, Black-crowned Monjita, Lesser Shrike-Tyrant, Patagonian and White-banded Mockingbirds, and Ringed Warbling-Finch. Turn north off the main road between San Antonio Oeste and Viedma in response to the 'Laguna del Monte' sign, written on a white tyre. The area around the windmill, the lake, and the farm, 1 to 1.5 km north of the main road, is the best.

Accommodation: San Antonio Oeste: Hotel Kandava (B).

The popular ski resort of **Bariloche**, on the southern shores of Lake Colhué-Huapí, near the Chile border, 1,600 km southwest of Buenos Aires, is surrounded by *Nothofagus* forest and Patagonian steppe. Goodies such as Torrent and Spectacled* Ducks, Andean Condor, Chilean Pigeon*, Green-backed Firecrown, Chilean Flicker, Magellanic Woodpecker, Des Murs' Wiretail, Thorn-tailed Rayadito, White-throated Treerunner, Black-throated Huet-huet, Chucao Tapaculo, Rufous-tailed Plantcutter, Patagonian Tyrant, Great Shrike-Tyrant and Austral Blackbird all occur here.

Bird the forest near Lake Colhué-Huapí and the steppe to the east, as well as three specific sites: (i) the **Cascada Los Alerces**, a waterfall which is reached by heading southwest on route 258, then west on route 254. Walk down the side road to Lake Fonck where Magellanic Woodpecker occurs; (ii) the viewpoint on Cerro Lopez 7 km from **Colonia Suiza** where Black-throated Huet-huet occurs; and (iii) the arid areas around **Estancia Perito Moreno** and beyond on **Route 23** where the rare Spectacled Duck* has been seen on the fast-flowing streams 10 km east of the estancia. Chilean Pigeon* occasionally occurs in Bariloche town.

CALAFATE

This small town on the shore of Lago Argentino is the gateway to Glacier NP. It is also one of the best bases from which to look for the rare Hooded Grebe*, which was only discovered in 1974 and was, until recently, considered to be an Argentine endemic (it has recently been reported from south Chile, see p. 175).

Specialities

Hooded Grebe*, Patagonian Tinamou, Andean and Spectacled* Ducks, Magellanic Plover*, Austral Parakeet, Magellanic Woodpecker, Des Murs' Wiretail, Austral Canastero*, Black-throated Huet-huet, Rufous-tailed Plantcutter, Chocolate-vented Tyrant, Great Shrike-Tyrant, Cinnamon-bellied Ground-Tyrant, Patagonian Yellow-Finch.

Although seemingly somewhat nomadic, persistent, well-prepared birders may see the very rare and beautiful Hooded Grebe near Calafate and/or on Strobel plateau lakes

Others

Lesser Rhea*, Silvery Grebe, Chilean Flamingo, Black-faced Ibis, Grey-breasted and Least Seedsnipes, Magellanic Oystercatcher, Rufous-chested and Tawny-throated Dotterels, Chilean Flicker, Andean Tapaculo, Fire-eyed Diucon, Austral Negrito, Grey-hooded Sierra-Finch.

Access

Hooded Grebe* was discovered at Lagunas de los Escarchados (Lagoons of the White Frost), on the south side of the road between Río Gallegos and Calafate, 70 km southeast of Calafate. Since its discovery this seemingly nomadic bird has not been present here every year. However, the lagoons are always worth checking, because (i) the other known sites for the grebe are a long way north, and (ii) Magellanic Plover* also occurs here. The roadside steppe east of here supports Chocolate-vented Tyrant.

Patagonian Tinamou occurs at Río Bote, 40 km east of Calafate. From the Hotel Rio Bote, on the south side of the Rio Gallegos–Calafate Road, take the 5-km track to the farm on the north side of the road, looking out for the tinamous crossing it.

Spectacled Duck* breeds (Sept–Jan) in the spectacular Glacier NP, 50 km west of Calafate, worth visiting just to see the world's biggest active glacier, the Perito Moreno, 5-km wide and 60 m high. Magellanic Woodpecker, Des Murs' Wiretail, Austral Canastero* and Black-throated Huet-huet also occur here.

The place to look for Hooded Grebe*, apart from near Calafate, is the **Strobel Plateau**. From Calafate head north on Route 40. This main road is 35 km east of Calafate. 113 km north of Tres Lagos turn west to **Lago Cardiel**. After 5 km, turn north on a track to an estancia and check the ponds along here. Alternatively continue west for 7 km to the lake, then turn north and explore the pools beyond the rock wall to the northwest. Shallow ponds with lots of Red Milfoil (*Myriophyllum elatinoides*) are the preferred habitat. These ponds often dry up hence the grebes are by no means guaranteed. The plateau is west of Gobernador Gregores.

Accommodation: Hotel Kaiken (A); private houses (C).

Patagonian Tinamou, Ruddy-headed Goose*, Austral Canastero* and Canary-winged Finch* all occur at Cabo Virgenes, 124 km southeast of Río Gallegos to the north of Tierra del Fuego across the Magellan Straits.

TIERRA DEL FUEGO

Map opposite

Ushuaia, the most southerly town in the world (55°S), is the best base from which to bird the windswept Tierra del Fuego NP with its steppe grasslands, wild shoreline, lakes and stunted *Nothofagus* forests, as well as a few other sites and the Beagle Channel. Many good birds restricted to the southernmost section of South America, the Falkland Islands and Antarctica, occur in this area, including King Penguin, Ruddy-headed Goose*, Magellanic Plover*, White-bellied Seedsnipe, Austral Canastero*, and Canary-winged* and Yellow-bridled Finches.

Specialities

Patagonian Tinamou, King and Gentoo Penguins, Magellanic Diving-Petrel, Upland, Kelp, Ashy-headed and Ruddy-headed* Geese, Flightless and Flying Steamerducks, Spectacled Duck*, White-throated Caracara, Snowy Sheathbill, Magellanic Plover*, White-bellied Seedsnipe, Austral Parakeet, Rufous-legged Owl, Magellanic Woodpecker, Short-billed Miner, Grey-flanked, Dark-bellied and Blackish Cinclodes, Austral Canastero*, Patagonian Tyrant, Cinnamon-bellied Ground-Tyrant, Patagonian Sierra-Finch, Canary-winged* and Yellow-bridled Finches, Patagonian Yellow-Finch.

Others

Great Grebe, Black-browed Albatross, Antarctic Giant Petrel, Southern Fulmar, Cape and White-chinned Petrels, Black-faced Ibis, Andean Condor, Bicoloured Hawk, Least Seedsnipe, Blackish and Magellanic Oystercatchers, Two-banded Plover, Rufous-chested and Tawny-throated Dotterels, Dolphin Gull, Southern Skua, Thorn-tailed Rayadito, White-throated Treerunner, Andean Tapaculo, Fire-eyed Diucon, Austral Negrito, Austral Thrush, Chilean Swallow, Black-chinned Siskin, Austral Blackbird.

Access

At **Ushuaia**, White-bellied Seedsnipe and Yellow-bridled Finch occur on the scree slopes around the Martial glacier, reached via the chairlift or a 3-km walk, 7 km behind the town. Yellow-bridled Finch also occurs in the fields south of Route 3, a few km west of town, west of the military base. Both the seedsnipe and the finch also occur on the scree slopes above Route 3 between Ushuaia and Rio Grande, at Garibaldi pass, 50 km northeast of Ushuaia. This latter site also supports Andean Condor and Ochre-naped Ground-Tyrant.

TIERRA DEL FUEGO

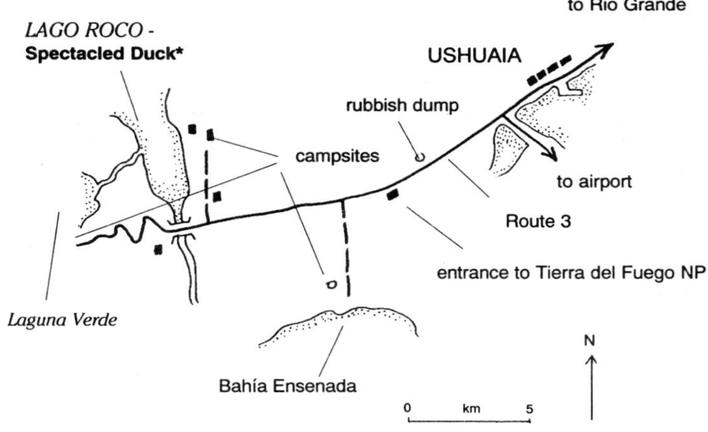

In Tierra del Fuego NP, the huge black Magellanic Woodpecker, complete with crazy red mohican crest, can be very confiding once found

White-throated Caracara occurs at the rubbish dump 7 km west of Ushuaia, *en route* to **Tierra del Fuego NP**, which is an excellent birding site, 10 km west of Ushuaia. Rufous-legged Owl, Magellanic Woodpecker and White-throated Treerunner occur around the Lago Roca campsite and café. Turn north before the bridge over the Lago Roca outlet past the NP HQ. Spectacled Duck* occurs on Lago Roca. Kelp Goose, White-throated Caracara and Andean Tapaculo occur around Bahia Ensenada, reached via the track south just inside the NP, before Lago Roca. This park is cold and snowy in the austral winter (April to Sept).

King Penguin occasionally appears in the Magellanic Penguin colony near Harberton at the eastern end of the Beagle Channel. To attempt to see this species contact La Agencia Caminante, 368 Deloqui, Ushuaia, who operate cruises. Gentoo Penguin and Blackish Cinclodes are also possible on islands and along the shoreline of the Beagle Channel.

Accommodation: Hotel Tolkeyen (A); private houses (B). TDF NP: camping (C).

At **Rio Grande** (236 km north of Ushuaia), it is possible to see Black-browed Albatross and Antarctic Giant Petrel from a dining table at the Las Ramblas restuarant, whilst the mudflats near the town centre support Two-banded Plover. Ruddy-headed Goose* occurs in the fields alongside Route 3 (15–20 km northwest of Rio Grande). Rufous-chested Dotterel, Short-billed Miner and Austral Canastero* occur at Cabo Peñas, 12 km south of Rio Grande. Head through the rubbish dump to the sea and continue along the coast until reaching an indistinct dirt track, 0.3 km before the low hills touch the dirt road. Head along this track through the hills and scour the the tussocky slopes next to a large lake. The roadside pools 1 km along the road to El Salvador, 3 km west of Rio Grande, occasionally hold Magellanic Plover* whilst Austral Canastero* occurs on the bushy slopes.

Accommodation: Hotel Isla del Mar (A); Hospedaje Miramar (B).

ADDITIONAL INFORMATION

Addresses

Asociacion Ornitological del Plata (AOP), 25 de Mayo 749, 2nd Floor, 1002 Buenos Aires. P. 312 8958. This is an active society which holds regular meetings.

Books

Birds of Argentina and Uruguay - A Field Guide, Narosky, T and Yzurieta, D. 1989.
Field Check-List to the Birds of Argentina, Straneck, R and Carrizo, G. 1991.

ARGENTINA ENDEMICS (16)	**BEST SITES**
Chubut Steamer-Duck*	South: Trelew
Bare-eyed Ground-Dove	Northwest: Tucuman–Salta area
Band-tailed Earthcreeper	South: Trelew and San Antonio Oeste
Comechingones Cinclodes	Northwest: Córdoba
Olrog's Cinclodes	Northwest: Córdoba
Steinbach's Canastero	Northwest: Mendoza
Patagonian Canastero	South: Trelew and San Antonio Oeste
Cordoba Canastero	Northwest: Córdoba and Tucumán–Salta area
White-throated Cachalote	South: Trelew and San Antonio Oeste
Sandy Gallito	Northwest and south: Tucuman–Salta area and San Antonio Oeste
Rusty-backed Monjita	South: Trelew and San Antonio Oeste
Salinas Monjita*	Northwest: Salinas Grandes and Santiago del Estero
Yellow-striped Brush-Finch	Northwest: Tucuman–Salta area
Carbonated Sierra-Finch	South: Trelew and San Antonio Oeste
Cinnamon Warbling-Finch	Northwest and south: Córdoba and San Antonio Oeste
Tucuman Mountain-Finch*	Northwest: Tucuman–Salta area

(Narosky's Seedeater*)	Recent DNA studies suggest this is a subspecies of Marsh Seedeater*

Northwest = Córdoba and Mendoza northwards.
South = Trelew southwards.

BOLIVIA

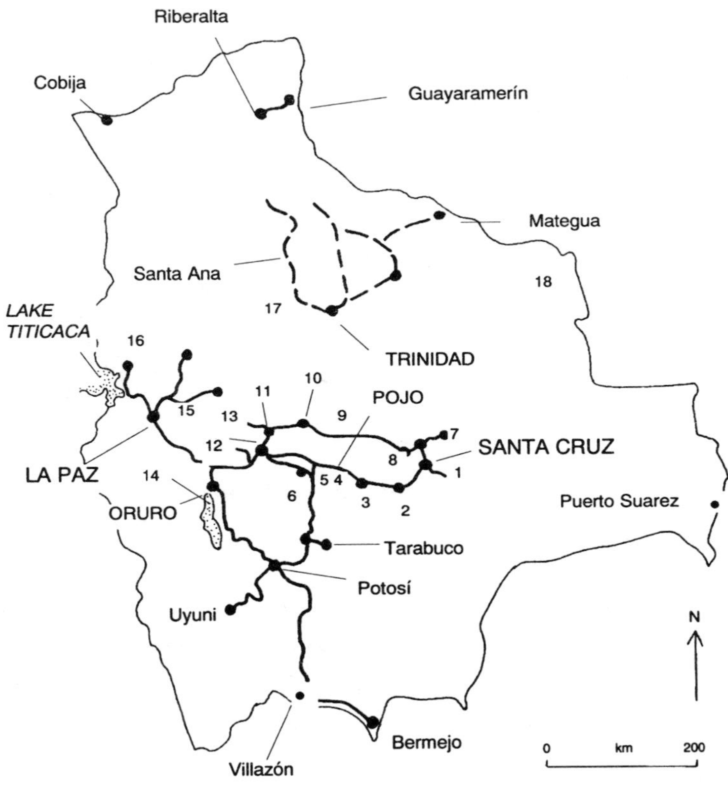

1	Santa Cruz	11	Villa Tunari–Cochabamba Road
2	Samaipata	12	Laguna Alalay
3	Comarapa and Tambo	13	Quillacollo Road
4	Siberia	14	Lake Uru-Uru
5	Siberia–Cochabamba Road	15	La Paz–Coroico Road
6	Incallajta ruins	16	Sorata
7	Okinawa 1 and 2	17	Beni Biological Station
8	Buena Vista	18	Noel Kempff Mercado NP
9	Sajta		
10	Villa Tunari		

INTRODUCTION

Summary

Although Bolivia is South America's poorest and least 'developed' country, it is one of the safest and friendliest. Fortunately, the few, somewhat rough roads pass through some of the continent's most pristine temperate and subtropical forests which support a number of rare species that otherwise occur only in remote areas of south Peru. This is a big country, but the best sites are relatively close together, enabling birders to see many of the star birds during a short trip.

Size

Bolivia is large (1,098,581 km^2), over eight times the size of England and 1.6 times larger than Texas, but still smaller than Peru and less than half the size of Argentina. Whilst much of the north remains inaccessible, most of the specialities occur at sites between Santa Cruz in the east and La Paz in the west, a distance of approximately 750 km.

Getting Around

Bolivia has few roads and only 5% of them are paved. Some of the best birding roads, such as that from La Paz to Coroico, are rather frightening, and others, such as the old road between Santa Cruz and Cochabamba, are in serious need of repairs. Four-wheel-drive is recommended but not absolutely essential. Modern buses run regularly along most routes but be sure to book in advance. The internal air network is quite extensive and railways cross the altiplano south of La Paz.

Accommodation and Food

There is plenty of accommodation to choose from, albeit mainly low-budget and basic. The trout and beer are recommended.

Health and Safety

It is advisable to start a trip to Bolivia at Santa Cruz in the eastern lowlands. It is then possible to acclimatise to the increasing altitude whilst travelling westwards towards the high Andes around La Paz. If arriving directly at La Paz by air, beware of altitude sickness.

Immunisation against yellow fever, malaria, typhoid and hepatitis is recommended. Also beware of Chagas disease spread by a hut-dwelling beetle, which is carried by half the Bolivian population and leads to sudden heart failure.

Bolivia is, arguably, the safest, friendliest country in South America. However, a very small minority may be less than helpful. This may be because of their connection with the drug trade, since coca is grown in many areas.

Climate and Timing

The varied topograpy creates a variable climate, although in general the main wet season is from November to March. It is wettest on the Andean massif and in the eastern lowlands from December to April. September is probably the best time to visit. This is the end of the southern winter when it is not too hot in the eastern lowlands and the Andean flowers are beginning to bloom. However, March, the end of the southern summer, is also a good time.

Visiting a number of sites involves considerable altitude changes so the full range of clothing is required, even during a single day. Be prepared for anything from damp, misty mornings and scorching sunshine to low temperatures at night.

Habitats

The Andes are at their widest (650 km) in Bolivia. The highest peaks are in the west on the border with Chile. To the east of these massive mountains is the cold, rugged, windswept altiplano which covers most of southwest Bolivia. Lake Titicaca, which lies northwest of La Paz, and the whole altiplano is encircled by puna, whilst to the east lie fertile valleys. Southeast of La Paz, near Cochabamba, semi-arid rainshadow valleys dominate the terrain.

Remnant groves of *Polylepis* woodland, as well as large, virtually untouched tracts of temperate forest and subtropical 'yungas' forest cover the northeast edge of the Andean region, breached by the spectacular La Paz–Coroico road which traverses 3,430 m (11,253 ft) in just 80 km, and the Villa Tunari–Cochabamba road. Beyond these green slopes to the north and northeast lies the largely inaccesible and vast mass of lowland Amazonian rainforest, interrupted here and there by wet savanna known as the Llanos de Mojos. This lowland wilderness covers some 60% of Bolivia's land surface.

Santa Cruz, Bolivia's second city, in the east, lies at the edge of the Gran Chaco, a land of scrub with scattered woodlands.

Conservation

Much of Bolivia's temperate and subtropical 'yungas' forests lie on the very steep northeastern slopes of the Andean massif, making them very difficult to exploit. They are relatively untouched and not under serious threat. However, *Polylepis* woodland, which the locals use for firewood, is disappearing fast, even here, and the Amazonian lowlands are slowly opening up, which is good for birders but bad for birds.

23 threatened species, of which seven are endemic, occur in Bolivia. A further two endemics are near-threatened.

Bird Families

Only 74 of the 92 families which regularly occur in South America are represented (due to the complete lack of coastal species), but this total includes 23 of the 25 Neotropical endemic families (no Sharpbill), and eight of the nine South American endemic families (no Magellanic Plover). Hence, Bolivia boasts the best range of families on the continent.

Well-represented families include tinamous (over 20), cracids (14), hummingbirds (over 70), furnariids (over 80) and cotingas (18).

Bird Species

Bolivia boasts the world's biggest list for a land-locked country, and yet the total of 1,274 species so far recorded is likely to increase, potentially to over 1,400. Indeed, once the remoter areas, especially in the Amazonian lowlands, have been fully explored, Bolivia may be found to support more species than Venezuela.

Non-endemic specialities and spectacular species include Short-winged Grebe, Andean* and Puna* Flamingos, Andean Condor, Horned Curassow*, Rufous-bellied Seedsnipe, Andean Avocet, Golden-collared Macaw, Red-tailed Comet, Blue-banded Toucanet, Hooded

Mountain-Toucan*, Olive-crowned Crescent-chest, Giant Antshrike, Stripe-headed and White-throated Antpittas, Slaty Gnateater, Chestnut-crested Cotinga, Band-tailed Fruiteater, Scimitar-winged Piha, Amazonian Umbrellabird, Andean Cock-of-the-Rock, Snow-capped Manakin, Hazel-fronted Pygmy-Tyrant, Bolivian Tyrannulet, White-eared Solitaire, Orange-browed Hemispingus, Straw-backed and Golden-collared Tanagers, Short-tailed Finch, Citron-headed Yellow-Finch*, Black-and-tawny Seedeater*, Moustached Flower-piercer and Rufous-bellied Saltator*.

Endemics

Bolivia has 18 endemic species, more than Argentina and Ecuador, but far less than Peru and Venezuela.

The endemic species include the rare and spectacular Blue-throated* and Red-fronted* Macaws, Wedge-tailed Hillstar*, two canasteros, Rufous-faced Antpitta, Cochabamba Mountain-Finch* and Bolivian Blackbird.

Bolivia is also the best country in which to look for the many species restricted to south Peru and west Bolivia.

Expectations

A three-week trip, which concentrates on the sites between Santa Cruz and La Paz, including the Villa Tunari–Cochabamba and the La Paz–Coroico roads, is likely to produce around 400 species including at least ten endemics. A longer trip, say six weeks, could push the list well past the 500 mark.

SANTA CRUZ

Santa Cruz (437 m/1,434 ft) is the best place to start a birding trip to Bolivia because it lies in the eastern lowlands and allows a gradual acclimatisation to increasingly high altitude whilst birding sites to the west *en route* to lofty La Paz, which is over 3,000 m (10,000 ft) higher up.

Bolivia's fastest growing city is surrounded by cerrado, chaco-like thorn scrub, gallery woodland and, at nearby Okinawa, wetlands. Hence the city environs are worth a look before heading west towards Cochabamba and much more exciting birding. Goodies here include a number of tyrants such as the rare Hudson's Black-Tyrant*, which has been recorded during the austral winter (April to September).

Specialities

Ringed Teal, Golden-collared Macaw, Blue-crowned Parakeet, Ashy-tailed Swift, White-eared Puffbird, White-rumped Monjita, Hudson's Black-Tyrant*, White-banded Mockingbird, Yellow-billed Cardinal, Black-capped Warbling-Finch, Grey-and-chestnut Seedeater*.

Others

Red-winged Tinamou, White-bellied Nothura, Greater Rhea*, Capped Heron, Red-legged Seriema, Collared Plover, Southern Lapwing, Picui Ground-Dove, Peach-fronted Parakeet, Blue-winged Parrotlet, Guira Cuckoo, Spot-backed Puffbird, Chestnut-eared Aracari, Toco Toucan, White Woodpecker, Campo Flicker, Narrow-billed Woodcreeper, Rufous Hornero, Chotoy Spinetail, Greater Thornbird, Mouse-coloured, White-bellied and Plain Tyrannulets, Pearly-vented Tody-Tyrant, Suiriri

Flycatcher, Tawny-crowned Pygmy-Tyrant, Austral Negrito, Spectacled Tyrant, Rufous Casiornis, Creamy-bellied Thrush, Chalk-browed Mockingbird, Thrush-like Wren, Masked Gnatcatcher, White-rumped Swallow, Yellowish Pipit, Hooded Siskin, Red-crested Cardinal, Sayaca Tanager, Red-crested Finch, Wedge-tailed Grass-Finch, Double-collared, White-bellied, Tawny-bellied and Dark-throated* Seedeaters, Black-backed Grosbeak, White-browed and Chopi Blackbirds, Shiny Cowbird.

Access

Head southeast out of Santa Cruz. After 8 km turn east towards Lomas de Arena. Red-legged Seriema and Blue-crowned Parakeet occur in the roadside chaco along the 12 km to Lomas de Arena. Alternatively, pass the Lomas de Arena turning, then turn west to a golf course (1–2 km away), where you need to ask permission, which is usually given, to bird the area. White-bellied Nothura, Ringed Teal, White-eared Puffbird, White-rumped Monjita and Grey-and-chestnut Seedeater* occur here.

There are two roads from Santa Cruz to Cochabamba. The **New Santa Cruz–Cochabamba road** heads north from Santa Cruz and west from Montero to Villa Tunari and Cochabamba, and the **Old Santa Cruz–Cochabamba road** heads southwest out of Santa Cruz to Samaipata, Comarapa and Cochabamba. There are some excellent birding sites along both of these roads and they are discussed separately below.

THE OLD SANTA CRUZ–COCHABAMBA ROAD

Golden-collared Macaw occurs 22 km west of San José, southwest of Santa Cruz. Follow and bird the stream south of the village to the west.

SAMAIPATA

Situated in the dry eastern Andean foothills, 120 km southwest of Santa Cruz, Samaipata has become virtually deserted since the opening of the new road between Santa Cruz and Cochabamba. For once, a new road has been good for the birds, since the excellent dry deciduous forest alongside the old road near Samaipata remains untouched. A number of localised species hard to see elsewhere occur here, including Ochre-cheeked Spinetail, Slaty Gnateater and Dull-coloured Grassquit.

Specialities

Tataupa Tinamou, White-faced Dove, Green-cheeked Parakeet, Buff-bellied Hermit, Slender-tailed Woodstar, Ocellated Piculet, Ochre-cheeked Spinetail, Slaty Gnateater, Grey-crested Finch, Ringed and Black-capped Warbling-Finches, Dull-coloured Grassquit.

Others

Andean Condor, Mitred Parakeet, Scaly-headed Parrot, Chestnut-eared Aracari, Olivaceous Woodcreeper, Buff-browed Foliage-gleaner,

Variable Antshrike, Black-capped Antwren, White-backed Fire-eye, Southern Beardless-Tyrannulet, Yellow-billed Tit-Tyrant, Cliff Flycatcher, Streak-throated Bush-Tyrant, Andean Tyrant, Swainson's Flycatcher, Purplish and Plush-crested Jays, Rufous-bellied Thrush, Two-banded Warbler, Saffron-billed Sparrow, Black-goggled Tanager, Purple-throated Euphonia, Black-and-rufous Warbling-Finch, Crested Oropendola, Epaulet Oriole.

Access

Bird the roadside forest 4–5 km east of Samaipata, especially where there is a small waterfall next to a bend in the road and the 'Pipeline' trail which leads east from Samaipata. The shrubs surrounding the small stone-walled fields alongside the road are also worth a prolonged look. Grey-crested Finch occurs alongside the road west to Comarapa.

Accommodation: Hotel Mily.

The southern HQ of **Amboro NP**, operated by the Fundacion Amigos de la Naturaleza (FAN), is situated in Samaipata. Large areas of untouched temperate and subtropical forest remain in the southern portion of this NP. Visits are permitted with ranger-guides (camping only). Solitary Eagle*, Dusky-legged Guan and Red-tailed Comet occur here.

COMARAPA AND TAMBO

Map opposite

Comarapa is just over 100 km west of Samaipata along the Cochabamba road. It is the centre of a farming community and much of the dry forest has been lost to agriculture. However, the rare endemic Red-fronted Macaw* still survives in the remnant forest patches near town, and this is the best place to look for it in Bolivia. The endemic Bolivian Earthcreeper* occurs at nearby Tambo.

Endemics

Red-fronted Macaw*, Bolivian Earthcreeper*.

Specialities

Dusky-legged Guan, Red-tailed Comet, Spot-backed Puffbird, Cream-backed Woodpecker, Brown-capped Tit-Spinetail, White-tipped Plantcutter, Greater Wagtail-Tyrant, Grey-crested Finch.

Others

Andean Tinamou, Andean Condor, Red-backed Hawk, Scissor-tailed Nightjar, Andean Swift, White-bellied Hummingbird, White-fronted and Striped Woodpeckers, Narrow-billed Woodcreeper, Stripe-crowned Spinetail, Streak-fronted Thornbird, Rufous-capped Antshrike, White-bellied Tyrannulet, White-winged Black-Tyrant, Saffron-billed Sparrow, Sayaca Tanager, Black-capped Warbling-Finch, Great Pampa-Finch, Golden-billed Saltator, Ultramarine Grosbeak, Bay-winged Cowbird.

Access

The tiny village of **Tambo**, 13 km east of Comarapa, lies in an arid cacti-

Look out for the snazzy Andean Swift around Comarapa and Tambo

clad canyon. Bolivian Earthcreeper* and Grey-crested Finch occur just north of the New Tribes Mission School here, in the dry river bed and around the fields past the piggery. The left fork in the track here continues through good habitat to a river.

The track north, 3 km west of Tambo on the main road to Comarapa, leads to a forested area worth exploring. Red-fronted Macaw* has been recorded here. This rarity may occur anywhere around Comarapa, and it has recently been seen just to the northwest between the two rivers.

Accommodation: Tambo: it is possible to stay at the Mission School, where the missionaries are very helpful with information on local birding sites and, on good days, make excellent tea. Comarapa: Hotel Central.

COMARAPA AND TAMBO

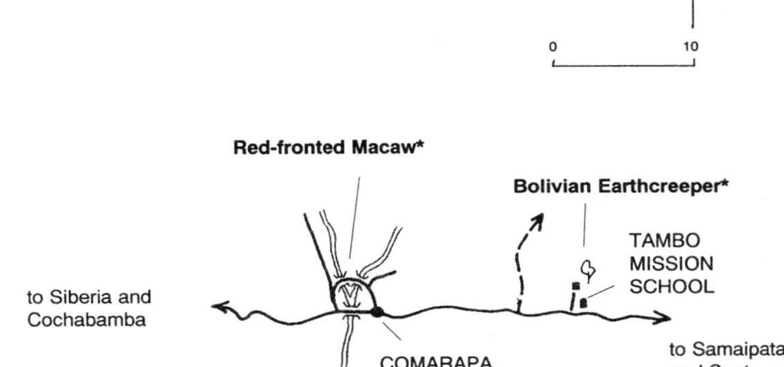

SIBERIA

The road from Comarapa to Pojo, 75 km to the west, traverses the Serrania de Siberia, where the southernmost humid temperate (cloud) forest in South America (2,438 m/8,000 ft to 2,743 m/9,000 ft) survives. This excellent forest supports three endemics, the rare Black-winged Parrot and two scarce cotingas: Chestnut-crested Cotinga and Band-tailed Fruiteater.

Endemics

Black-hooded Sunbeam, Rufous-faced Antpitta, Grey-bellied Flower-piercer.

Specialities

White-throated Hawk, Mountain Caracara, Black-winged Parrot, Violet-throated Starfrontlet, Blue-capped Puffleg, Light-crowned Spinetail, Giant Antshrike, Chestnut-crested Cotinga, Band-tailed Fruiteater, Brown-capped Redstart, Pale-legged Warbler, White-browed Conebill, Chestnut-bellied Mountain-Tanager, Rufous-sided and Rusty-browed Warbling-Finches.

Others

Andean Condor, Puna Hawk, Andean Guan, Scaly-naped Parrots, Andean Swift, Green Violet-ear, Bar-bellied Woodpecker, Variable Antshrike, Rufous Antpitta, Unicoloured Tapaculo, Red-crested Cotinga, Streak-necked Flycatcher, Tawny-rumped and White-throated Tyrannulets, Highland Elaenia, Crowned Chat-Tyrant, White-winged Black-Tyrant, Glossy-black Thrush, Rufous-naped and Stripe-headed Brush-Finches, Blue-backed Conebill, Blue-winged Mountain-Tanager, Plush-capped Finch, Great Pampa-Finch, Masked Flower-piercer.

Access

The roadside shrubs a few km west of Comarapa are worth checking for hummers. The forest begins 23 km west of Comarapa and continues for 20 km, just before the village of Siberia. Bird from the roadside and along any trails you may find. One good trail is 55 km west of Comarapa.

SIBERIA–COCHABAMBA ROAD

This stretch of road, west of Siberia, climbs even further, before descending again towards Cochabamba. It passes through the best area in Bolivia for the endemic Iquico Canastero. Three more endemics and a number of localised species, notably the near-endemic Citron-headed Yellow-Finch, also occur along here.

Endemics

Wedge-tailed Hillstar, Iquico Canastero, Bolivian Warbling-Finch, Cochabamba Mountain-Finch*.

Specialities

Red-tailed Comet, Rock Earthcreeper, Freckle-breasted Thornbird, Olive-crowned Crescent-chest, Grey-bellied Shrike-Tyrant, Cinereous Ground-Tyrant, Brown-backed Mockingbird, Fulvous-headed Brush-Finch, Giant Conebill*, Black-hooded Sierra-Finch, Rufous-sided Warbling-Finch, Citron-headed Yellow-Finch*, Rufous-bellied Saltator*.

Others

Andean Tinamou, Darwin's Nothura, Andean Condor, Black-chested Buzzard-Eagle, Red-backed Hawk, Spot-winged Pigeon, Bare-faced Ground-Dove, Andean Swift, Creamy-breasted Canastero, Streak-fronted Thornbird, Andean Tapaculo, Tufted Tit-Tyrant, D'Orbigny's Chat-Tyrant, Rufous-webbed Tyrant, Spot-billed Ground-Tyrant, Great Thrush, Brown-bellied Swallow, Band-tailed Sierra-Finch, Greenish Yellow-Finch.

Access

Bird the roadside, especially any *Polylepis* woodland. Wedge-tailed Hillstar, Bolivian Warbling-Finch and Citron-headed Yellow-Finch* have been seen recently 78 km west of Comarapa, and Iquico Canastero 90 km west of Comarapa. A grove of *Polylepis* may still exist 2 km east of the 340 km post, from Santa Cruz. Fulvous-headed Brush-Finch, Bolivian Warbling-Finch, Cochabamba Mountain-Finch* and Citron-headed Yellow-Finch* occur along the mountain track, which leads west just before the river, in the tiny village of **Puente Lope Mendosa**, 156 km west of Comarapa. Brown-backed Mockingbird also occurs, around the village.

INCALLAJTA RUINS

These ruins, near Pocona, some 130 km southeast of Cochabamba, are a good site for the rare endemic Wedge-tailed Hillstar*, as well as other hummers and warbling-finches.

Endemics

Wedge-tailed Hillstar*.

Specialities

Ornate Tinamou, Red-tailed Comet, Tawny Tit-Spinetail*, Fulvous-headed Brush-Finch, Rufous-sided, Rusty-browed and Ringed Warbling-Finches.

Others

Darwin's Nothura, Andean Condor, Spot-winged Pigeon, Bare-faced Ground-Dove, Andean Swift, Andean Hillstar, Giant Hummingbird, Streak-fronted Thornbird, Rufous-capped Antshrike, Yellow-billed and Tufted Tit-Tyrants, White-browed Chat-Tyrant, Brown-capped Redstart, Golden-billed Saltator.

Access

Turn south 7 km west of Epizana towards Pocona (22 km south of barrier). Turn west 3 km before Pocona to reach the ruins, 10 km away. Wedge-tailed Hillstar* occurs 2 km beyond the ruins car park on the steep slopes across the river *en route* to the Mirador viewpoint.

Accommodation: there is no accommodation here, although the church in Pocona has a dormitory which they may let you use on request. Otherwise, camping is possible near the ruins.

THE NEW SANTA CRUZ–COCHABAMBA ROAD

There are a number of excellent birding sites along this road.

White-bellied Nothura, Greater Rhea*, and Tawny-bellied and Dark-throated* Seedeaters occur around Viru Viru airport, 16 km north of Santa Cruz *en route* to Montero. Bird the airport entrance road, the road north from the airport including the first track east after the toll house and another road east, which leads to a farm, further north along the main road to Montero.

OKINAWA 1 AND 2

These two Japanese rice-growing villages lie at the edge of a big wetland, where Ringed Teal, Yellow-breasted Crake and Golden-collared Macaw occur as well as a number of wetland species, widespread and common in South America, but more localised in Bolivia.

Specialities

Ringed Teal, Yellow-breasted Crake, Golden-collared Macaw, Blue-fronted Parrot, White-rumped Monjita.

Others

Southern Screamer, Brazilian Teal, Whistling Heron, Rufescent Tiger-Heron, Plumbeous Ibis, Wood and Maguari Storks, Snail Kite, Savanna and Black-collared Hawks, Wattled Jacana, Picazuro Pigeon, Nacunda Nighthawk, Gilded Sapphire, Little Woodpecker, Greater Thornbird, Yellow-browed Tyrant, Red-crested Cardinal, Crested Oropendola, Solitary Cacique, Troupial, Chopi Blackbird.

Access

Turn east just north of Montero (37 km north of Santa Cruz) towards Okinawa. The farther east you go the wetter and better (for birds) it gets.

Heading west from Montero towards Buena Vista there is a good wet, wooded area some 25 km east of Buena Vista where Maguari Stork, Pied Lapwing, Chestnut-fronted Macaw, Ashy-tailed Swift, Chestnut-eared Aracari, and Black-capped Donacobius occur. 11 km east of Buena Vista there is a remnant patch of chaco on the north side of the road where Red-legged Seriema and Toco Toucan occur.

BUENA VISTA

Map below

Buena Vista is a three-hour drive northwest of Santa Cruz. Birding near this small town may produce such goodies as Black-capped Tinamou and Band-tailed Manakin. The Amboro NP northern HQ is situated within this town and those with time to spare or a mind to look for the two rare endemics, Bolivian Recurvebill* and Ashy Antwren*, as well as Horned Curassow*, may be able to arrange a mini-expedition here.

Specialities

Black-capped Tinamou, White-bellied Nothura, Speckled Chachalaca, Little Nightjar, Band-tailed Manakin.

Others

Muscovy Duck, Capped Heron, Green Ibis, Grey-necked Wood-Rail, Sungrebe, Collared Plover, Chestnut-fronted Macaw, White-eyed and Dusky-headed Parakeets, Blue-winged Parrotlet, Blue-headed Parrot, Hoatzin, Guira Cuckoo, Tropical Screech-Owl, Grey Potoo, Scissor-tailed Nightjar, Rufous-throated Sapphire, White-tailed Goldenthroat, Blue-crowned Trogon, Blue-crowned Motmot, White-necked Puffbird, Black-fronted Nunbird, Lettered and Chestnut-eared Aracaris, Crimson-crested Woodpecker, Greater Thornbird, Small-billed Elaenia, Rufous Casiornis, Black-capped Donacobius, Orange-headed and Guira Tanagers, Russet-backed Oropendola, Solitary Cacique, White-browed and Chopi Blackbirds.

Access

The road leading southeast from the Buena Vista plaza to El Cairo passes through wet fields where Muscovy Duck occurs, but the best birding

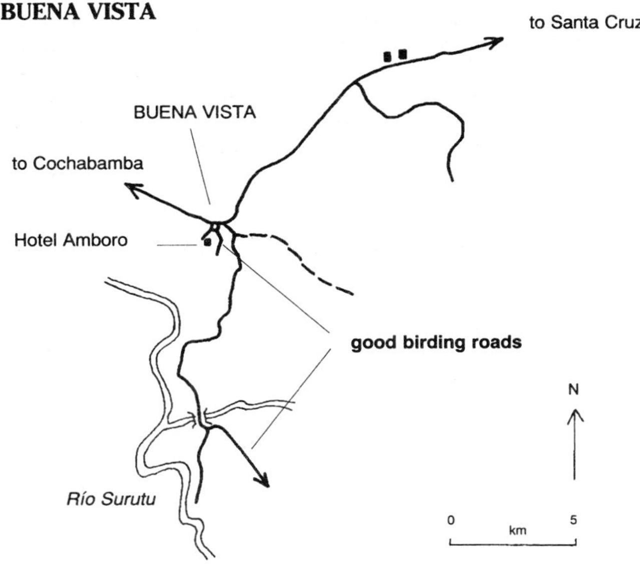

area around Buena Vista is near the Río Surutu, along the road to Amboro NP (see below). Head east then southeast out of town and take the left fork after the second river. There are a few side-trails and tracks worth exploring along the stretch of road after the fork.

The northern HQ of **Amboro NP** is 200 m along the road leading southwest from the plaza. Crested Eagle*, Razor-billed and Horned* Curassows, and the rare endemics, Bolivian Recurvebill* and Ashy Antwren* occur in this NP. Obtain a permit, directions and, hopefully, a guide, at the HQ. There is no accommodation in the park but it is possible to sleep in one of the huts. It takes 1.5 hours by road and 1.5 hours by foot to reach one of the best areas. There are no trails, but the part of the forest where Ashy Antwren* occurs can be worked via a stream which has to be crossed a few times.

400 m further down the road southwest from the plaza, on the right-hand side, is the Hotel Amboro (A), which has excellent grounds where Band-tailed Manakin occurs. The owner is a keen conservationist and may be willing to assist with local directions or visits to Amboro NP.

SAJTA

Further west along the new Santa Cruz–Cochabamba road there is a good patch of lowland rainforest just west of Sajta. The rare Grey-bellied Goshawk* has been recorded here.

Specialities

Black-capped Tinamou, Grey-bellied Goshawk*, Slate-coloured Hawk, Round-tailed Manakin, Flammulated Bamboo-Tyrant.

Others

King Vulture, Spix's Guan, Blue-and-yellow Macaw, Cobalt-winged Parakeet, Blue-headed and Mealy Parrots, Short-tailed and Ashy-tailed Swifts, White-bearded and Reddish Hermits, Lettered and Curl-crested Aracaris, Channel-billed Toucan, Red-necked Woodpecker, Ocellated and Buff-throated Woodcreepers, Red-billed Scythebill, Plain-winged Antshrike, Black-faced, Chestnut-tailed, Black-throated and Spot-backed Antbirds, Black-faced Antthrush, Screaming Piha, McConnell's Flycatcher, Short-tailed Pygmy-Tyrant, Rufous-tailed Flatbill, Ruddy-tailed Flycatcher, Pink-throated Becard, Moustached Wren, Southern Nightingale-Wren, White-banded Swallow, Magpie, Yellow-backed and Paradise Tanagers, Amazonian Oropendola.

Access

Turn north just west of Sajta in response to a 'University of Cochabamba' sign, to a gate. Permission is usually granted to enter and bird the forest beyond the sawmill and nursery. The road north from Sajta itself skirts the eastern edge of this forest and is also worth birding.

VILLA TUNARI

The new Santa Cruz–Cochabamba road eventually reaches Villa Tunari, a small resort where Bolivians enjoy swimming in the local rivers which flow through forest in the upper tropical zone. This forest supports some good birds, most of which are widespread throughout South America but not present at many sites in Bolivia. Oilbird occurs in nearby caves.

Specialities

Oilbird, White-throated Woodpecker, Round-tailed and Fiery-capped Manakins.

Others

Little Tinamou, Fasciated Tiger-Heron*, Speckled Chachalaca, Spix's Guan, Plumbeous Pigeon, Blue-headed Parrot, White-necked Jacobin, Fork-tailed Woodnymph, Rufous Motmot, White-necked Puffbird, Black Nunbird, Cuvier's Toucan, Buff-throated and Ocellated Woodcreepers, Chestnut-tailed Antbird, Screaming Piha, Sepia-capped Flycatcher, Southern Nightingale-Wren, Andean Slaty Thrush, White-winged Shrike-Tanager, Thick-billed Euphonia, Paradise Tanager, Black-faced Dacnis, Yellow-rumped Cacique, Giant Cowbird.

Access

The riverine forest a few km west of Hotel el Puente (A), which is reached by turning south after crossing the second river east of Villa Tunari, supports White-throated Woodpecker, Round-tailed and Fiery-capped Manakins, White-winged Shrike-Tanager and Paradise Tanager. West of the hotel turn left (south) over a small river, past a football pitch and through a small village to the best forest. The Dutch owner of the Hotel el Puente (which has excellent grounds well worth birding) can arrange a trip to a nearby Oilbird cave. This involves a rugged trek and crossing a river (where Fasciated Tiger-Heron* occurs) on a chair hanging from a rope.

VILLA TUNARI–COCHABAMBA ROAD **Map p. 91**

The new road from Santa Cruz heads west and south out of Villa Tunari to Cochabamba, through superb subtropical 'yungas', temperate, and elfin forests on the northeast Andean slope. It then passes over a high ridge (3,500 m/11,483 ft) before descending to Cochabamba.

This stretch of road is one of the best birding sites in South America. As well as four endemics, it supports a number of species restricted to southeast Peru and west Bolivia which are virtually impossible to see in Peru and remain rare or unrecorded elsewhere in Bolivia itself (especially along the similar La Paz–Coroico road). These star birds include Stripe-faced Wood-Quail, Black-winged Parrot, Hazel-fronted Pygmy-Tyrant and Slaty Tanager. Furthermore, this site is also one of the best, if not *the* best in Bolivia (and elsewhere) for Hooded Tinamou*, Blue-banded Toucanet, Hooded Mountain-Toucan*, Upland Antshrike, White-throated Antpitta, Chestnut-crested Cotinga, Band-tailed Fruiteater, Bolivian Tyrannulet and the rare Straw-backed Tanager.

Endemics

Black-hooded Sunbeam, Black-throated Thistletail, Rufous-faced Antpitta, Yungas Tody-Tyrant, Grey-bellied Flower-piercer.

Specialities

Hooded Tinamou*, Stripe-faced Wood-Quail, Black-winged and Speckle-faced Parrots, Swallow-tailed Nightjar, Violet-throated Starfrontlet, Versicoloured Barbet, Blue-banded Toucanet, Hooded Mountain-Toucan*, Light-crowned Spinetail, Upland Antshrike, White-throated Antpitta, Chestnut-crested Cotinga, Band-tailed Fruiteater, Hazel-fronted Pygmy-Tyrant, Bolivian and Buff-banded Tyrannulets, Unadorned Flycatcher, Rufous-bellied Bush-Tyrant*, White-eared Solitaire, Three-striped Hemispingus, Rust-and-yellow and Slaty Tanagers, Chestnut-bellied Mountain-Tanager, Straw-backed Tanager, Moustached Flower-piercer.

*The Villa Tunari–Cochabamba road is the best site in South America for Hooded Mountain-Toucan**

Others

Brown Tinamou, Sickle-winged Guan, Plumbeous Pigeon, Red-billed Parrot, White-tipped Swift, Collared Inca, Amethyst-throated Sunangel, Long-tailed Sylph, Golden-headed Quetzal, Masked Trogon, Red-necked Woodpecker, Buff-browed Foliage-gleaner, Amazonian Umbrellabird, Andean Cock-of-the-Rock, Blue-backed Manakin, Ochre-faced Tody-Flycatcher, Sclater's and White-banded Tyrannulets, Slaty-backed and Brown-backed Chat-Tyrants, Spotted Nightingale-Thrush,

Andean Slaty-Thrush, Spectacled Redstart, Three-striped Warbler, Superciliared and Black-eared Hemispinguses, Hooded and Scarlet-bellied Mountain-Tanagers, Orange-eared, Saffron-crowned and Spotted Tanagers, Deep-blue Flower-piercer, Dusky-green and Russet-backed Oropendolas, Mountain Cacique.

Access

Anywhere along this road is worth birding, but there are some particularly good tracks and trails; Blue-banded Toucanet, White-throated Antpitta, Hazel-fronted Pygmy-Tyrant and Slaty Tanager occur along the trail west, just north of Miguelito (95 km from Cochabamba). Chestnut-crested Cotinga occurs along the track which goes northwest just north of Corani, to Tablas Montes. Black-winged Parrot, Hooded Mountain-Toucan*, and Rufous-faced Antpitta occur along the track south from the clearing in the village of Corani (72 km from Cochabamba).

VILLA TUNARI–COCHABAMBA ROAD

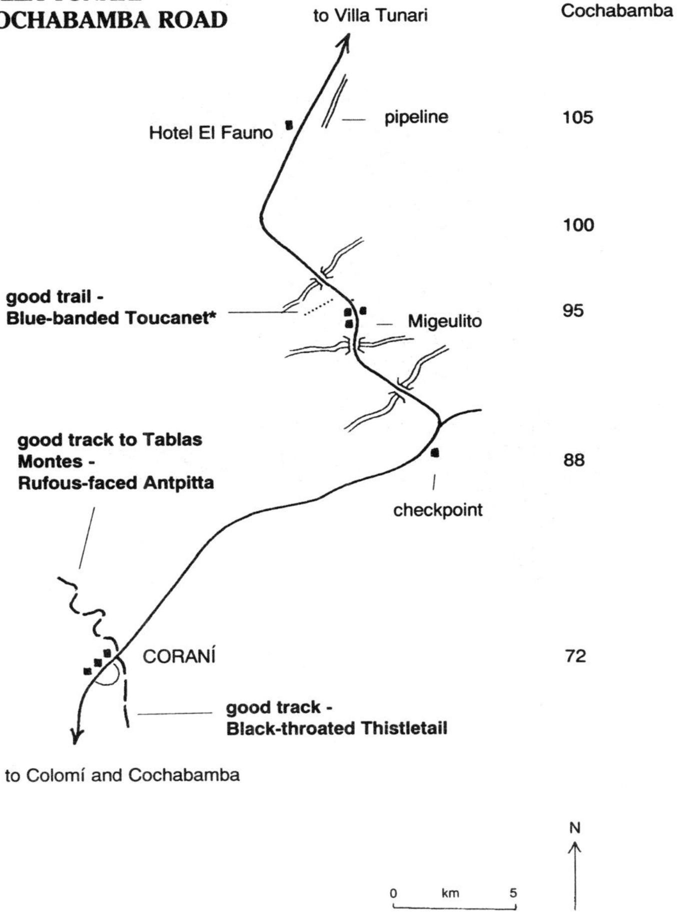

Km from Cochabamba

to Villa Tunari

Hotel El Fauno

pipeline

105

100

good trail - Blue-banded Toucanet*

Migeulito

95

good track to Tablas Montes - Rufous-faced Antpitta

88

checkpoint

CORANÍ

72

good track - Black-throated Thistletail

to Colomí and Cochabamba

N

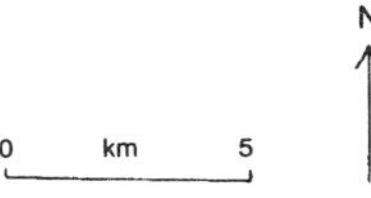

COCHABAMBA

The city of Cochabamba (2,570 m/8,414 ft) lies in a 10 km-wide natural arena of rolling hills below snow-capped Cerro Tunari (5,180 m/16,995 ft). The dry and sunny climate is very pleasant. The city is near some great birding sites, not least the road to Villa Tunari (see previous site) and the Quillacollo road.

LAGUNA ALALAY

This shallow high-altitude lake, just a short walk southeast of Cochabamba city, is well worth a look since it supports a wide variety of waterfowl, the stunning Many-coloured Rush-Tyrant and Short-billed Pipit.

Specialities

Short-billed Pipit.

Others

White-tufted and Silvery Grebes, Andean Duck, Speckled Teal, Yellow-billed and White-cheeked Pintails, Puna Teal, Red Shoveler, Puna Ibis, Cinereous Harrier, Black-chested Buzzard-Eagle, Plumbeous Rail, Wattled Jacana, Andean Lapwing, Andean Gull, Wren-like Rushbird, Many-coloured Rush-Tyrant, Grey-bellied Shrike-Tyrant, Cinereous Ground-Tyrant, Andean Negrito, White-winged Black-Tyrant, Andean Swallow, Blue-and-yellow Tanager, Greenish Yellow-Finch, White-browed Blackbird, Bay-winged Cowbird.

Access

The east side of the lake is the least disturbed, most vegetated, and hence, best for birding. Rare visitors to this lake have included Andean Avocet.

Accommodation: Residencial Florida, Av. 25e Mayo.

QUILLACOLLO ROAD

The road north from Quillacollo, 10 km west of Cochabamba, rises precipitously through shrub-covered slopes (excellent for hummingbirds) and some remnant groves of *Polylepis* woodland to puna grassland just below a high pass (4,000 m/13,123 ft). Four endemics, as well as Olive-crowned Crescent-chest, Short-tailed Finch and Rufous-bellied Saltator* are the birds worth making an extra effort for at this excellent site.

Endemics

Wedge-tailed Hillstar*, Bolivian Warbling-Finch, Cochabamba Mountain-Finch*, Grey-bellied Flower-piercer.

Specialities

Andean Parakeet, Red-tailed Comet, Rock and Plain-breasted

Earthcreepers, Tawny Tit-Spinetail*, Streak-throated Canastero, Olive-crowned Crescent-chest, Cinnamon-bellied (April-Sept), Cinereous and White-fronted Ground-Tyrants, Giant Conebill*, Black-hooded Sierra-Finch, Short-tailed Finch, Rufous-bellied Saltator*.

Others

Darwin's Nothura, Torrent Duck, Mountain Caracara, Andean Lapwing, Black-winged Ground-Dove, Grey-hooded Parakeet, Green and Sparkling Violet-ears, Andean Hillstar, Giant Hummingbird, Andean Flicker, White-winged Cinclodes, Puna Canastero, Rufous-webbed Tyrant, Puna Ground-Tyrant, Plumbeous and Ash-breasted Sierra-Finches, White-winged Diuca-Finch, Band-tailed Seedeater.

Access

Polylepis is disappearing fast. If the areas mentioned below no longer exist, and that is quite possible, some exploration will be necessary. Head west out of Cochabamba to Quillacollo and turn north at the second roundabout (the one with a statue hidden by trees) towards Liriuni. Wedge-tailed Hillstar* occurs at the first series of hairpin bends. 3–4 km on from the first bends there is a fork. Turn right to **Liriuni** hot springs. The *Polylepis* and scrub north of the road a few km east of here supports Cochabamba Mountain-Finch* and Rufous-bellied Saltator*, whilst Olive-crowned Crescent-chest has been seen above here. The left fork climbs to **Cerro Tunari** and there are more patches of *Polylepis* along here. Short-tailed Finch, amongst the boulder fields near the pass, and Bolivian Warbling-Finch occur along here.

The highly localised endemic Bolivian Blackbird occurs west of Cochabamba alongside the road to Oruro. A good place to look is in introduced *Eucalyptus* trees. The area between 34 and 60 km west of the city is particularly good.

LAKE URU-URU

The localised Andean Avocet is usually present on Lake Uru-Uru, just south of Oruro

The 204-km road southwest from Cochabamba to Oruro ascends to puna and the scruffy town of Oruro (3,700 m/12,139 ft), near Lake Uru-Uru where, when the water levels are suitable, the specialities include Short-winged Grebe, the two rare Andean flamingos, and Andean Avocet.

Specialities

Short-winged Grebe, Andean* and Puna* Flamingos, Andean Avocet, Puna Plover, Brown-backed Mockingbird, Puna Yellow-Finch.

Others

White-tufted Grebe, Andean Duck, Andean Goose, Crested Duck, Puna Teal, Chilean Flamingo, Cinereous Harrier, Least Seedsnipe, Andean Lapwing, Andean Gull, Bare-faced Ground-Dove, Mountain Parakeet, Andean Flicker, Common, Rufous-banded and Slender-billed Miners, Cordilleran Canastero, Wren-like Rushbird, Many-coloured Rush-Tyrant, Andean Negrito, Yellow-winged Blackbird.

Access

The lake is just five minutes drive south of Oruro.

15 km southeast of **Chaillapata**, *en route* south from Oruro to Potosí, the road passes through an excellent wetland where Andean Flamingo*, Andean Avocet, Puna Plover, Golden-spotted Ground-Dove and Puna Miner occur.

Rufous-bellied Seedsnipe, Puna Miner and White-throated Sierra-Finch occur near **Potosí.** Head south out of town towards Tupiza and after a few km (when the eucalypts cease to dot the hillside on the left) take the track to the left to the end. Explore this area and upwards.

Citron-headed Yellow-Finch*, almost endemic to Bolivia, occurs around the village of **Betanzos**, two hours from Potosí on the road north to Sucre. Bolivian Blackbird occurs around **Tarabuco** village, two hours east of Sucre.

The southernmost state of Bolivia, **Tarija**, is still relatively little explored by birders but does support three very special birds: the endemic Red-fronted Macaw* and near-endemic Slender-tailed Woodstar, (both occur near Bermejo on the Argentinian border), and Rufous-throated Dipper* which has been recorded 25 km northwest of Entre Ríos.

LA PAZ

La Paz, the highest capital city in the world, lies at 3,636 m (11,929 ft) in a 5-km wide canyon below the snow-capped Mount Illimani (6,402 m/21,004 ft).

White-tipped Plantcutter is fairly common in gardens in lower parts of the city. Streak-throated Canastero and Peruvian Sierra-Finch occur at the Valley of the Moon, a local tourist spot. It is possible to go on a number of treks from La Paz. One such trek, the Takesi, is good for birds such as the endemic Black-hooded Sunbeam, Scaled Metaltail, Stripe-headed Antpitta, White-eared Solitaire, White-browed Conebill and Short-tailed Finch.

LA PAZ–COROICO ROAD

Map p. 96

This is one of South America's most spectacular roads, descending 4,300 m (14,108 ft) in just 80 km. The road's steepness is the main reason why so much superb forest remains.

There are four major zones to concentrate on: puna grassland and bogs above the treeline at El Cumbre (4,600 m/15,092 ft) (beware of altitude sickness); shrubby precipitous canyons above the Pongo turn-off; untouched stunted temperate forest with bamboo below the Pongo turn-off; and the subtropical 'yungas' forest, characterised by silver-leafed *cecropia* trees, beyond Chuspipata. Roadside birding is brilliant, but there are also a number of trails.

The roadside forests here support a number of species restricted to southeast Peru and west Bolivia, which are very rarely seen in Peru or elsewhere in Bolivia (especially along the Villa Tunari–Cochabamba road, see p. 89), notably Stripe-headed Antpitta, White-crowned Tapaculo, Scimitar-winged Piha, Yungas Manakin, Orange-browed Hemispingus and Golden-collared Tanager.

Endemics

Black-throated Thistletail, Rufous-faced Antpitta, Yungas Tody-Tyrant.

Specialities

Rufous-bellied Seedsnipe, Diademed Sandpiper-Plover*, Violet-throated Starfrontlet, Scaled Metaltail, Black-streaked Puffbird, Versicoloured Barbet, Blue-banded Toucanet, Hooded Mountain-Toucan*, Plain-breasted Earthcreeper, Tawny Tit-Spinetail*, Light-crowned Spinetail, Streak-throated, Line-fronted* and Scribble-tailed Canasteros, Barred Antthrush, Stripe-headed Antpitta, White-crowned Tapaculo, Scimitar-winged Piha, Yungas Manakin, Bolivian Tyrannulet, Ochraceous-breasted Flycatcher, Rufous-bellied Bush-Tyrant*, White-collared Jay, White-eared Solitaire, Orange-browed, Drab and Three-striped Hemispinguses, Golden-collared Tanager, Chestnut-bellied Mountain-Tanager, Peruvian Sierra-Finch, Short-tailed Finch, Moustached Flower-piercer.

Others

Mountain Caracara, Andean Guan, Puna and Andean Snipes, Grey-breasted Seedsnipe, White-throated Quail-Dove, Great Sapphirewing, Sword-billed Hummingbird, Booted Racket-tail, Rufous-capped Thornbill, Smoky-brown and Crimson-mantled Woodpeckers, Slender-billed Miner, Andean Tit-Spinetail, Pearled Treerunner, Streaked Tuftedcheek, Striped Treehunter, Undulated and Rufous Antpittas, Barred Fruiteater, Blue-backed Manakin, Marble-faced Bristle-Tyrant, D'Orbigny's Chat-Tyrant, Black-billed Shrike-Tyrant, Rufous-naped and Puna Ground-Tyrants, Andean Solitaire, Sepia-brown and Mountain Wrens, Black Siskin, Citrine and Three-striped Warblers, Grass-green Tanager, Superciliared and Black-eared Hemispinguses, Scarlet-bellied Mountain-Tanager, Orange-bellied Euphonia, Blue-naped Chlorophonia, White-winged Diuca-Finch, Bright-rumped Yellow-Finch, Rusty Flower-piercer.

Access

The road to Coroico begins at Villa Fatima. Roadside birding is excellent and there are also some superb trails. Puna Snipe, Rufous-bellied Seedsnipe and Peruvian Sierra-Finch occur around the bog on the right side of the road, and around the lake left of the statue, at El Cumbre pass. Diademed Sandpiper Plover* has also been recorded here. Bird the valley to the west before the Pongo turn-off (km 29 from Villa Fatima) (Line-fronted Canastero* and Stripe-headed Antpitta), the track west from Cotapata, which begins just before the garage, and the trail west beyond Cotapata (km 42.5 from Villa Fatima) (Hooded Mountain-Toucan*, Black-throated Thistletail, Orange-browed Hemispingus, Golden-collared Tanager and Moustached Flower-piercer). The 6-km stretch of road to Chuspipata, the closed road to Chulimani, and the road on to Coroico are also well worth birding.

LA PAZ–COROICO ROAD

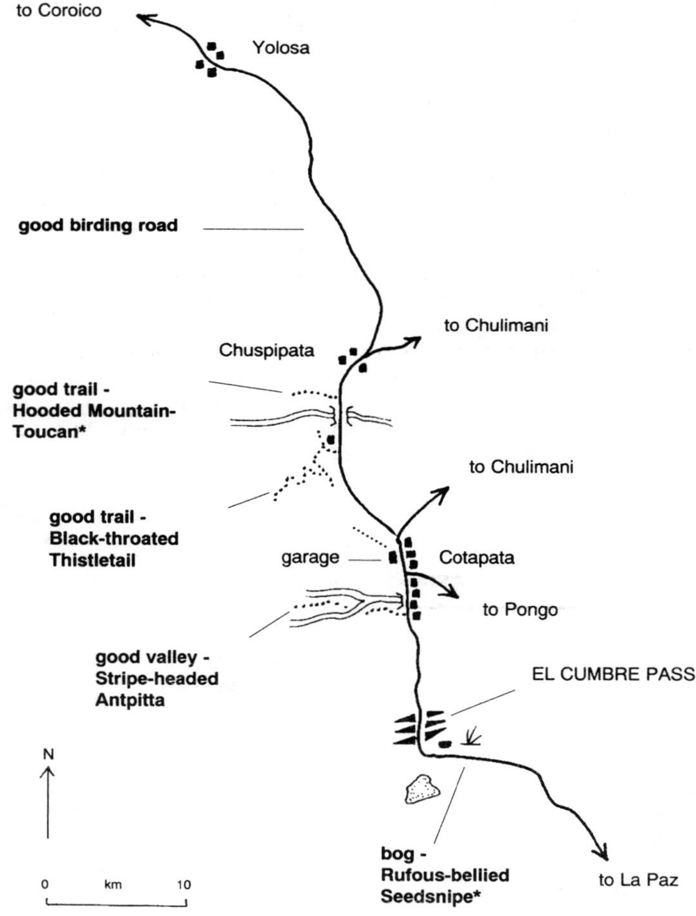

The black, white and chestnut White-eared Solitaire is an elusive species, but is seen with some regularity alongside the La Paz–Coroico road

SORATA

This small village, some six hours northwest of La Paz, situated beneath Mt Illampu, is the only known site for the endemic Berlepsch's Canastero*.

Endemics

Berlepsch's Canastero*.

Specialities

Andean Avocet, Rust-and-yellow Tanager.

Others

Black-chested Buzzard-Eagle, Mitred and Grey-hooded Parakeets, Andean Swift, Sparkling Violet-ear, White-bellied Hummingbird, Giant Hummingbird, Green-tailed Trainbearer, Bar-bellied Woodpecker, Tufted Tit-Tyrant, Rufous-breasted Chat-Tyrant, Black-billed Shrike-Tyrant, White-winged Black-Tyrant, Yellow-bellied Siskin, Band-tailed Sierra-Finch, Band-tailed Seedeater, Black-backed Grosbeak, Golden-billed Saltator, Mountain Cacique.

Access

En route to Sorata from La Paz look out for Ornate Tinamou and Short-billed Pipit around the edge of **Lake Titicaca**, as well as Short-winged Grebe on the lake itself.

The endemic Berlepsch's Canastero* occurs on the slopes at the back of the cemetery in **Sorata**, reached from Huarina on Lake Titicaca by heading inland to Achacachi (where roadside bogs support Andean Avocet) and up. Introduced eucalypts are a good place to look for the canastero.

Accommodation: Residencial Sorata; Hotel Prefectural.

The large, remote reserve known as **Beni Biological Station**, at the base of the Andean foothills in north Bolivia near San Borja, was established to protect lowland rainforest where the rare Wattled Curassow*and Unicoloured Thrush* occur. The station is a long, arduous seven-hour truck trip from Trinidad.

The rare endemic Blue-throated Macaw* was first seen in the wild near Tinidad in 1992.

The forests and savanna around **Riberalta**, a cattle centre, and **Guayaramerín**, a gold-prospector's town, in far north Bolivia on the Brazilian border, support a number of rare and spectacular Amazonian species such as Long-tailed Potoo, Rusty-necked Piculet, White-bellied Spinetail, Amazonian Umbrellabird, Fiery-capped Manakin, Pale-bellied Tyrant-Manakin, Zimmer's* and Stripe-necked Tody-Tyrants, Rufous-sided Pygmy-Tyrant*, White-eyed Attila, and Slate-coloured and Tawny-bellied Seedeaters.

The forests and savanna around **Cobija**, on the border with Brazil in extreme northwest Bolivia, support Brazilian Tinamou, Chestnut-headed and Black-banded Crakes, Scarlet-shouldered Parrotlet, Amazonian Pygmy-Owl, Black-bellied Thorntail, White-throated Jacamar, Brown-banded Puffbird, Fulvous-chinned Nunlet and Chestnut-shouldered Antwren.

NOEL KEMPFF MERCADO NP

This huge (607,050 ha) NP, situated in far east Bolivia next to the Brazilian border, was established in the late 1980s to protect cerrado, gallery forest, rainforest (igapo and terra firme), savanna and marshes. This is one of the continent's remotest wildernesses, where even indigenous people are absent.

Such pristine untouched habitats are very rare in South America so the park's list is one of the best on the continent. The 700 or so species recorded include such rarities as Zigzag Heron*, White-browed Hawk, Long-tailed Potoo, Snow-capped Manakin and Zimmer's Tody-Tyrant*.

This is a remote site only accessible via chartered flights from Santa Cruz.

Specialities

Zigzag Heron*, White-browed Hawk, Bare-faced Curassow, Starred Wood-Quail, Ocellated Crake*, Hyacinth Macaw*, Crimson-bellied Parakeet, Long-tailed Potoo, Horned Sungem, Rufous-necked Puffbird, Black-girdled Barbet, Rusty-necked Piculet, Saturnine Antshrike, Band-tailed Antbird, Amazonian Antpitta, Collared Crescent-chest, Snow-capped and Flame-crested Manakins, Pale-bellied Tyrant-Manakin, Snethlage's and Zimmer's* Tody-Tyrants, Rufous-sided Pygmy-Tyrant*, Curl-crested Jay, Tooth-billed Wren, White-banded and White-rumped Tanagers, Coal-crested Finch*, Black-and-tawny* and Grey-and-chestnut* Seedeaters, Black-throated Saltator, Yellow-billed Blue Finch*.

Others

Grey and Brazilian Tinamous, Southern Screamer, Blue-throated Piping-Guan, Razor-billed Curassow, Sungrebe, Sunbittern, Blue-and-yellow,

Scarlet and Red-and-green Macaws, Hoatzin, Spectacled Owl, Scissor-tailed Nightjar, Great Dusky Swift, Pied Puffbird, Red-necked Aracari, Channel-billed and Toco Toucans, Cream-coloured Woodpecker, White-chinned and Red-billed Woodcreepers, White-shouldered and Amazonian Antshrikes, Rusty-backed Antwren, White-backed Fire-eye, Black-spotted Bare-eye, Amazonian Umbrellabird, Band-tailed and Fiery-capped Manakins, Grey Monjita, Grey-chested Greenlet, Flavescent Warbler, Rusty-collared and Dark-throated* Seedeaters.

Other Wildlife

Giant Otter, Coatimundi, Night Monkey, Maned Wolf, Tapir, Marsh Deer.

Access

Scour the tourist agencies in Santa Cruz for the latest details on how to visit this NP.

The comfortable Flor de Oro Lodge on the banks of the Río Itenze/Guaporé is accessible by air from Santa Cruz (3 hours). Zigzag Heron* and Flame-crested Manakin occur here. It is surrounded by cerrado, gallery forest and lowland (igapo) forest. Bird the lodge area, the trail up the north face of the Caparus plateau to semi-deciduous woodland, and the series of oxbows surrounded by forest (terra firme), an hours' boat ride up river.

It is possible to fly up to the pristine cerrado on top of the Caparus plateau (camping equipment required). Horned Sungem, Collared Crescent-chest and Rufous-sided Pygmy-Tyrant* occur here.

It is also possible to fly to the wetter rainforest at Los Fierros (also connected to Santa Cruz by air) on the west side of the Caparus plateau, an area even richer than that around Flor de Oro. The forest here is accessible, ironically, via logging tracks. Extensive savanna and marshes lie close by. Starred Wood-Quail, Ocellated Crake*, Long-tailed Potoo, Saturnine Antshrike and Snow-capped Manakin occur here.

The Reserva Florestal Bajo Paraguá in northeast Santa Cruz is another remote site where Bolivian rarities such as Zimmer's Tody-Tyrant* and Tooth-billed Wren occur.

ADDITIONAL INFORMATION

Addresses

Fundacion Armonia, Casilla (Box) 3081, Santa Cruz, Bolivia (tel: 522919; fax 324971)

Asociacion Boliviana para la Proteccion de las Aves (ABPA), Casilla (Box) 3257, Cochabamba, Bolivia

Books

An Annotated List of the Birds of Bolivia, Remsen, J and Traylor, M. 1989. Buteo Books.

The Birds of the High Andes, Fjeldsa, J and Krabbe, N. 1990. Apollo. (This book covers most birds found in Bolivia).

BOLIVIAN ENDEMICS (18)	BEST SITES
Blue-throated Macaw*	North: very rare, found for the first time in the wild in 1992 in Beni dept
Red-fronted Macaw*	Central: Comarapa
Coppery Thorntail*	Unknown, possibly northeast: known only from two specimens collected around 1852
Wedge-tailed Hillstar*	Central: Quillacollo road, Pojo–Cochabamba road and Pocona
Black-hooded Sunbeam	Northwest and Central: Siberia, Chapare road and La Paz area
Bolivian Earthcreeper*	Central: Comarapa
Black-throated Thistletail	Central and northwest: Coroico road and Villa Tunari–Cochabamba road
Iquico Canastero	Central: Epizana–Cochabamba road
Berlepsch's Canastero*	Northwest: Sorata
Bolivian Recurvebill*	Northwest and central: very rare, previously known from only four specimens, recently rediscovered in upper Río Saguayo valley, Amboro NP
Ashy Antwren*	Northwest and central: very rare, Amboro NP
Rufous-faced Antpitta	Northwest and central: Siberia, Chapare road and Coroico road
Yungas Tody-Tyrant	Northwest and central: rare, Chapare road and Coroico road
Unicoloured Thrush*	North: Beni Biological Station
Bolivian Warbling-Finch	Central: Puente Lopé Mendosa and Quillacollo road
Cochabamba Mountain-Finch*	Central: Puente Lopé Mendosa and Quillacollo road
Grey-bellied Flower-piercer	Central: Siberia, Chapare and Quillacollo roads (This species has recently been reported from Argentina).
Bolivian Blackbird	Central: Cochabamba and Tarabuco

North = north of old Cochabamba–Santa Cruz road.
Northwest = north and northwest of La Paz.
Central = new road between Cochabamba and Santa Cruz, and south to Potosí.

BRAZIL

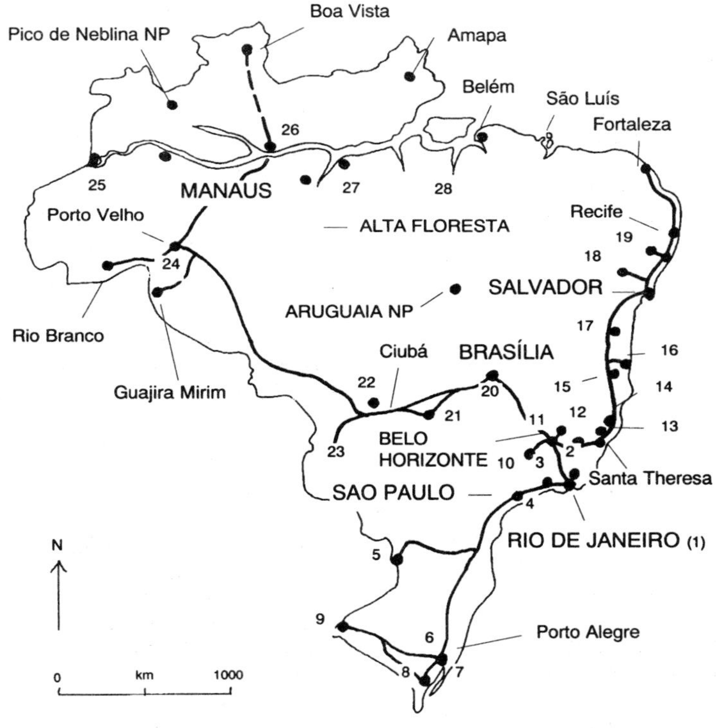

1 Rio de Janeiro
2 Serra dos Orgãos NP
3 Itatiaia NP
4 Ubatuba area
5 Iguaçu falls
6 Aparados da Serra NP
7 Mostardas Peninsula
8 Rio Grande
9 Uruguaiana
10 Serra da Canastra NP
11 Serra do Cipo NP
12 Rio Doce State Park
13 Nova Lombardia BR
14 Sooretama BR
15 Monte Pascoal NP
16 Porto Seguro
17 Boa Nova
18 Canudos
19 Parque Estadual da Pedra Talhada
20 Brasília NP
21 Emas NP
22 Chapada dos Guimarães NP
23 The Pantanal
24 Rondonia
25 Tabatinga
26 Manaus
27 Amazonia (Tapajós NP)
28 Carajás

INTRODUCTION

Summary

Brazil is very, very large, and so is the 'field guide', but the infrastructure, especially the internal air network, is excellent. However, despite the fact that getting to all the major habitats, from Amazonian rainforest, to the Pantanal and the Atlantic forests of the east coast, is easy, it would be impossible to do justice to all the sites on a short trip. A trip that concentrates on the east coast will provide the chance to see many of the endemics though. One drawback is obtaining permission to visit many of the best sites (a basic knowledge of Portuguese is recommended), but patient and persistent birders who overcome such hurdles may be rewarded with some of South America's most threatened and beautiful birds.

Size

Brazil is the fifth largest country in the world, and by far the largest in South America (8,511,965 km^2). It is larger than the whole of Western Europe, almost as large as the USA and takes up nearly half of the South American continent, covering over six times the area of Peru and 30 times the area of Ecuador. Hence the distances to be covered are huge: 4,320 km north to south, and 4,328 km west to east.

Getting Around

The excellent internal air network, which connects the larger cities several times a day and is capable of reaching even the remotest parts of the country, overcomes the problem of great distances between major birding sites. Although the best roads are still concentrated in the southeast, those in the interior are being paved, enhancing the already wide-reaching, comfortable and efficient bus network.

A basic knowledge of Portuguese would be a tremendous asset in Brazil, not only to help getting around but also to pave the way to gaining permission to visit many of the reserves.

Health and Safety

Mosquitoes are ubiquitous and protection against malaria and dengue fever is essential, even if just visiting Rio. Visitors intending to bird the Amazon basin need immunisation against yellow fever, cholera, hepatitis and typhoid. Sandflies abound here, and the water needs boiling and/or purifying.

Safety is a major problem, especially in Rio and Salvador, where it is advisable not to wear jewellery, travel alone or carry one cruzeiro more than you need. Muggers are numerous in Rio, although some have been good enough to leave their victims with the fare back to their hotel! These problems are a legacy of mass poverty, which also means overcharging is common. Expect minor rip-offs, but be generous to those who give you a good deal.

Climate and Timing

The best time for birding in Brazil is May to October, especially September–October. The early part of the year is hot and holiday time for the Brazilians, which means hotels, planes and buses tend to be booked. This is also the rainy season in the southeast and north. May to

October is the dry season in the Pantanal, thus concentrating the birds, and the least humid time in the southeast.

Habitats

The Amazon basin of west and north Brazil accounts for more than a third of the country. This is a hot, wet and humid region which was once an endless carpet of lowland rainforest, the most avian-rich forest on earth (in the west). At times the depressing destruction of this great wilderness was measured at one football pitch-sized area per second. Vast areas are now devoid of vegetation altogether or covered in poor secondary growth.

North of the Amazon basin the Guiana highlands are, for the most part, still forested. The Brazilian highlands (also known as the central plateau) lie to the southeast of the Amazon basin. This plateau, ranging in height from 300 m (984 ft) to 900 m (2,953 ft) is mostly covered in dry, rolling grasslands known as the 'campo' and scrubby woodlands known as 'cerrado'. The world's biggest seasonally flooded freshwater wetland, the Pantanal, lies at the southwest base of the central plateau. To the northeast, the land is arid and covered in patchy, scrubby woodland known as 'caatinga'.

The Serro do Mar, an isolated, narrow range of mountains, rising to 2,898 m (9, 508 ft), runs parallel to the southeast coast. Less than 5% of the cloud and subtropical ('Atlantic') forests that once covered the slopes of these mountains, remains, and most of the birds, many of them endemic, occur only in the few reserves set up to protect the remnants. Subtropical forest also surrounds the world's most impressive waterfall at Iguaçu in the southwest, on the border with Argentina.

The state of Rio Grande do Sul in far south Brazil is a land of wet 'pampa' grasslands with pockets of *Araucaria* (Monkey-puzzle) forest.

Conservation

Brazil has suffered heavily at the hand's of the mad axeman. Much of the Amazonian rainforest and 95% of the Atlantic forest have been lost, so far. These incredibly rich habitats and the birds inhabiting them are now amongst the most endangered on earth. Many of the other habitats, such as the cerrado, have been badly degraded, and even the Pantanal is threatened by crazy hydrological schemes. Cynical or, some say, realistic birders are rushing to Brazil, for obvious reasons. Where will it all end? Will tour companies be offering trips to the 'Great Brazilian Desert' in years to come?

Almost 100 (97) threatened species, of which no less than 58 are endemic, occur in Brazil. A further 35 endemics are near-threatened. The habitats these species occur in need to be conserved now.

Bird Families

Brazil is second only to Peru in terms of total family diversity, and second only to Bolivia in terms of family endemism. 84 of the 92 families which regularly occur in South America are represented, including 23 of the 25 Neotropical endemic families and seven of the nine South American endemic families. Only Magellanic Plover and seedsnipes are absent.

Well-represented families include trumpeters, parrots, toucans, antbirds, gnateaters and cotingas.

Bird Species

Despite its huge size, Brazil's list of 1,661 species is only the third largest in South America. Both Colombia and Peru boast more birds, but they have lowland Amazonian rainforest *and* temperate and subtropical Andean forests within their boundaries. The Andes do not cross Brazil's borders.

Non-endemic specialities and spectacular species include Brazilian Merganser*, Crowned Eagle*, Bare-faced Curassow, Hyacinth Macaw*, Rufous Potoo, Fiery-tailed Awlbill*, Plovercrest, Racket-tailed Coquette, Rufous-capped Motmot, Helmeted Woodpecker*, Giant Antshrike, Black-bellied Gnateater, Collared Crescent-chest, Spotted Bamboowren*, Black-necked and Guianan Red-Cotingas, Swallow-tailed Cotinga*, Purple-breasted and Pompadour Cotingas, Crimson Fruitcrow, Bare-throated Bellbird*, Guianan Cock-of-the-Rock, Black-and-white Monjita*, Cock-tailed Tyrant*, Guianan Gnatcatcher, Yellow-faced Siskin*, and a number of fabulous tanagers.

Many specialities include species confined to southeast Brazil, north-east Argentina and east Paraguay.

Endemics

Brazil has more endemic species than any other South American country and one of the highest totals for any country in the world, 179 in all. Of these, 76 (43%) occur only at various sites near the coast north from São Paulo to Salvador. A further 23 occur from São Paulo to the north-east, and 25 are restricted to sites only north of Salvador, making a total for the east 'coast' of 124 (69%). Another 30 (17%) are restricted to the Amazon basin, especially the east.

Brazil's endemics include the rarest bird in the world, Spix's Macaw*, of which only one lonely individual remains in the wild. The rest of the endemics, of which many are endangered, include six cracids, 15 more parrots including Lear's Macaw*, White-winged Potoo, 12 humming-birds, Three-toed Jacamar*, four woodcreepers, 16 furnariids, 32 antbirds, two gnateaters, ten cotingas including Grey-winged Cotinga*, Hooded* and Black-headed* Berryeaters, the possibly extinct Kinglet Calyptura*, Banded*, White-tailed and White-winged* Cotingas, 19 tyrants, and 22 'tanagers' including Seven-coloured Tanager* and Black-legged Dacnis*.

Expectations

Although Brazil is a huge country, it is possible to see 500 species including up to 50 endemics on a three-week trip that includes the southeast and east, and up to 700 on more prolonged trips of 40 days or so. Whilst it may be possible to squeeze in a brief visit to the Pantanal or the Amazon area on such a trip, it is advisable to make a long, long trip, or several short ones, in order to see Brazil's best birds.

RIO DE JANEIRO

A small sample of Brazil's southeastern endemics can be seen in the vicinity of Rio de Janeiro, on the southeast coast. Tijuca NP, where over 200 species have been recorded, is the best site, but most of the birds here, and a lot more, occur in nearby Serra dos Orgãos NP (55 km north), and Itatiaia NP (174 km west).

Endemics
Plain Parakeet, Scaled Antbird, Black-cheeked Gnateater, Pin-tailed Manakin, Eye-ringed Tody-Tyrant*, Yellow-lored Tody-Flycatcher, Long-billed Wren, Rufous-headed Tanager.

Specialities
Southern Pochard, Maroon-bellied Parakeet, Violet-capped Woodnymph, Scaled and Lesser Woodcreepers, Spot-breasted Antvireo*, Streak-capped Antwren, White-shouldered Fire-eye, Bare-throated Bellbird*, Blue Manakin, Crested Doradito, Ruby-crowned, Brazilian and Green-headed Tanagers, Capped Seedeater, Unicoloured Blackbird.

Others
Magnificent Frigatebird, Blackish Rail, Wattled Jacana, Picazuro Pigeon, Plain-breasted Ground-Dove, Scaly-headed Parrot, Guira Cuckoo, Dusky-throated Hermit, Channel-billed Toucan, White-barred Piculet, Campo Flicker, Sharp-tailed Streamcreeper, Streaked Xenops, Eastern Slaty Antshrike, Plain Antvireo, White-flanked Antwren, Masked Water-Tyrant, Lemon-chested Greenlet, Yellow-legged, Rufous-bellied and Creamy-bellied Thrushes, White-rumped Swallow, Masked Yellowthroat, Red-crowned Ant-Tanager, Purple-throated Euphonia, Red-necked Tanager, Blue Dacnis, Double-collared Seedeater, Green-winged Saltator, Chestnut-capped and White-browed Blackbirds.

Access
The **Botanical Gardens** (140 ha), 8 km south of the city centre near Lagoa Rodrigo, are worth a look if you have some time to spare. The best birding near the city centre though is in **Tijuca NP**, southwest of the city centre. Bird along the Estrada do Redentor road, which runs east from the road bisecting the park, to Corcovado. There is a Dusky-throated Hermit lek midway between the toll gate and Corcovado.

Crested Doradito and Capped Seedeater occur at the small marsh in Reserva Biologica da Barra west of Lagoa da Tijuca, southwest of Tijuca NP. The scarce Southern Pochard has been recorded on Lagoa Piratininga, southeast of Niteroi, which is on the east side of Rio harbour.

Accommodation: Ipanema Beach: Arpoador Inn. Rio: Hotel De Bret (A), Hotel Nacional (A).

It may be possible to fly from Rio to **Ilha de Trindade**, an island over 1,000 km off the coast of east Brazil, where Herald Petrel, Red-footed Booby and Common White-Tern occur.

SERRA DOS ORGÃOS NP

Although only 55 km north of Rio, this spectacular, mountainous NP (30,000 ha) supports more endemics than any other reserve in Brazil (although the Boa Nova area (see p.128), near Salvador, supports more endemics per sq km).

Many of the birds are similar to those that occur at Itatiaia NP. They both support Itatiaia Thistletail, Rufous-tailed Antbird*, Black-and-gold Cotinga*, Brown Tanager* and Bay-chested Warbling-Finch, but the rare Grey-winged Cotinga* occurs only in Serra dos Orgãos NP. This is *the* species to concentrate on here, although Three-toed Jacamar*, Rio de Janeiro Antwren* and Swallow-tailed Cotinga*, three birds which are very scarce, all occur in the vicinity.

Many species which occur only in southeast Brazil, east Paraguay and northeast Argentina are also found in this NP. These include Mantled Hawk*, Rufous-capped Motmot and Spot-billed Toucanet.

Endemics

Plain Parakeet, Brown-backed Parrotlet*, Sombre Hummingbird, Brazilian Ruby, Three-toed Jacamar*, Crescent-chested Puffbird, Yellow-eared Woodpecker, Itatiaia Thistletail, Pallid Spinetail, Red-eyed Thornbird, Pale-browed Treehunter, White-collared Foliage-gleaner, Rufous-backed Antvireo, Star-throated and Rio de Janeiro* Antwrens, Ferruginous, Rufous-tailed*, Ochre-rumped*, Scaled, Rio de Janeiro* and White-bibbed Antbirds, Such's Antthrush, Black-cheeked Gnateater, Black-and-gold* and Grey-winged* Cotingas, Hooded Berryeater*, Cinnamon-vented Piha*, Pin-tailed Manakin, Eye-ringed* and Hangnest* Tody-Tyrants, Yellow-lored Tody-Flycatcher, Grey-capped* and Serra do Mar* Tyrannulets, Velvety Black-Tyrant, Grey-hooded Attila, Long-billed Wren, Brown*, Cinnamon, Rufous-headed, Azure-shouldered*, Golden-chevroned and Brassy-breasted Tanagers, Bay-chested Warbling-Finch.

Specialities

Tataupa Tinamou, Rufous-thighed Kite, Mantled Hawk*, Slaty-breasted Wood-Rail, Blue-winged Macaw*, Pileated Parrot*, Sooty Swift, Scale-throated Hermit, Plovercrest, Violet-capped Woodnymph, Rufous-capped Motmot, Spot-billed Toucanet, Yellow-browed Woodpecker*, Black-billed Scythebill, Rufous-capped and Chicli Spinetails, Black-capped Foliage-gleaner, Sharp-billed Treehunter, Giant, Tufted, Large-tailed and White-bearded* Antshrikes, Bertoni's Antbird, Streak-capped Antwren, White-shouldered Fire-eye, Speckle-breasted Antpitta, Mouse-coloured Tapaculo, Swallow-tailed Cotinga*, Bare-throated Bellbird*, Blue Manakin, Drab-breasted Bamboo-Tyrant, Blue-billed Black-Tyrant, Shear-tailed Grey-Tyrant*, Ruby-crowned, Diademed and Red-necked Tanagers, Uniform Finch, Red-rumped Warbling-Finch, Yellow-billed Blue Finch*.

Others

Brown Tinamou, Barred Forest-Falcon, Spot-winged Wood-Quail, Plumbeous Pigeon, Mottled Owl, White-collared Swift, Planalto and Reddish Hermits, Black Jacobin, Glittering-bellied Emerald, Fork-tailed Woodnymph, White-throated Hummingbird, Sapphire-spangled Emerald, Surucua Trogon, White-eared Puffbird, White-barred Piculet, Yellow-throated Woodpecker, White-throated, Planalto and Lesser Woodcreepers, White-eyed Foliage-gleaners, Rufous-breasted Leaftosser, Rufous-capped Antshrike, Short-tailed Antthrush, Variegated Antpitta, Rufous Gnateater, Sharpbill, Planalto, Rough-legged and Mottle-cheeked Tyrannulets, Cliff Flycatcher, Chestnut-crowned and White-winged Becards, Rufous-crowned Greenlet, Yellow-legged, Pale-breasted and White-necked Thrushes, Hellmayr's Pipit, White-rimmed

Warbler, Black-goggled Tanager, Chestnut-bellied Euphonia, Blue-naped Chlorophonia, Pileated Finch.

Access

The NP entrance is just south of Teresópolis, 55 km northeast of Rio. Giant Antshrike occurs near the dam, 3 km west of the entrance, and White-bearded Antshrike* a little above it. Black-and-Gold Cotinga* occurs above the dam. Grey-winged Cotinga* occurs 3–4 hours' walk up from the dam, in the vicinity of the two demolished shelters, on the Da Pedro do Sino trail. Hooded Berryeater* also occurs along here.

Swallow-tailed Cotinga* occurs along the northern edge of the park alongside the road between Teresópolis and Petrópolis. 22 km west of Teresópolis there is a 3-km track south to a garden where the cotinga has been recorded. Three-toed Jacamar* and Rio de Janeiro Antbird* occur in the remnant forest patches north of Nova Fribugo (northeast of Teresópolis). Rio de Janeiro Antwren* was discovered in 1988 at Santo Aleixo, south of Teresópolis, but has yet to be relocated.

Accommodation: Hotel Teresópolis. It is possible to stay in the park with prior permission from the National Forestry Institute, Serra dos Orgãos, Casa 8, CEP 25.950, Teresópolis.

ITATIAIA NP

Map p. 109

This superb park (12,000 ha), 174 km west of Rio (257 km east of São Paulo) protects subtropical and temperate forests, with large stands of giant bamboo, as well as high grasslands on the slopes of some of Brazil's highest mountains. Mount Itatiaia rises to 2,787 m (8,490 ft).

The NP is a must for any birder in search of the Brazil's endemics. Situated in the midst of the Serra do Mar, its list of over 270 includes many of them. Restricted endemics shared with Serra dos Orgãos NP include Itatiaia Thistletail, Rufous-tailed Antbird*, Black-and-gold Cotinga*, Brown Tanager* and Bay-chested Warbling-Finch, whilst the specialities include Slaty Bristlefront*, Swallow-tailed Cotinga* and Olive-green Tanager. Many of the species shared with Serra dos Orgãos are somewhat easier to find here, and a week long stay may produce over 150 species, including over 20 endemics.

Many species restricted to southeast Brazil, east Paraguay and northeast Argentina also occur here. These include Tawny-browed Owl, Plovercrest, Giant Antshrike, Black-capped Manakin*, Shear-tailed Grey-Tyrant* and Thick-billed Saltator*.

Endemics

Plain Parakeet, Brown-backed Parrotlet*, Sombre Hummingbird, Brazilian Ruby, Itatiaia Thistletail, Pallid Spinetail, Pale-browed Treehunter, White-collared Foliage-gleaner, Rufous-backed Antvireo, Star-throated Antwren, Ferruginous, Rufous-tailed*, Ochre-rumped*, Rio de Janeiro* and White-bibbed Antbirds, Such's Antthrush, Slaty Bristlefront*, White-breasted Tapaculo, Black-and-Gold Cotinga*, Pin-tailed Manakin, Brown-breasted Bamboo-Tyrant, Yellow-lored Tody-Flycatcher, Serra do Mar Tyrannulet*, Velvety Black-Tyrant, Grey-hooded Attila, Brown*, Cinnamon, Rufous-headed, Olive-green, Golden-chevroned, Brassy-breasted and Gilt-edged Tanagers, Bay-chested Warbling-Finch.

The rare black and yellow Swallow-tailed Cotinga is seen occasionally at Itatiaia NP*

Specialities

Mantled Hawk*, Dusky-legged Guan, Slaty-breasted Wood-Rail, Ash-throated Crake, Maroon-bellied Parakeet, Pileated Parrot*, Tawny-browed Owl, Sooty and Ashy-tailed Swifts, Scale-throated Hermit, White-vented Violet-ear, Plovercrest, Violet-capped Woodnymph, Rufous-capped Motmot, Saffron Toucanet*, Red-breasted Toucan, White-spotted and Yellow-browed* Woodpeckers, Scaled Woodcreeper, Rufous-capped Spinetail, White-browed Foliage-gleaner*, Sharp-billed Treehunter, Giant, Tufted, Large-tailed and White-bearded* Antshrikes, White-shouldered Fire-eye, Speckle-breasted Antpitta, Mouse-coloured Tapaculo, Shrike-like* and Swallow-tailed* Cotingas, Bare-throated Bellbird*, Blue and Black-capped* Manakins, Drab-breasted Bamboo-Tyrant, Planalto Tyrannulet, Olivaceous Elaenia, Eared Pygmy-Tyrant, White-rumped Monjita, Blue-billed and Crested Black-Tyrants, Streamer-tailed Tyrant, Shear-tailed Grey-Tyrant*, Rufous-tailed Attila, Curl-crested Jay, Chestnut-headed, Ruby-crowned, Brazilian, Diademed and Green-headed Tanagers, Uniform Finch, Red-rumped Warbling-Finch, Thick-billed Saltator*, Golden-winged Cacique.

Others

Brown Tinamou, Black-and-white and Ornate Hawk-Eagles, Barred Forest-Falcon, Spot-winged Wood-Quail, Plumbeous Pigeon, Blue-winged Parrotlet, Guira Cuckoo, White-collared Swift, Black Jacobin, Festive Coquette, White-throated Hummingbird, Sapphire-spangled Emerald, Surucua Trogon, White-eared Puffbird, White-barred Piculet, Green-barred Woodpecker, White-throated, Planalto and Lesser Woodcreepers, Firewood-gatherer, Buff-fronted Foliage-gleaner, Rufous-breasted Leaftosser, Sharp-tailed Streamcreeper, Streaked Xenops, Rufous-capped Antshrike, Spot-breasted Antvireo*, Variegated Antpitta, Rufous Gnateater, Sharpbill, Ochre-faced Tody-Flycatcher, Rough-legged, Yellow, Sooty and Mottle-cheeked Tyrannulets, Large-headed Flatbill, White-throated Spadebill, Black-tailed and Cliff Flycatchers, Masked Water-Tyrant, Rufous-crowned Greenlet, Yellow-legged and Pale-breasted Thrushes, Black-capped Donacobius, Tawny-headed Swallow, White-rimmed Warbler, Magpie, Orange-headed, Black-goggled and Fawn-breasted Tanagers, Purple-throated and Chestnut-bellied Euphonias, Blue-naped Chlorophonia, Blue Dacnis, Swallow-Tanager, Long-tailed Reed-Finch, Great Pampa-Finch, White-bellied Seedeater, Black-throated Grosbeak, Chestnut-capped and Chopi Blackbirds.

ITATIAIA NP

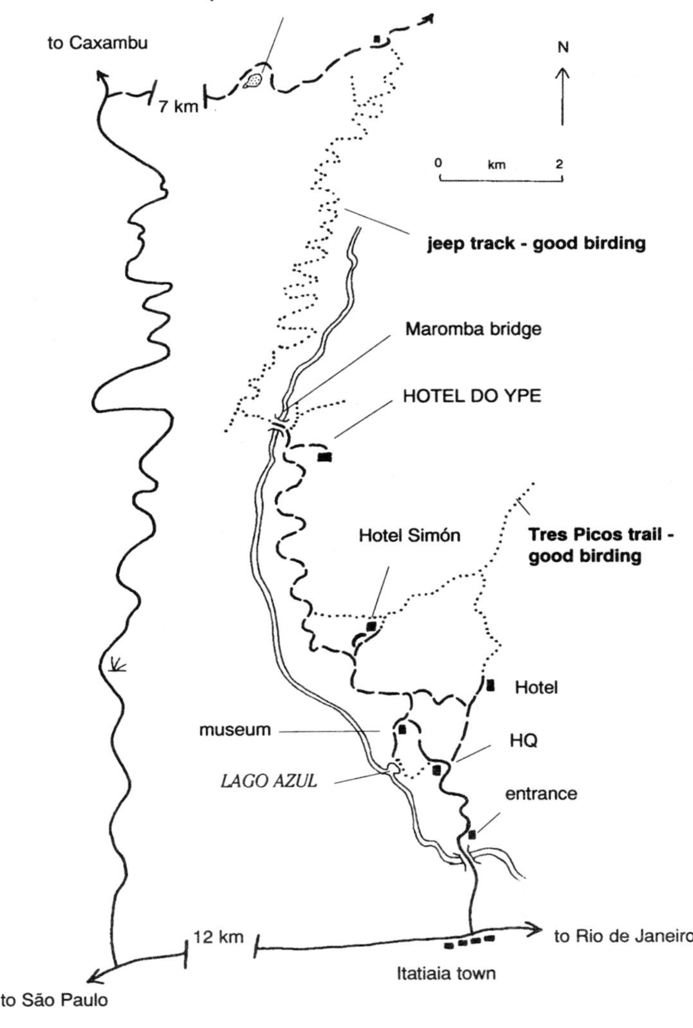

Other Wildlife

Black-faced Titi Monkey.

Access

The entrance to the NP is north of the town of Itatiaia, 174 km west of Rio on the road to São Paulo. From the HQ walk west to bird Lago Azul and bird the road up from the HQ, past the Museum, to Hotel Simón, where the Tres Picos trail is worth a prolonged look. The lower sections of the park support White-bearded Antshrike*, the very rare Shrike-like Cotinga* (do not *expect* to see this mythical creature), and Swallow-tailed* Cotinga. The vicinity of the Hotel do Ype (the best place to stay) is an excellent birding area. Bird tables and feeders in the grounds of the hotel attract hummers and tanagers. There is also a broken water pipe here which forms a miniature marsh where Slaty-breasted Wood-Rail and Sharp-tailed Streamcreeper sometimes appear. Giant Antshrike also occurs around this hotel. Above the hotel, bird the road up to Maromba bridge, and the jeep-track beyond. Black and-Gold Cotinga*, Black-capped Manakin*, Brown-breasted Bamboo-Tyrant, Serra do Mar Tyrannulet* and Bay-chested Warbling-Finch all occur on the higher reaches of the jeep-track and along the Tres Picos trail.

High-altitude species include Plovercrest, Itatiaia Thistletail, Slaty Bristlefront*, Planalto Tyrannulet, Velvety and Crested Black-Tyrants, Shear-tailed Grey-Tyrant* and Diademed Tanager. The Caxambu road is a good way to reach the high temperate forest and grasslands which support these species. To reach this road return to the main road between Rio and São Paulo, head west for 12km then turn north. Bird the marsh 6 km from here then turn east onto a track after a further 20 km. Bird the pool 8 km up from here (Itatiaia Thistletail) and the area beyond.

Rio de Janeiro Antbird* occurs east of the park near Penedo.

Accommodation: Hotel do Ype (A), one of the best hotels for birders in the world; Hotel Simón; Hotel Repouso.

UBATUBA AREA

The expensive beach resort of Ubatuba, 367 km south of Rio, lies near remnant lowland Atlantic forest, which supports some very rare and restricted endemics, including Red-eyed Thornbird, Black-hooded Antwren*, rediscovered in 1987, Squamate Antbird, Slaty Bristlefront*, Buff-throated Purpletuft*, Fork-tailed Tody-Tyrant*, Oustalet's Tyrannulet*, Olive-green Tanager and Black-legged Dacnis*.

Endemics

Plain Parakeet, Brown-backed Parrotlet*, Saw-billed Hermit*, Brazilian Ruby, Crescent-chested Puffbird, Pallid Spinetail, Red-eyed Thornbird, Pale-browed Treehunter, Rufous-backed Antvireo, Salvadori's*, Unicoloured* and Black-hooded* Antwrens, Ochre-rumped, Scaled and Squamate Antbirds, Such's Antthrush, Slaty Bristlefront*, White-breasted Tapaculo, Hooded Berryeater*, Buff-throated Purpletuft*, Pin-tailed Manakin, Brown-breasted Bamboo-Tyrant, Eye-ringed*, Hangnest* and Fork-tailed* Tody-Tyrants, Yellow-lored Tody-

Flycatcher, Oustalet's Tyrannulet*, Grey-hooded Attila, Long-billed Wren, Rufous-headed, Olive-green, Azure-shouldered*, Golden-chevroned and Brassy-breasted Tanagers, Black-legged Dacnis*.

Specialities

Solitary* and Tataupa Tinamous, Mantled Hawk*, Slaty-breasted Wood-Rail, Pileated Parrot*, Scale-throated Hermit, Rufous-capped Motmot, Red-breasted Toucan, Yellow-fronted and White-spotted Woodpeckers, Rufous-capped and Grey-bellied Spinetails, Black-capped Foliage-gleaner, Giant, Tufted and Chestnut-backed Antshrikes, Bertoni's Antbird, Speckle-breasted Antpitta, Spotted Bamboowren*, Mouse-coloured Tapaculo, Bare-throated Bellbird*, Blue Manakin, Grey-hooded Flycatcher, Bay-ringed Tyrannulet*, Eared Pygmy-Tyrant, White-rumped Monjita, Brazilian, Green-headed and Red-necked Tanagers, Uniform Finch, Buffy-fronted Seedeater*, Sooty Grassquit, Golden-winged Cacique.

Others

Brown Tinamou, Magnificent Frigatebird, Black Hawk-Eagle, Spot-winged Wood-Quail, Blue-winged Parrotlet, Lesser Swallow-tailed Swift, Planalto and Reddish Hermits, Swallow-tailed Hummingbird, Versicoloured and Glittering-throated Emeralds, White-necked Puffbird, Channel-billed Toucan, Yellow-throated, Blond-crested and Lineated Woodpeckers, Rufous-breasted Leaftosser, Plain Xenops, Spot-backed Antshrike, Short-tailed Antthrush, Rufous Gnateater, Sharpbill, White-bearded Manakin, Large-headed Flatbill, Sulphur-rumped Flycatcher, Greyish Mourner, Chestnut-crowned and White-winged Becards, Rufous-crowned and Lemon-chested Greenlets, Long-billed Gnatwren, White-thighed Swallow, Neotropical River Warbler, Orange-headed, Black-goggled and Fawn-breasted Tanagers, Black-throated Grosbeak, Crested Oropendola, Red-rumped Cacique, Giant Cowbird.

Access

Red-eyed Thornbird, Unicoloured* and Black-hooded* Antwrens, and Black-legged Dacnis* occur in the mangroves between Angra dos Reis, 151 km west of Rio, and Parati, 100 km further west. The Serra da Bocaina NP lies inland from Paratí but has been little explored by birders. West of Paratí, just east of km post 45, east of Ubatuba, turn north on to a track to **Fazenda Capricornio**. Permission is usually given to bird here, a good site for the rare Buff-throated Purpletuft*. Spotted Bamboowren*, Fork-tailed Pygmy-Tyrant* and Buffy-fronted Seedeater* occur at a second fazenda, reached by turning right before reaching Capricornio. Ask permission to bird here also.

Purple-winged Ground-Dove*, Blue-bellied Parrot*, Canebrake Groundcreeper* and Black-backed Tanager* have been recorded at **Boraceia Forest Reserve**, three hours east of São Paulo. For permission to visit and details of birding contact: Dra. Francesca do Val, Museu de Zoologia, Avenida Nazare 481, São Paulo, Caixa Postal 7172-01051.

Some 200 km due southwest of São Paulo, some forest still remains along the coast from Iguape south to Ariri, and on the **Ilha do Cardoso**. This island, accessible from Cananéia, supports the endemic Red-tailed Parrot*. The most reliable site for this is the Research Station run by Centro de Estudos e Pesquisas Aplicadas de Recursos Naturals da Ilha do Cardoso (CEPARNIC), Caixa Postal 43, 11990 Cananéia, São Paulo, P. 0138 51 1163. There are many other birds here including Unicoloured Antwren*, Black-headed Berryeater* and Black-backed Tanager*.

Mantled Hawk* and Festive Coquette occur in **Marumbi NP**, situated between São Joao da Graciosa (14 km north of Morretes) and the Curitaba–São Paulo road. A road traverses this well-forested mountainous park, enabling excellent access, whilst the nearby Morretes-Matinhos road has several tracks leading into lowland forest. Other species present include Saw-billed Hermit*, Squamate Antbird, Sharpbill, Bay-ringed Tyrannulet* and Chestnut-backed Tanager. Morretes makes a good base.

IGUAÇU FALLS

Map p. 63

Iguacu falls, the world's biggest waterfall, in south Brazil on the border with Argentina, is surrounded by subtropical forest which supports some rare birds restricted to south Brazil, northeast Argentina and east Paraguay. Particular specialities of this site include Black-fronted Piping-Guan*, Ochre-collared Piculet, Helmeted Woodpecker*, Spotted Bamboowren*, Sao Paulo Tyrannulet* and Green-chinned Euphonia*.

Whilst the Brazilian side of the falls is the best place to look for Helmeted Woodpecker*, the Argentinian side is much better overall (see p. 62).

Specialities

Solitary Tinamou*, Dusky-legged Guan, Black-fronted Piping-Guan*, Maroon-bellied Parakeet, Pileated Parrot*, Great Dusky Swift, Saffron Toucanet*, Red-breasted Toucan, Ochre-collared Piculet, Yellow-fronted, White-spotted, Helmeted* and Robust Woodpeckers, Streak-capped Antwren, Spotted Bamboowren*, Blue Manakin, Grey-hooded Flycatcher, Southern Antpipit, Southern Bristle-Tyrant*, Sao Paulo* and Bay-ringed* Tyrannulets, Eared Pygmy-Tyrant, Russet-winged Spadebill*, Plush-crested Jay, Creamy-bellied Gnatcatcher*, Chestnut-headed and Ruby-crowned Tanagers, Green-chinned Euphonia*, Green-headed Tanager, Uniform Finch.

Others

Collared Forest-Falcon, Spot-winged Wood-Quail, Sungrebe, White-eyed Parakeet, Versicoloured Emerald, Black-throated and Surucua Trogons, Toco Toucan, Blond-crested Woodpecker, White-throated, Planalto and Lesser Woodcreepers, Ochre-breasted and White-eyed Foliage-gleaners, Rufous-winged Antwren, Short-tailed Antthrush, Variegated Antpitta, Red-ruffed Fruitcrow, Wing-barred Manakin, Grey Elaenia, Three-striped Flycatcher, Rufous-crowned Greenlet, Pale-breasted Thrush, Golden-crowned Warbler, Chestnut-vented Conebill, Orange-headed, Guira, Black-goggled and Fawn-breasted Tanagers, Blue-naped Chlorophonia, Red-rumped Cacique, Epaulet Oriole, Giant Cowbird.

Other Wildlife

Jaguar. Especially good for butterflies.

Access

The Poco Preto trail, 2.5 km from the Iguaçu NP entrance, near the falls, is the best place to bird. Helmeted Woodpecker*, Sao Paulo Tyrannulet* and Creamy-bellied Gnatcatcher* all occur along here, as well as Jaguar. Permission must be obtained from the director at the park entrance, but this is not usually a problem. The falls, at their most impressive in January and February, are good for Great Dusky Swift at dusk. Another good site, north of Foz do Iguaçu, is the Itaipu dam. However, it would be wiser to cross to the more extensive and more accessible forest on the Argentina side (see p. 62).

Accommodation: Falls: Hotel das Cataratas (A). Foz do Iguaçu: many.

APARADOS DA SERRA NP

Maps p. 114 and p. 116

Situated 187 km northeast of Porto Alegre in Rio Grande do Sul, Brazil's southernmost state, this NP protects high plateau grasslands dotted with marshes and stands of *Araucaria* forest, around the spectacular Itaimbezinho canyon. As well as the endemics, which include the localised Long-tailed Cinclodes and Striolated Tit-Spinetail, specialities of this site include Giant Snipe, Vinaceous Parrot*, Straight-billed Reedhaunter* and Speckle-breasted Antpitta.

Endemics

Long-tailed Cinclodes, Striolated Tit-Spinetail, Brown-breasted Bamboo-Tyrant, Serro do Mar Tyrannulet*, Velvety Black-Tyrant, Azure-shouldered Tanager*.

Specialities

Spotted Nothura, Dusky-legged Guan, Slaty-breasted Wood-Rail, Giant Snipe, Vinaceous Parrot*, Rusty-barred Owl, Sooty and Biscutate Swifts, Plovercrest, Mottled Piculet*, White-spotted Woodpecker, Scaled Woodcreeper, Araucaria Tit-Spinetail*, Chicli, Grey-bellied and Olive Spinetails, Straight-billed Reedhaunter*, Sharp-billed Treehunter, Giant and Large-tailed Antshrikes, Dusky-tailed Antbird, Speckle-breasted Antpitta, Mouse-coloured Tapaculo, Black-capped Manakin*, Rough-legged and Greenish Tyrannulets, Olivaceous Elaenia, Sooty Tyrannulet, Black-and-white Monjita*, Blue-billed Black-Tyrant, Azure Jay*, Chestnut-headed, Diademed and Chestnut-backed Tanagers, Red-rumped Warbling-Finch, Lesser Grass-Finch*, Thick-billed Saltator*, Golden-winged Cacique, Saffron-cowled Blackbird*.

Others

Brown Tinamou, White-tufted Grebe, Buff-necked Ibis, Barred and Collared Forest-Falcons, Spot-winged Wood-Quail, Blackish and Plumbeous Rails, Spot-flanked Gallinule, Red-legged Seriema, Variable Screech-Owl, Rufous-throated Sapphire, Campo Flicker, Planalto, White-throated and Lesser Woodcreepers, Firewood-gatherer, Buff-browed Foliage-gleaner, Variable Antshrike, Variegated Antpitta, Rufous Gnateater, Grey Monjita, Yellow-browed Tyrant, Hellmayr's Pipit, White-rimmed Warbler, Long-tailed Reed-Finch, Black-and-rufous Warbling-Finch, Great Pampa-Finch, Yellow-rumped Marshbird.

Access

Head north out of Porto Alegre towards Taquará and São Francisco de Paula. Turn north off the RS 020 3 km west of São Francisco de Paula, on the road to Canela (a good birding area). A little way along this road (good for Speckle-breasted Antpitta), on the left, is the Hotel Veraneio Hampel (which would make a good base), complete with a nature trail well worth birding.

From São Francisco de Paula head north towards Tainhas on an excellent birding road to the park itself. Long-tailed Cinclodes and Straight-billed Reedhaunter* occur along the track to the east after 18 km. Back on the main dirt road, continue past Tainhas and then turn east in response to the 'Itaimbezinho' sign. Straight-billed Reedhaunter* occurs in the marsh to the east of this road, before the NP entrance gate. Long-tailed Cinclodes occurs around the warden's house. Vinaceous Parrot* can usually be seen from the café overlooking the canyon.

APARADOS DA SERRA NP

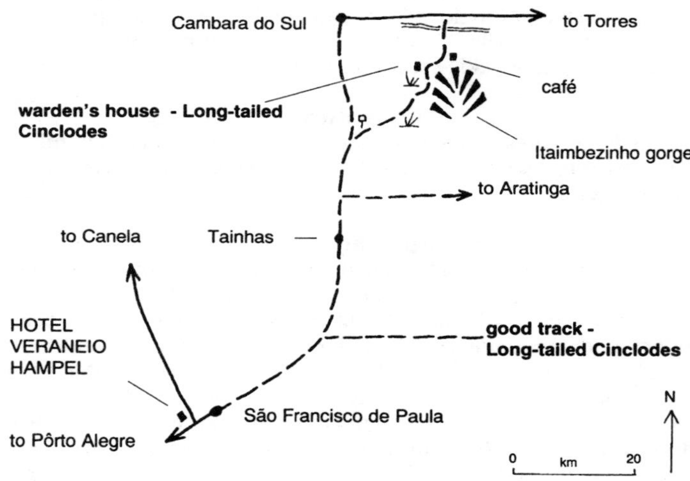

Returning to the main road between Sao Francisco de Paula and Cambara do Sul, head north to Cambara do Sul, then east towards Torres. 17 km east of Cambara do Sul turn south to the NP. Striolated Tit-Spinetail occurs along here, between the ford and the NP entrance. Giant Snipe occurs over the high ground 16 km east of Cambara do Sul.

Accommodation: near São Francisco de Paula: Hotel Veraneio Hampel.

Another good site, 250 km northwest of Aparados da Serra, is the **Aracuri-Esmeralda Ecological Station**, north of Vacaria. Red-spectacled Parrot*, Araucaria Tit-Spinetail* and Saffron-cowled Blackbird* occur here.

MOSTARDAS PENINSULA

Map p. 116

The 250-km long finger of marshy land, southeast of Porto Alegre on the east side of Lagoa dos Patos, supports a number of birds which are common further south but are otherwise scarce in Brazil. The species are similar to those which occur at the more easily accessible site at Taim near Rio Grande (see p. 116). However, seawatching here in suitable conditions may produce Magellanic Penguin and Yellow-nosed Albatross, and Short-billed Pipit is a particular speciality.

Specialities

Spotted Nothura, Magellanic Penguin, Yellow-nosed Albatross, Ringed Teal, Snowy-crowned Tern, Curve-billed Reedhaunter, Black-and-white Monjita*, Short-billed Pipit, Red-crested Cardinal, Diademed Tanager, Brown-and-yellow Marshbird, Scarlet-headed Blackbird.

Others

Greater Rhea*, White-tufted and Great Grebes, Black-browed Albatross, White-chinned Petrel, Wilson's Storm-Petrel, Southern Screamer, Fulvous and White-faced Whistling-Ducks, Black-necked and Coscoroba Swans, Rosy-billed Pochard, Chilean Flamingo, Bare-faced and White-faced Ibises, Maguari Stork, Snail Kite, Long-winged Harrier, Savanna Hawk, Aplomado Falcon, Grey-necked Wood-Rail, Spot-flanked Gallinule, White-winged and Red-gartered Coots, Wattled Jacana, South American Snipe, Collared and Two-banded Plovers, Rufous-chested and Tawny-throated Dotterels, Grey-headed and Brown-hooded Gulls, Large-billed, South American and Yellow-billed Terns, Southern Skua, Common Miner, Wren-like Rushbird, Sooty and White-crested Tyrannulets, White Monjita, Austral Negrito, Spectacled Tyrant, Masked Gnatcatcher, White-rumped and Tawny-headed Swallows, Yellowish Pipit, Hooded Siskin, Golden-crowned and Neotropical River Warblers, Blue-and-Yellow Tanager, Red-crested Finch, Long-tailed Reed-Finch, Grassland Yellow-Finch, Great Pampa-Finch, Yellow-winged, Chestnut-capped and White-browed Blackbirds, Bay-winged Cowbird.

Access

Seawatching can be excellent from the pier at the south end of **Capão da Canoa**, 138 km east of Porto Alegre on the coast road to Torres. Magellanic Penguin, hundreds of albatrosses and Snowy-crowned Tern have been seen from here in July. Curve-billed Reedhaunter occurs on the north side of the lake just west of Capão da Canoa.

The track that traverses the Mostardas peninsula is rough, and may be impassable during the winter months, especially after rain. Beware! The track starts 78 km east of Porto Alegre. From here it is 43 km to **Solidao**, another good seawatching point. Greater Rhea* and Common Miner occur alongside the 65 km stretch of track south from here to Mostardas. Waterfowl abounds on Lagoa **Capão da Fuma** just west of Mostardas. Chilean Flamingo, Black-necked and Coscoroba Swans, Rufous-chested and Tawny-throated Dotterels, Short-billed Pipit and Brown-and-yellow Marshbird occur at **Lagoa do Peixe NP**, south of Mostardas. Turn east 29 km south of Mostardas just south of Tavares, to reach this excellent area.

RIO GRANDE

Map below

Rio Grande, in far south Brazil, is an ideal base from which to explore the wetlands of the extreme south, where the birds are similar to those found on the Mostardas Peninsula (see p. 115). However, species occuring more reliably here include Sulphur-bearded Spinetail, Freckle-breasted Thornbird and Bay-capped Wren-Spinetail*.

Goodies for Brazil shared with the Mostardas peninsula include Two-banded Plover, Rufous-chested and Tawny-throated Dotterels, Snowy-crowned Tern, Common Miner, Curve-billed Reedhaunter, and Black-and-white Monjita*.

Specialities

Spotted Nothura, Ringed Teal, Giant Wood-Rail, Snowy-crowned Tern, Chicli and Sulphur-bearded Spinetails, Freckle-breasted Thornbird, Bay-capped Wren-Spinetail*, Curve-billed Reedhaunter, Crested and Warbling Doraditos, Black-and-white Monjita*, Red-crested Cardinal, Brown-and-yellow Marshbird, Scarlet-headed Blackbird.

SOUTH BRAZIL

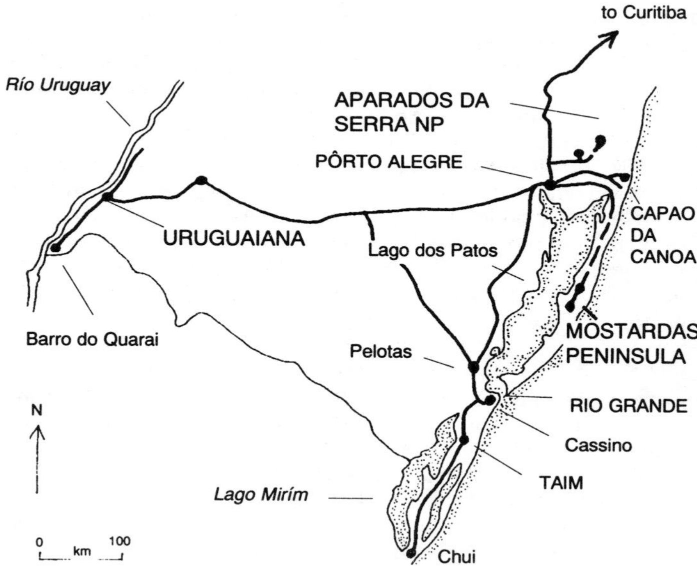

Others

Greater Rhea*, White-tufted and Great Grebes, Southern Screamer, Black-necked and Coscoroba Swans, Rosy-billed Pochard, Bare-faced and White-faced Ibises, Roseate Spoonbill, Maguari Stork, Snail Kite, Long-winged and Cinereous Harriers, Plumbeous Rail, Spot-flanked Gallinule, White-winged, Red-gartered and Red-fronted Coots, Limpkin, Wattled Jacana, South American Snipe, Collared and Two-banded

Plovers, Rufous-chested and Tawny-throated Dotterels, Grey-headed and Brown-hooded Gulls, Large-billed, South American and Yellow-billed Terns, Picui Ground-Dove, Monk Parakeet, Guira Cuckoo, Gilded Sapphire, Green-barred Woodpecker, Common Miner, Bar-winged Cinclodes, Stripe-crowned Spinetail, Wren-like Rushbird, Rufous-capped Antshrike, Small-billed Elaenia, Many-coloured Rush-Tyrant, White Monjita, Spectacled and Yellow-browed Tyrants, White-rumped and Tawny-headed Swallows, Long-tailed Reed-Finch, Black-and-rufous Warbling-Finch, Great Pampa-Finch, Rusty-collared Seedeater, Yellow-winged, Chestnut-capped and White-browed Blackbirds.

Access

Rio Grande is 274 km south of Porto Alegre. Seawatching can be good off Cassino, 24 km southeast of Rio Grande. Head north 6 km along the beach (suitable for vehicles) to the jetty. Common Miner, Freckle-breasted Thornbird and Bay-capped Wren-Spinetail* occur behind the sand dunes here. The best marshes are south of **Taim**, 82 km southwest of Cassino, where the whole area is worth exploring. There is an Ecological Station here where it is possible to stay.

Accommodation: Rio Grande: Hotel Charrua (A).

URUGUAIANA

Map p. 116

This cattle centre, 772 km west of Porto Alegre, in the extreme southwest corner of Brazil on the Argentina border, lies near remnant *Espinilho* parkland. Most of the specialities here are at the edge of their restricted ranges (centred on Argentina and Paraguay), and in Brazil are likely to be seen only here.

Specialities

Spotted Nothura, Ringed Teal, Giant Wood-Rail, Short-billed Canastero, Lark-like Brushrunner, Checkered Woodpecker, Scimitar-billed Woodcreeper, Black-crowned Monjita, Plush-crested Jay, White-banded Mockingbird, Red-crested Cardinal, Screaming Cowbird.

Others

Greater Rhea*, Great Grebe, Southern Screamer, Masked Duck, Maguari Stork, Snail Kite, Long-winged and Cinereous Harriers, Great Black-Hawk, Aplomado Falcon, South American Snipe, Large-billed and Yellow-billed Terns, Spot-winged Pigeon, Guira Cuckoo, Campo Flicker, Narrow-billed Woodcreeper, Tufted Tit-Spinetail, Brown Cachalote, Sooty-fronted and Stripe-crowned Spinetails, Suiriri Flycatcher, Tawny-crowned Pygmy-Tyrant, Grey and White Monjitas, Masked Gnatcatcher, Yellowish Pipit, Black-and-Rufous and Black-capped Warbling-Finches, Great Pampa-Finch, Golden-billed Saltator, Ultramarine Grosbeak, Bay-winged Cowbird.

Access

Bird the road south from Uruguaiana to Barro do Quaraí on the Uruguay border, especially the parkland, between kms 62 and 69 from Uruguaiana.

SERRA DA CANASTRA NP

Maps opposite and p. 121

The cerrado, gallery forest and wild rivers of Serra da Canastra NP, 330 km west of Belo Horizonte (and some 500 km northwest of Rio as the Brazilian Merganser*, for which this park is famous, flies) support two very rare birds: Brazilian Merganser*, which has a depressing estimated world population of just 250, and Brasilia Tapaculo*, both of which are extremely difficult to see.

This NP is also one of the few sites where the endemics, Stripe-breasted Starthroat, Grey-backed Tachuri*, and Grey-eyed Greenlet, are known to occur.

Endemics

Golden-capped Parakeet*, Stripe-breasted Starthroat, Brasilia Tapaculo*, Yellow-lored Tody-Flycatcher, Grey-backed Tachuri*, Velvety Black-Tyrant, Grey-eyed Greenlet, White-striped Warbler, Cinnamon Tanager, Dubois' Seedeater.

Serra da Canastra NP is about the only site in Brazil where it is possible to see the endangered Brazilian Merganser, although it might take more than one visit*

Specialities

Spotted Nothura, Brazilian Merganser*, Giant Snipe, Blue-winged Macaw*, Long-trained Nightjar*, Great Dusky Swift, White-vented Violet-ear, Campo Miner, Chicli Spinetail, Rufous-winged Antshrike, White-shouldered Fire-eye, Helmeted Manakin, Sharp-tailed Tyrant*, White-rumped Monjita, Crested Black-Tyrant, Cock-tailed Tyrant*, Curl-crested Jay, Ruby-crowned Tanager, Black-masked* and Pileated Finches, Stripe-tailed Yellow-Finch, Black-throated Saltator, Yellow-billed Blue Finch*.

Others

Brown Tinamou, Greater Rhea*, Black-chested Buzzard-Eagle, Collared Forest-Falcon, Red-legged Seriema, Blue-winged Parrotlet, Striped Cuckoo, Scissor-tailed Nightjar, Planalto Hermit, Swallow-tailed Hummingbird, White-eared Puffbird, Toco Toucan, White and Blond-crested Woodpeckers, Pale-breasted Spinetail, Firewood-gatherer, Buff-fronted Foliage-Gleaner, Sharp-tailed Streamcreeper, Eastern Slaty

Antshrike, Rufous Gnateater, Red-ruffed Fruitcrow, Cliff Flycatcher, Grey Monjita, Rufous Casiornis, Swainson's Flycatcher, Rufous-crowned Greenlet, White-rumped and Tawny-headed Swallows, Hellmayr's Pipit, Rufous-crowned Greenlet, White-rimmed Warbler, Wedge-tailed Grass-Finch, Yellow-bellied Seedeater, Crested Oropendola, Yellow-rumped Marshbird.

Other Wildlife

Giant Anteater, Giant Armadillo, Maned Wolf.

Access

The park entrance is 7 km west of São Roque de Miñas, northwest of Piumhi, which is on the road between Belo Horizonte and Ribeirão Prêto.

SERRA DA CANASTRA NP

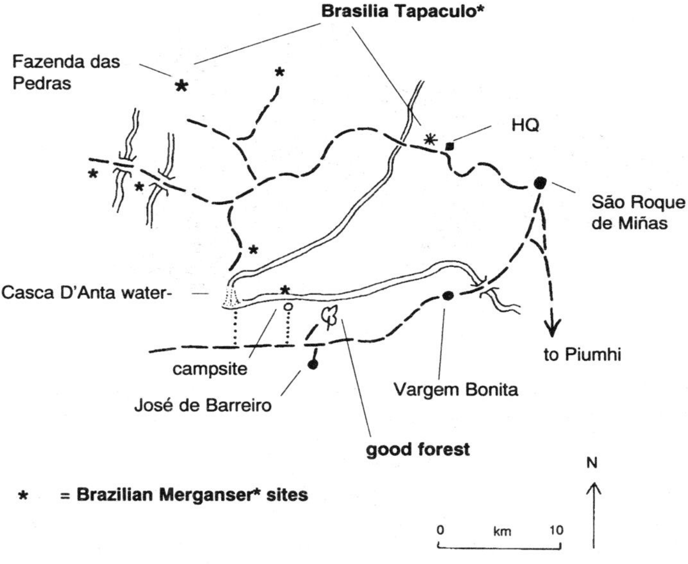

Brasilia Tapaculo* occurs in the thick scrub by the stream 4 km from the entrance. As befits a tapaculo, this bird is a hyper-skulker and very difficult to see. 16 km further on the turning northwest goes to Fazenda das Pedras, 10 km away. Campo Miner occurs along this side road. Brazilian Merganser*, Brasilia Tapaculo* and Dubois' Seedeater have all been recorded around this fazenda. The turning north 6 km along the road to this fazenda reaches a river after 4 km, where the merganser has also been recorded. 2 km south of the Fazenda das Pedras turning there is another side-road, to the west, which crosses two rivers where the merganser also occurs. The best place for the elusive duck, though, is above and below the Casca D'Anta waterfall, 29 km from the entrance. The best chance of seeing it is probably by scouring the river whilst walking alongside it, from the campsite 6 km to the east of the waterfall base to the waterfall. The best time to look is dawn and dusk, from July to August.

Golden-capped Parakeet* and Red-ruffed Fruitcrow occur in the patch of forest 29 km southwest of São Roque de Miñas. Walk the trail north of the road, 10 km east of the Casca D'Anta waterfall.

Accommodation: São Roque de Miñas: Hotel Faria (basic).

Hyacinth Visorbearer*, Serra Antwren*, Dusky-tailed Antbird, Shrike-like Cotinga*, Grey-backed Tachuri* and Pale-throated Serra-Finch* occur in the picturesque **Parque Natural do Caraca**, 123 km east of Belo Horizonte. Bird the woods beside the path to Tanque Grande and the picnic sites below the monastery. It is also possible to see the rare Maned Wolf here. They are attracted to food put out for them on the monastery steps in the evenings.

Red-eyed Thornbird, Serra Antwren*, Ochre-rumped Antbird*, and Hangnest Tody-Tyrant* occur at nearby **Estacao Ambiental de Peti**, 15 km from Santa Barbara. For permission to visit contact: CEMIG - Reserva de Peti, Sr. Antonio Procopio Sampaio Rezende, Av. Barbacena, 1200-20, Belo Horizonte, Minas Gerais 30190.

SERRA DO CIPO NP

Map p. 121

The cerrado and gallery forest of Serra do Cipo NP, 120 km north of Belo Horizonte, supports seven endemics including Cipo Canastero* which is only known from this site, where it was discovered in 1985. This is also the most reliable site for Hyacinth Visorbearer* and Grey-backed Tachuri*.

Endemics

Lesser Nothura*, Hyacinth Visorbearer*, Cipo Canastero*, Grey-backed Tachuri*, White-naped Jay, Grey-eyed Greenlet, Pale-throated Serra-Finch*.

Specialities

Spotted Nothura, Giant Snipe, Long-trained Nightjar*, White-vented Violet-ear, Horned Sungem, Checkered Woodpecker, Black-capped Antwren*, Collared Crescent-chest, Plain-crested Elaenia, Crested Black-Tyrant, White-bellied Warbler, Cinereous Warbling-Finch*.

Others

Blackish Rail, Red-legged Seriema, Peach-fronted Parakeet, Blue-winged Parrotlet, Planalto Hermit, Swallow-tailed Hummingbird, Sapphire-spangled Emerald, White-barred Piculet, Campo Flicker, Little, Green-barred and Lineated Woodpeckers, Narrow-billed Woodcreeper, Sooty-fronted Spinetail, Eastern Slaty Antshrike, Grey Elaenia, Fuscous Flycatcher, Grey and White-rumped Monjitas, White-winged Black-Tyrant, Hooded Tanager, Swallow-Tanager, Yellow-rumped Marshbird.

EAST BRAZIL

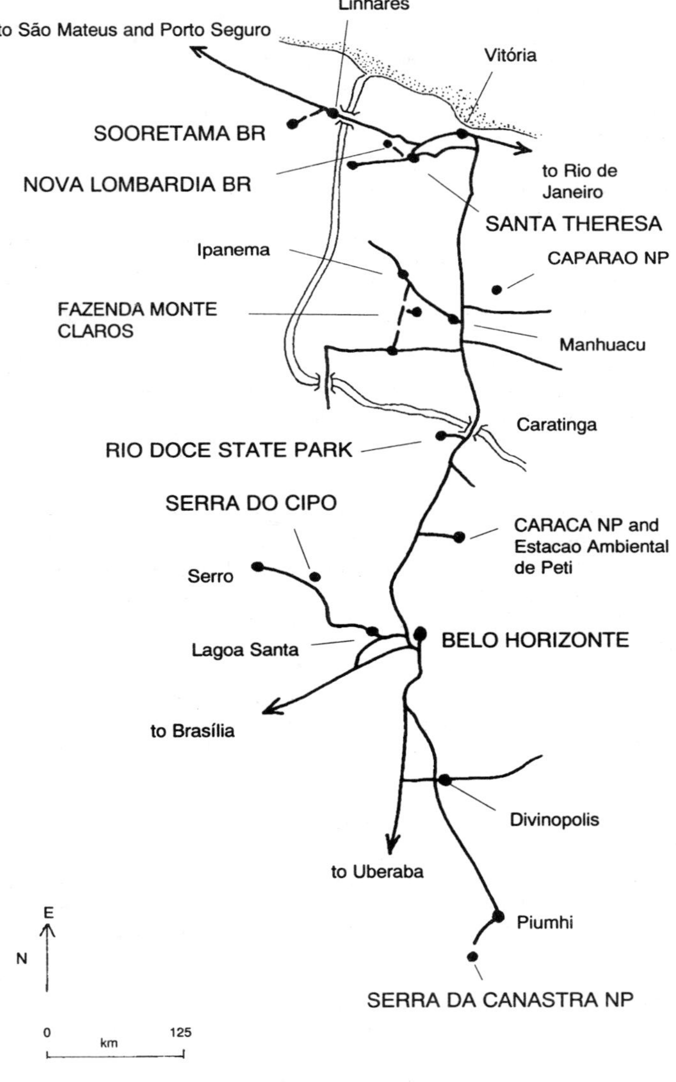
Linhares
to São Mateus and Porto Seguro
Vitória
SOORETAMA BR
NOVA LOMBARDIA BR
to Rio de Janeiro
SANTA THERESA
Ipanema
CAPARAO NP
FAZENDA MONTE CLAROS
Manhuacu
Caratinga
RIO DOCE STATE PARK
SERRA DO CIPO
CARACA NP and Estacao Ambiental de Peti
Serro
Lagoa Santa
BELO HORIZONTE
to Brasília
Divinopolis
to Uberaba
E
N
Piumhi
SERRA DA CANASTRA NP
0
km
125

Access

From Belo Horizonte head north to Lagoa Santa from where the park is signposted northwards. Sapphire-spangled Emerald occurs in the roadside forest just before the administration building, 1 km before the Hotel Veraneio and the bridge. Check flowering trees, especially at the end of July. From the Hotel Veraneio, a good base, it is 9 km uphill to the Chapeu de Sol Hotel. Giant Snipe and Cinereous Warbling-Finch* occur along the rough track to the northwest 3 km beyond the Chapeu de Sol Hotel. Hyacinth Visorbearer*, Cipo Canastero*, Grey-backed Tachuri* and Pale-throated Serra-Finch* all occur 5 km above the Chapeu do Sol Hotel, in the rocky outcrops on the right-hand side of the road.

Accommodation: Hotel Veraneio. Chapeu de Sol Hotel (basic).

RIO DOCE STATE PARK

Map p. 121

This immensely important reserve, 210 km east of Belo Horizonte, protects one of the largest remaining tracts of Atlantic forest in southeast Brazil. Although permission is usually required to visit, access is difficult and the birds are similar to Sooretama, it may well be worth birding if time is available since many more birds than have already been recorded are probably present. The park's specialities include the endemics, Minute Hermit and Forbe's Blackbird*.

Endemics

Golden-capped Parakeet*, Minute Hermit, Crescent-chested Puffbird, Tail-banded Hornero, Black-cheeked Gnateater, Cinnamon-vented Piha*, Yellow-lored Tody-Flycatcher, Grey-hooded Attila, Forbes' Blackbird*.

Specialities

Dusky-legged Guan, Rufous-sided Crake, Slaty-breasted Wood-Rail, Ash-throated Crake, Blue-winged Macaw*, Rufous-vented Ground-Cuckoo, Rusty-barred Owl, Rufous-capped Motmot, Yellow-fronted Woodpecker, Chestnut-backed Antshrike, Thrush-like Schiffornis, Eared Pygmy-Tyrant, Pileated Finch.

Others

Capped Heron, Mealy Parrot, Black-necked Aracari, Channel-billed Toucan, White-barred Piculet, Campo Flicker, Blond-crested Woodpecker, Lesser Woodcreeper, Rufous Hornero, Ochre-breasted and White-eyed Foliage-gleaners, Streaked Xenops, Eastern Slaty and Cinereous Antshrikes, White-flanked and Rufous-winged Antwrens, Red-ruffed Fruitcrow, White-bearded Manakin, Ochre-faced Tody-Flycatcher, Masked Water-Tyrant, Sirystes, Green-backed and Crested Becards, Yellow-legged and Pale-breasted Thrushes, Moustached Wren, Chestnut-bellied Euphonia, Green-winged Saltator.

Other Wildlife

Brazilian Tapir, Capybara.

The Black Jacobin is one of many hummingbirds attracted to the feeders at the late Dr Augusto Ruschi's house in Santa Theresa

Access

Head east from Belo Horizonte for 168 km then turn north towards Timotei. The reserve entrance is 42 km north along the road to Timotei. Unfortunately there are few access points into the impressive forest here. However, a 21-km road does bisect the reserve from the northwest (19 km north of the reserve entrance) to the southeast. Also bird the 'Primary Forest Footpath' to the east, 3 km northwest of the bridge at the southeast end of the road. Forbes' Blackbird* occurs around the administration buildings. The pools around here and the 'Secondary Forest Footpath' near the Casa de Tabua Lodge are also worth exploring fully.

Obtain permission from the IEF.

Accommodation: Hotel Pousada; Casa de Tabua Lodge within the park (permission and food required, although there is a café 14 km away).

The private **Estacao Biologica de Caratinga Reserve**, 392 km east of Belo Horizonte, has been established within the Fazenda Monte Claros (Map p.121). The forest here supports some scarce endemics, notably Yellow-legged Tinamou* and Three-toed Jacamar* (which has not been recorded for some time). Golden-capped Parakeet* and Vinaceous Parrot* also occur here. Otherwise, the birds are similar to those found at Nova Lombardia (see p. 124). The reserve was actually established to protect the habitat of the rare Woolly Spider-monkey (also known as the Muriqui), and it also supports Brown Howler, Tufted-ear Marmoset and Brown Capuchin. It is situated 52 km east of Caratinga, but Ipanema is probably a better access point, 22 km to the east. Turn north off the BR 262, 284 km east of Belo Horizonte to Manhuaçu. Tail-banded Hornero occurs along the 86-km road to Ipanema from here. There is a maze of trails to explore.

Accommodation: on site (A); contact the Director, Estacao Biologica de Caratinga, CP 82 36 950 Ipanema, Minas Gerais.

The small hill town of **Santa Theresa**, 76 km northwest of Vitória, (Map p. 121), is the home of the late Dr Augusto Ruschi, a hummingbird expert who, on finding it virtually impossible to study these hyperactive birds in the wild, resorted, with great success, to putting up sugar-solution feeders inside and outside his house. Fortunately this practice has been continued, providing birders (on Saturdays) with a unique opportunity to experience one of the ornithological wonders of the world: perhaps over a hundred hummers, of up to 15 species at any one time, performing at close quarters. It is an experience not to be missed. The endemic hummingbirds to look out for are Saw-billed Hermit*, the fabulous Frilled Coquette, Sombre Hummingbird and Brazilian Ruby. Others include Rufous-breasted, Scale-throated, Dusky-throated and Planalto Hermits, the huge Swallow-tailed Hummingbird, the striking Black Jacobin, White-vented Violet-ear, Glittering-bellied Emerald, Violet-capped Woodnymph, White-chinned Sapphire, White-throated Hummingbird, Versicoloured and Glittering-throated Emeralds and Amethyst Woodstar. The estate is at the east end of town. Tail-banded Hornero occurs in the town plaza.

Accommodation: Hotel Pierazzo (A).

NOVA LOMBARDIA BIOLOGICAL RESERVE

Map p. 121

This reserve, 9 km north of Santa Theresa and 90 km from Vitória, protects montane Atlantic forest where over 200 species have been recorded, including many endemics. The habitat and birds are akin to those of Serra dos Orgãos NP (see p. 105) and Itatiaia NP (see p. 107), but particular specialities include Weid's Tyrant-Manakin, and Grey-capped* and Oustalet's* Tyrannulets. Although permission is required to visit, keen birders will be well rewarded for this extra effort.

Endemics

Frilled Coquette, Brazilian Ruby, Crescent-chested Puffbird, Yellow-eared Woodpecker, Pallid Spinetail, Pale-browed Treehunter, White-collared Foliage-gleaner, Plumbeous Antshrike*, Salvadori's Antwren*, Ferruginous and White-bibbed Antbirds, Such's Antthrush, Black-cheeked Gnateater, White-breasted Tapaculo, Hooded Berryeater*, Cinnamon-vented Piha*, Pin-tailed Manakin, Weid's Tyrant-Manakin, Yellow-lored Tody-Flycatcher, Grey-capped* and Oustalet's* Tyrannulets, Grey-hooded Attila, Cinnamon, Rufous-headed, Azure-shouldered*, Golden-chevroned and Gilt-edged Tanagers, Dubois' Seedeater.

Specialities

Maroon-bellied Parakeet, Ashy-tailed Swift, Scale-throated Hermit, Spot-billed Toucanet, Black-billed Scythebill, Rufous-capped and Chicli Spinetails, Black-capped Foliage-gleaner, Shrike-like Cotinga*, Bare-throated Bellbird*, Blue Manakin, Eared Pygmy-Tyrant, Russet-winged Spadebill*, Rufous-brown Solitaire*, Ruby-crowned, Sayaca, Green-headed and Red-necked Tanagers, Buffy-fronted Seedeater*, Sooty Grassquit.

Others

Brown Tinamou, Harpy Eagle*, Barred Forest-Falcon, Blackish Rail, Plumbeous Pigeon, Scissor-tailed Nightjar, Channel-billed Toucan, White-barred Piculet, White-throated, Scaled and Lesser Woodcreepers, Rufous Hornero, Ochre-breasted Foliage-gleaner, Spot-backed and Variable Antshrikes, Spot-breasted Antvireo*, Red-ruffed Fruitcrow, Sharpbill, Masked Water-Tyrant, Pale-breasted and Creamy-bellied Thrushes, Rufous-crowned Greenlet, Black-goggled Tanager, Red-crowned Ant-Tanager.

Other Wildlife

Maned Sloth.

Access

Head west out of Santa Theresa towards Colatina and look for an unmarked track up a steep hill to the northeast edge of town. Turn right here and right again at the fork after 6 km. The roadside at the start of this track to the reserve is good for seedeaters. Bird the two roads which traverse the reserve and the paths which lead east and west from near the warden's house. Harpy Eagle* has been seen from the junction of the roads to Lombardy and Fundao.

Obtain permission to visit Nova Lombardia from IBAMA, who have an office in Vitória.

Accommodation: Santa Theresa: Hotel Pierazzo (A).

SOORETAMA BIOLOGICAL RESERVE **Map p. 121**

The lowland Atlantic forest protected by this reserve, 50 km north of Linhares (136 km north of Vitoria), is one of only three sites where the world's rarest cracid, Red-billed Curassow*, occurs. Over 300 species have been recorded including a superb selection of rare endemics, notably White-necked Hawk*, Minute and Hook-billed* Hermits, Striated Softtail*, and Scalloped Antbird*, as well as unusually large numbers of woodpeckers and cotingas. Although permission is required to visit, this site is a must for the serious endemic hunters.

Endemics

White-necked Hawk*, Red-billed Curassow*, Blue-throated* and Plain Parakeets, Red-browed Parrot*, Minute and Hook-billed* Hermits, Crescent-chested Puffbird, Yellow-eared Woodpecker, Striated Softtail*, Pale-browed Treehunter, Plumbeous Antshrike*, Star-throated and Band-tailed* Antwrens, Scaled and Scalloped* Antbirds, Black-cheeked Gnateater, Black-headed Berryeater*, Banded* and White-winged* Cotingas.

Specialities

Solitary* and Small-billed Tinamous, Rufous-vented Ground-Cuckoo, Tawny-browed Owl, Rufous-capped Motmot, Spot-billed Toucanet, Black-capped Foliage-gleaner, Chestnut-backed Antshrike, Bare-throated Bellbird*, Brazilian Tanager.

Others

Rusty-margined Guan, Mealy Parrot, Ocellated Poorwill, Black-necked Aracari, White-barred Piculet, Campo Flicker, White, Red-stained, Yellow-throated and Blond-crested Woodpeckers, Buff-throated Woodcreeper, Eastern Slaty and Cinereous Antshrikes, Rufous-capped Antthrush, White-crowned and White-bearded Manakins, Yellow-bellied Elaenia, Sulphur-rumped Flycatcher, Sirystes, Greyish Mourner, Green-backed, Chestnut-crowned and Black-capped Becards, Cocoa Thrush, Moustached Wren, Long-billed Gnatwren, Pectoral Sparrow, Turquoise Tanager, White-bellied Seedeater, Yellow-green Grosbeak, Green-winged Saltator.

Access

To reach the reserve head north out of Linhares on the BR 101 towards São Mateus, and turn northwest after 14 km on to a track to the park HQ. Bird the track northwest of the HQ, where Red-billed Curassow* has been known to wander out of the forest after heavy rain. Also bird the 6-km trail which leads east before reaching the HQ and the small trail behind the HQ.

Obtain permission to visit this reserve from IBAMA, who have an office in Vitória.

Accommodation: Linhares: Hotel Pratti Park (A); Hotel Linhatur. It also possible to stay at the park HQ, with permission.

MONTE PASCOAL NP

Map opposite

The forest protected in this NP, roughly halfway between Vitória and Salvador, near Porto Seguro, on Brazil's east coast, supports a number of spectacular endemics, of which White-necked Hawk*, Black-headed Berryeater*, and Banded* and White-winged* Cotingas are particularly noteworthy. Much of the park is inaccessible to all but those on major expeditions, and Red-billed Curassow* is not a realistic possibility here, since it occurs in the remote areas.

Endemics

White-necked Hawk*, Red-billed Curassow*, Blue-throated* and Plain Parakeets, Red-browed Parrot*, Long-tailed Woodnymph, Salvadori's* and Band-tailed* Antwrens, Scaled Antbird, Black-cheeked Gnateater, Black-headed Berryeater*, Cinnamon-vented Piha*, Banded* and White-winged* Cotingas.

Specialities

Black-fronted Piping-Guan*, Ash-throated Crake, White-eared Parakeet, Racket-tailed Coquette, Bare-throated Bellbird*, Bay-ringed Tyrannulet*.

Others

Harpy Eagle*, Black-and-white* and Ornate Hawk-Eagles, Spot-winged Wood-Quail, Ash-throated Crake, Black-throated Trogon, Campo Flicker, Eastern Slaty Antshrike, Screaming Piha, White-crowned and Striped Manakins, Cocoa Thrush, Opal-rumped Tanager, Green-winged Saltator.

MONTE PASCOAL NP

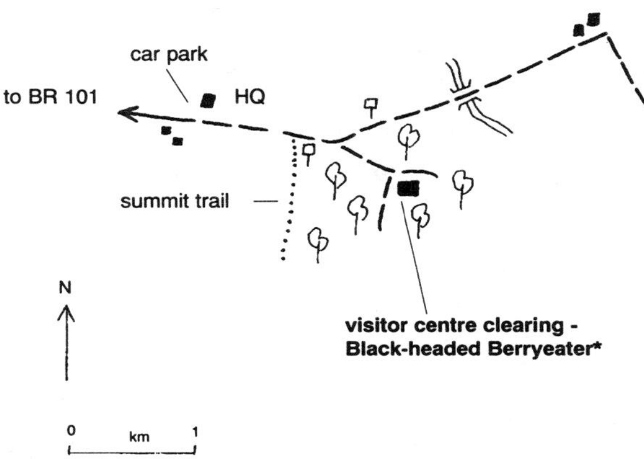

Access

Access is restricted to the western side of the NP, 32 km north of Itamaraju on Route BR 101. The Visitor's Centre is 1 km inside the park. The clearing around here is good for Black-headed Berryeater*. White-necked hawk* occurs along the trail to the right, between the entrance and the visitor centre.

PORTO SEGURO

Some rare endemics can be found near this small seaside resort roughly midway between Vitória and Salvador on Brazil's east coast. The habitat and birds are akin to those at Monte Pascoal (p. 126) and Sooretama (p. 125), although Banded* and White-winged* Cotingas are probably easier to see near here, if suitable habitat can be located. Hence this site is a must for those birders in search of these two superb birds.

Endemics

Plain Parakeet, Red-browed Parrot*, Crescent-chested Puffbird, Yellow-eared Woodpecker, Band-tailed Antwren*, Scaled Antbird, Black-cheeked Gnateater, Black-headed Berryeater*, Banded* and White-winged* Cotingas, Dubois's Seedeater.

Specialities

Plain-bellied Emerald, Black-capped Foliage-gleaner, Bare-throated Bellbird*, Eared Pygmy-Tyrant, Green-headed Tanager.

The lovely maroon White-winged Cotinga, a rare Brazilian endemic, is best looked for near Porto Seguro*

Others

Reddish Hermit, Swallow-tailed Hummingbird, Blue-chinned Sapphire, White-fronted Nunbird, Black-necked Aracari, Channel-billed Toucan, Yellow-throated Woodpecker, Lesser Woodcreeper, Screaming Piha, White-crowned and Striped Manakins, Black-capped Becard, Long-billed Gnatwren, Turquoise and Opal-rumped Tanagers.

Access

The Porto Seguro Reserve and Pau Brasil Ecological Station lie on the north side of the road 14 km west of Porto Seguro. They are owned by a mining company and some persuasion may be needed to enter. However, roadside birding along here can be very productive once suitable habitat has been located. Patience and perseverence, the top two qualities of the best birders, are required at this site.

BOA NOVA

Map opposite

The Boa Nova area, 230 km inland from Ilhéus on Brazil's east coast, supports more endemics per sq km than any other area in Brazil (Serra dos Orgãos (see p. 105) boasts the most within a NP). This is owing to the variety of habitats: caatinga to the west of town, dry and wet forest to the east. The really rare and restricted endemics occuring here include Broad-tipped Hermit, Spotted Piculet, Striated Softtail*, Silvery-cheeked Antshrike, Pileated* and Narrow-billed* Antwrens, Slender Antbird*, and White-browed Antpitta*.

Endemics

Caatinga and Plain Parakeets, Broad-tipped Hermit, Frilled Coquette, Spotted Piculet, Yellow-eared Woodpecker, Tail-banded Hornero, Pallid and Grey-headed Spinetails, Striated Softtail*, Silvery-cheeked Antshrike, Pileated* and Narrow-billed* Antwrens, Ferruginous and Ochre-rumped* Antbirds, Rio de Janiero*, Slender* and White-bibbed Antbirds, White-browed Antpitta*, Black-cheeked Gnateater, Pin-tailed

Manakin, Hangnest Tody-Tyrant*, Yellow-lored Tody-Flycatcher, Grey-hooded Attila, Red-cowled Cardinal, Cinnamon, Golden-chevroned and Gilt-edged Tanagers, Dubois' Seedeater.

Specialities

Small-billed Tinamou, Mantled Hawk*, Rufous-sided and Ash-throated Crakes, Maroon-bellied Parakeet, Pileated Parrot*, Scale-throated Hermit, Violet-capped Woodnymph, Rufous-capped and Ochre-cheeked Spinetails, Tufted and Rufous-winged Antshrikes, Black-bellied Antwren, Mouse-coloured Tapaculo, Blue Manakin, Greenish Schiffornis, Brazilian Tanager.

Others

Red-winged Tinamou, Planalto and Reddish Hermits, Swallow-tailed Hummingbird, Amethyst Woodstar, Golden-spangled Piculet, Yellow-throated Woodpecker, Scaled and Lesser Woodcreepers, Spot-backed, Great and Eastern Slaty Antshrikes, Spot-breasted Antvireo*, Rufous-winged Antwren, Rufous Gnateater, Sharpbill, Planalto Tyrannulet, Grey and White Monjitas, Chestnut-crowned Becard, Rufous-crowned Greenlet, Black-capped Donacobius, Long-billed Gnatwren, Orange-headed and Black-goggled Tanagers, Chestnut-bellied Euphonia, Green-winged Saltator.

BOA NOVA

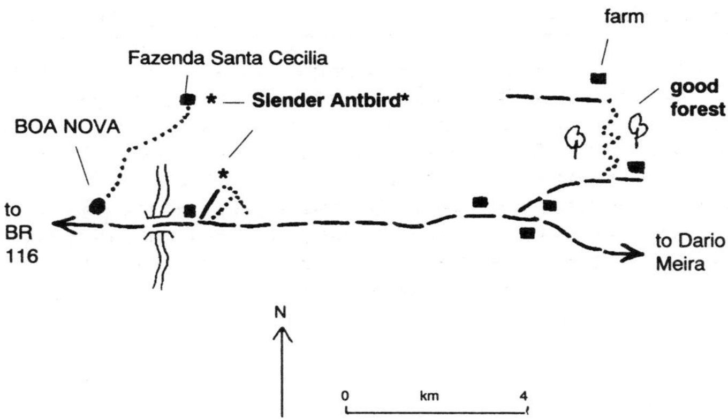

Access

Boa Nova is 230 km west of Ilhéus, via Ibicarai and Poçoes. Head north from Poçoes towards Jequié on the BR 116. Turn right along here to Boa Nova and Dario Meira. Silvery-cheeked Antshrike occurs south of this road after 1 km. It is a further 20 km to the village of Boa Nova, which lies to the north of the road. From here head northeast 4 km to the Fazenda Santa Cecilia, where Slender Antbird* occurs. Returning to the Dario Meira road, head east for 2 km to a bridge. Slender Antbird* occurs in the dry forest to the north of the road past this bridge. Further east along the road to Dario take the left fork which runs north from the road 10 km east of Boa Nova, then take the second left along an excellent trail through wet forest, where Rio de Janeiro Antbird* occurs.

Accommodation: Boa Nova: Hotel Central (C).

The endemics, Little Wood-Rail, Hooded Visorbearer*, Fringe-backed Fire-eye* and Bahia Tapaculo*, as well as Southern Pochard, Black-billed Scythebill, Rufous-capped Spinetail, Tufted and Rufous-winged Antshrikes, Stripe-necked Tody-Tyrant, Yellow-faced Siskin*, which may be an endemic, and Bicoloured Conebill occur around **Salvador** on Brazil's east coast. Little Wood-Rail and Bicoloured Conebill occur in the mangroves at the mouth of the Río Traripe. Take the BA 026 road west to Santo Amaro and turn south after 6 km, towards São Francisco do Conde, then take the minor road straight on at the crossroads after 8 km, into the mangroves (best at low tide). The very rare Fringe-backed Fire-eye* occurs in the remnant forest 8 km west of Santo Amaro. The recently discovered Bahia Tapaculo* occurs around Valença, south of Salvador, where some good forest still remains.

NORTHEAST BRAZIL

The vast interior of Bahia state in northeast Brazil is arid country, covered with thorny woodland and scrub characterised by cacti and terrestrial bromeliads - a habitat known as 'caatinga'. Owing to burning, grazing and clearance for agriculture, it is one of the most threatened habitats in South America and hence supports an endangered avifauna, which includes the rarest bird in the world, Spix's Macaw*, only one specimen of which remains. Naturally the site is top secret.

CANUDOS

The remote red-rock escarpment of Raso da Catarina supports the world's sole population of Lear's Macaw*, which was only discovered in the wild in 1978. It is a sad fact, but there are only 60 individuals left, and they can be seen near Canudos, also known as Cocorobo, 320 km northwest of Aracaju on the rough road to Juazeiro. You must arrange to see them through local wardens. This is a good site for many other northeastern endemics and most of the endemics listed below occur only here and at a few other sites.

Endemics

White-browed Guan*, Lear's Macaw*, Caatinga Parakeet, Broad-tipped Hermit, Stripe-breasted Starthroat, Red-shouldered Spinetail, Great Xenops*, Silvery-cheeked Antshrike, Pectoral Antwren*, Hang-nest Tody-Tyrant*, Ash-throated Casiornis, White-naped Jay, Long-billed Wren, Red-cowled Cardinal, Scarlet-throated Tanager, White-throated Seedeater.

Specialities

Small-billed Tinamou, White-bellied Nothura, Blue-winged Macaw*, Least Nighthawk, Rufous Cachalote, Rufous-winged Antshrike, Black-bellied Antwren, Lesser Wagtail-Tyrant, Capped Seedeater.

Others

Lesser Yellow-headed Vulture, Blue-crowned Parakeet, Dark-billed Cuckoo, Swallow-tailed and Ruby-topaz Hummingbirds, Spot-backed Puffbird, Little and Green-barred Woodpeckers, Red-billed Scythebill, Pearly-vented Tody-Tyrant, Tawny-crowned Pygmy-Tyrant, Masked Water-Tyrant, Rufous Casiornis, Green-backed and Crested Becards, Rufous-crowned Greenlet, Chestnut-vented Conebill, Wedge-tailed Grass-Finch, White-bellied Seedeater, White-browed Blackbird.

Access

Pectoral Antwren* and Scarlet-throated Tanager occur in the gallery forest 21 km west of Jeremoabo, east of Canudos. In Canudos arrange for the SEMA (Secretaria Especial do Meio Ambiente) wardens to accompany you to the roosting canyons of Lear's Macaw, a long, long walk. All around Canudos some good caatinga can be found. Caatinga Parakeet, Broad-tipped Hermit, Red-shouldered Spinetail, Black-bellied Antwren, Lesser Wagtail-Tyrant, White-naped Jay and Long-billed Wren occur here. A particularly good birding area is 2 km north of Cem Cinquenta village, 5 km to the east of Canudos.

(Information for this site has been minimised to help protect Lear's Macaw* from cagebird collectors)

Accommodation: Canudos: Hotel Brazil (C).

The scarce Southern Pochard occurs on roadside lakes along Route BR 407 from Capim Grosso to **Senhor do Bonfim**, some 200 km south of Juazeiro, and the road to Campo Formoso. Scarlet-throated Tanager occurs near the rubbish dump near Filadelfia and Great Xenops* has been recorded near Serrinha to the southeast.

Hooded Visorbearer* occurs around **Cachoeira** waterfall, 19 km east of Morro do Chapeu on Route BA 052. Ask at the restaurant opposite the Palace Hotel in Morro for precise directions and check low vegetation once on site. Spotted Piculet occurs in the roadside forest south towards **Utingá**. Caatinga Parakeet, Grey-backed Tachuri* and Pale-throated Serra-Finch* also occur around Morro do Chapéu.

Southern Pochard and Wied's Tyrant-Manakin occur in **Diamantina NP**, near Lençóis. A major expedition would be necessary to cover this mountain range fully, but there are some roads which skirt the edge of the NP.

PARQUE ESTADUAL DA PEDRA TALHADA

Map p. 133

This excellent forested reserve, 160 km west of Maceió, in the state of Alagoas, supports many of the northeastern endemics, and most of those listed below, notably Jandaya Parakeet, Pinto's Spinetail*, Alagoas Foliage-gleaner*, Alagoas Antwren*, Alagoas Tyrannulet*, the spectacular Seven-coloured Tanager* and Forbe's Blackbird*, which are known only from this site and a few others. This is also one of the few known sites for Yellow-faced Siskin*, which may be endemic to Brazil, since specimens taken in Venezuela may have been escaped cagebirds.

Endemics

White-collared Kite*, Jandaya and Plain Parakeets, Pygmy Nightjar*, Frilled Coquette, Long-tailed Woodnymph, Sombre Hummingbird, Tawny Piculet, Pinto's* and Grey-headed Spinetails, Alagoas Foliage-gleaner*, Unicoloured Antwren*, Scaled Antbird, Alagoas Antwren*, Fringe-backed Fire-eye*, Scalloped Antbird*, Black-cheeked Gnateater, Black-headed Berryeater*, Buff-throated Purpletuft*, Buff-breasted Tody-Tyrant, Alagoas Tyrannulet*, Red-cowled Cardinal, Seven-coloured Tanager*, White-throated Seedeater, Forbes' Blackbird*.

Specialities

Small-billed Tinamou, Mantled Hawk*, Rufous-sided Crake, Cinereous-breasted and Ochre-cheeked Spinetails, Rufous-winged Antshrike, White-backed Fire-eye, Bearded Bellbird*, Smoky-fronted Tody-Flycatcher, Yellow-faced Siskin*, Capped Seedeater.

Others

Lesser Yellow-headed Vulture, Speckled Chachalaca, Tropical Screech-Owl, Lesser Swallow-tailed Swift, Black-eared Fairy, Spot-backed Puffbird, Black-necked Aracari, Golden-spangled Piculet, Red-stained Woodpecker, Wedge-billed Woodcreeper, Sooty-fronted Spinetail, Great Antshrike, Short-tailed Antthrush, Sharpbill, Red-headed and Blue-backed Manakins, Masked Water-Tyrant, Cocoa Thrush, Tropical Gnatcatcher, Yellowish Pipit, Flavescent Warbler, Pectoral Sparrow, Orange-headed and Guira Tanagers, Chestnut-bellied Euphonia, Wedge-tailed Grass-Finch, White-bellied Seedeater, Yellow-green and Black-throated Grosbeaks, White-browed Blackbird.

Access

The park is 25 km northeast of Quebrangulo, a small town 150 km west of Maceio.

Forbes' Blackbird* occurs between the visitor's accommodation next to the school and the warden's house at the park entrance 2 km away. There are a number of side-trails off the main trail from the warden's house worth exploring. The best one goes left at the first major 'cross'. Alagoas Antwren* occurs along here, high in the canopy. After crossing the stream along here, turn left to a dilapidated house where Russet-crowned Crake occurs, and walk on to a clearing where you may find Yellow-faced Siskin*.

All forest habitats are worth exploring inland from Maceió, 259 km south of Recife.

White-collared Kite* and Alagoas Foliage-gleaner* occur at **Fazenda Bananeira**, 70 km northwest of Maceió. 10 km north of the Murici junction on the BR 101 there is a rough track known as Usina Bititinga II, which leads west for 10 km to the village of Bititinga. From here, a very rough track leads 11 km north into excellent forest.

Alagoas Curassow*, if it is not extinct, may occur around São Miguel dos Campos, south of Maceió. But there is little forest left here, and by the time you read this there may be no trees and no curassows.

Recife is the departure point for flights to **Isla Fernando de Noronha**, a tiny archipelago 345 km east of Brazil's northeast coast. The six islets are covered in cacti-scrub which supports two endemics:

Noronha Elaenia and Noronha Vireo. South American rarities, such as Squacco Heron, also turn up here.

The dry evergreen woodland on the isolated Serra de Baturité, near Fortaleza, on Brazil's north coast, supports some rare endemics including

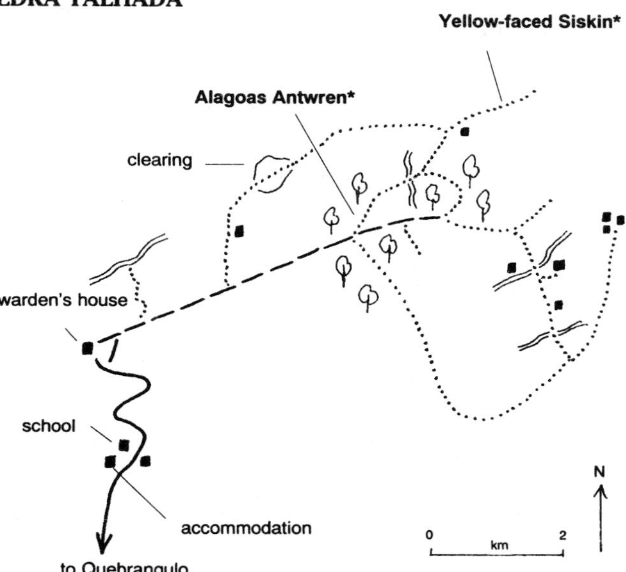

Jandaya Parakeet, Tawny Piculet, Tail-banded Hornero, Glossy Antshrike, Pectoral Antwren*, Hooded Gnateater*, Buff-breasted Tody-Tyrant and Ash-throated Casiornis, as well as White-eared Parakeet, Spot-billed Toucanet, Ochre-cheeked Spinetail, Band-tailed Manakin.

The city of **São Luís**, almost midway between Belém and Fortaleza on Brazil's north coast, lies near mangroves where a few specialities occur. Scarlet Ibis, Rufous Crab-Hawk and the endemic Little Wood-Rail occur in the mangroves around the village of Iguaiba, 30 km east of São Luís, as well as Mangrove Cuckoo, Plain-bellied Emerald and Bicoloured Conebill. The best way to see these species is to hire a boat, with guide, from the local villagers. Buff-browed Chachalaca and Zimmer's Woodcreeper have also been recorded near São Luís.

BRASILÍA NP

Situated just 9 km north of Brazil's capital, Brasília, this small NP (28,000 ha) protects some of the best remaining cerrado in Brazil. Although permission is required to visit some areas of the NP, most of the best birds can be seen near the entrance, where access is unrestricted. This is one of the few sites where the endemic Black-bellied Seedeater* has been recorded, and it is a reliable place to look for Bare-faced Curassow and the snazzy Collared Crescent-chest.

Endemics

Lesser Nothura*, Brasilia Tapaculo*, White-striped Warbler, Black-bellied Seedeater*.

Specialities

Small-billed and Dwarf* Tinamous, Southern Pochard, Bare-faced Curassow, Ocellated* and Rufous-faced* Crakes, Yellow-faced Parrot*, Least Nighthawk, Sickle-winged Nightjar*, White-vented Violet-ear, Horned Sungem, White-eared Puffbird, White-wedged Piculet, Checkered Woodpecker, Campo Miner, Russet-mantled Foliage-gleaner*, Black-capped Antwren, Collared Crescent-chest, Cock-tailed* and Streamer-tailed Tyrants, Curl-crested Jay, White-bellied Warbler, White-banded and White-rumped Tanagers, Coal-crested Finch*.

Others

Red-winged Tinamou, Green Ibis, Short-tailed Hawk, Rusty-margined Guan, Grey-necked Wood-Rail, Red-legged Seriema, Southern Lapwing, Picazuro Pigeon, Grey-fronted Dove, Peach-fronted and Canary-winged Parakeets, Nacunda Nighthawk, Fork-tailed Palm-Swift, Swallow-tailed Hummingbird, Fork-tailed Woodnymph, Long-billed Starthroat, Rufous-tailed Jacamar, Toco Toucan, Green-barred Woodpecker, Campo Flicker, Variable Antshrike, Rufous Gnateater, Euler's Flycatcher, Grey and White-rumped Monjitas, Tawny-headed Swallow, Moustached Wren, Yellowish Pipit, Masked Yellowthroat, Saffron-billed Sparrow, Orange-headed and Black-goggled Tanagers, Blue Dacnis, Wedge-tailed Grass Finch, Plumbeous Seedeater, Green-winged Saltator, Chopi Blackbird.

Access

Public access is restricted to the entrance area but the the forest surrounding the swimming pool here is excellent. Birding at dawn and dusk may produce Small-billed Tinamou, Bare-faced Curassow and Curl-crested Jay. Collared Crescent-chest occurs in the scrub around the entrance. Rusty-margined Guan occurs along the road around the northern perimeter, reached via Colorado Garage on the Formosa road.

Permission to visit restricted areas of the park should be sought from the Delegacia Estadual do IBDF, Av. W-3 N, Q513, Edif. Imperador. These areas include Lake Barragem Santa María where the rare Rufous-faced Crake* and Southern Pochard have been recorded.

Accommodation: Planalto Hotel.

The remote **Araguaia NP**, situated 750 km northwest of Brasília, more or less in the centre of the country, protects cerrado, gallery forest and part of the seasonally-flooded Ilha do Bananal, a huge wetland. Glossy Antshrike, Crimson-fronted Cardinal and Scarlet-throated Tanager occur along the river, Chestnut-bellied Guan* and Bananal Antbird occur in the gallery forest; Brazilian Tinamou, Rufous Nightjar, Cinnamon-throated Hermit and Grey-headed Spinetail occur in the dry woodland, and Long-tailed Ground-Dove, Hyacinth Macaw* and Yellow-faced Parrot*,in the cerrado. It is possible to fly to Santa Teresinha from Brasília and stay at the NP HQ, across the Rio Araguaia, with permission from IBAMA.

EMAS NP

This large (130,000-ha) but somewhat remote NP, roughly halfway between Brasília and Cuiabá (via Goiânia), protects some of Brazil's best remaining cerrado, as well as gallery forest. Although permission is required to visit and stay here, the birder with time to spare could be well rewarded for the extra effort involved. Rarities such as Crowned Eagle*, White-winged Nightjar*, Bearded Tachuri*, Cinereous Warbling-Finch*, and three seedeaters occur here as well as what is probably the best variety of large mammals in South America. Indeed, the rolling tree-dotted savanna grasslands are somewhat reminescent of Africa.

Endemics

Lesser Nothura*, White-striped Warbler.

Emas NP is named after the Greater Rhea

Specialities

Spotted Nothura, Dwarf Tinamou*, Crowned Eagle*, Bare-faced Curassow, Grey-breasted Crake, Yellow-faced* and Blue-fronted Parrots, White-winged Nightjar*, White-eared Puffbird, Pale-crested Woodpecker, Campo Miner, Russet-mantled Foliage-gleaner*, Large-billed Antwren, Collared Crescent-chest, Helmeted Manakin, Sharp-tailed Tyrant*, White-rumped Monjita, Cock-tailed* and Streamer-tailed Tyrants, Bearded Tachuri*, Curl-crested Jay, White-bellied Warbler, White-banded and White-rumped Tanagers, Coal-crested* and Black-masked* Finches, Cinereous Warbling-Finch*, Marsh*, Grey-and-chestnut* and Chestnut* Seedeaters, Black-throated Saltator.

Others

Red-winged Tinamou, Greater Rhea*, Long-winged Harrier, Red-legged Seriema, Pale-vented Pigeon, Blue-and-yellow Macaw, Peach-fronted Parakeet, Scissor-tailed Nightjar, Fork-tailed Palm-Swift, Gilded Sapphire, White-tailed Goldenthroat, Toco Toucan, White Woodpecker, Sooty Tyrannulet, Grey Monjita, Rufous Casiornis, Masked Gnatcatcher, Tawny-headed Swallow, Chopi Blackbird.

Other Wildlife

Giant Anteater, Puma, Pampas and Marsh Deer, Capybara, Giant Otter, Giant and Six-banded Armadillos, Pampas Fox, Maned Wolf, Brazilian Tapir, Anaconda.

Access

The park entrance is 130 km southwest of Mineiros, off the BR 359, a long, long drive from Brasília or Cuiabá. However, it is possible to fly in from Goiânia or Campo Grande (south Pantanal). Bare-faced Curassow and Pale-crested Woodpecker occur along the walk by the Río Formosa.

Permission to visit and stay should be sought from IBAMA.

Accommodation: It is possible to stay in the park. Take enough food for yourself and the potential cooks.

CHAPADA DOS GUIMARÃES NP

Chapada dos Guimarães NP, 68 km northeast of Cuiabá, lies at the western edge of the central plateau. The campo and (Amazonian) gallery forest around the Veu de Noiva waterfall support a number of species not found in the nearby Pantanal, including Dot-eared Coquette and Fiery-capped Manakin.

Specialities

Tataupa Tinamou, Crowned Eagle*, Blue-winged Macaw*, Pheasant Cuckoo, Biscutate Swift, Cinnamon-throated Hermit, Dot-eared Coquette, Horned Sungem, White-eared Puffbird, Yellow-ridged Toucan, Pale-crested Woodpecker, Rufous-winged Antshrike, Large-billed Antwren, White-backed Fire-eye, Band-tailed, Helmeted and Fiery-capped Manakins, Crested Black-Tyrant, Purplish and Curl-crested Jays, White-bellied Warbler, White-banded and White-rumped Tanagers, Coal-crested* and Red-crested Finches, Black-throated

Saltator, Yellow-billed Blue Finch*.

Others

Undulated and Red-winged Tinamous, Greater Rhea*, King Vulture, Pearl Kite, Tiny Hawk, Bat Falcon, Rusty-margined Guan, Red-legged Seriema, Pale-vented Pigeon, Red-and-green Macaw, White-eyed and Peach-fronted Parakeets, Little and Guira Cuckoos, Glittering-bellied Emerald, Blue-crowned Motmot, Brown Jacamar, Little Woodpecker, Sooty-fronted Spinetail, Rusty-backed Antwren, Planalto Tyrannulet, Suiriri Flycatcher, Plain-crested Elaenia, Grey Monjita, Crested Becard, Pectoral Sparrow, Guira Tanager, Swallow-Tanager, Epaulet Oriole.

Other Wildlife

Black-capped Capuchin.

Access

Head north out of Cuiabá. Blue-winged Macaw* occurs at Portao do Inferno after 52 km. After 57 km the road reaches the 60-m high Veu de Noiva waterfall. Helmeted Manakin occurs in the forest at the bottom of this waterfall, reached by a steep trail behind the café, and along the trails by the next waterfall along the main road. 9 km beyond the Veu de Noiva, the road south to a radar station runs through good forest. Further along the main road, turn north on the track to Agua Fria. Point-tailed Palmcreeper occurs along here. Further still towards Chapada dos Guimarães turn south to the Portao da Fe religious centre. Horned Sungem occurs along this road. Behind the centre is more excellent forest where Cinnamon-throated Hermit, Dot-eared Coquette, Pale-crested Woodpecker, and Band-tailed Manakin and Fiery-capped Manakins occur.

Accommodation: Pousada da Chapada (A); Hotel Turismo (B).

The rare, endemic Blue-eyed Ground-Dove* has been recorded at the **Serra das Araras Ecological Station**, northeast of Cáceres, southwest of Cuiabá. Birds similar to those found in the Pantanal occur around the **Hotel Fazenda Barranquinho**, 60 km southwest of Cáceres.

THE PANTANAL

Map p. 139

The Pantanal is the biggest freshwater wetland in the world and, therefore, one of the world's best birding areas. The numbers of birds, especially at the end of the dry season in September, has to be seen to be believed. Even the Llanos of Venezuela cannot compete with the Pantanal at its best.

The area is usually flooded by the rising Río Paraguay from October to March, with the waters receding to their lowest in September when masses of waterbirds throng the remaining pools. A dead-end road, known as the 'Transpantaneira', pierces the Pantanal. It runs south from Poconé, 93 km south of Cuiabá, through extensive forest and open-water areas, for a distance of 148 km, to Porto Jofre. Birding from this road can produce over 100 species in a day, including many herons, ibises and raptors, the endemic Chestnut-bellied Guan* and last, but by no means least, the beautiful Hyacinth Macaw*.

*One of the Pantanal's star birds is the Hyacinth Macaw**

Endemics

Chestnut-bellied Guan*.

Specialities

Ringed Teal, Crowned Eagle*, Chaco Chachalaca, Bare-faced Curassow, Long-tailed Ground-Dove, Hyacinth* and Golden-collared Macaws, Nanday and Yellow-chevroned Parakeets, Blue-fronted Parrot, Buff-bellied Hermit, White-wedged Piculet, Pale-crested Woodpecker, Great Rufous Woodcreeper, Chicli, Cinereous-breasted and White-lored Spinetails, Rufous Cachalote, Mato Grosso and Band-tailed Antbirds, Helmeted Manakin, Stripe-necked Tody-Tyrant, Plain Tyrannulet, Bearded Tachuri*, White-rumped Monjita, Dull-capped Attila, White-naped Xenopsaris, Purplish and Curl-crested Jays, Fawn-breasted Wren, Red-crested and Yellow-billed Cardinals, Coal-crested* and Red-crested Finches, Black-and-tawny Seedeater*, Black-backed Grosbeak, Scarlet-headed Blackbird.

Others

Greater Rhea*, Southern Screamer, Fulvous, White-faced and Black-bellied Whistling-Ducks, Muscovy and Comb Ducks, Capped Heron, Rufescent Tiger-Heron, Bare-faced, Plumbeous, Buff-necked and Green Ibises, Roseate Spoonbill, Wood and Maguari Storks, Jabiru, Lesser Yellow-headed and King Vultures, Pearl and Snail Kites, Crane Hawk, Great Black-Hawk, Savanna and Black-collared Hawks, Laughing Falcon, Blue-throated Piping-Guan, Grey-necked Wood-Rail, Sungrebe, Sunbittern, Limpkin, Red-legged Seriema, Wattled Jacana, Pied and Southern Lapwings, Large-billed and Yellow-billed Terns, Picui Ground-Dove, Blue-and-yellow Macaw, Peach-fronted, Monk and Canary-winged Parakeets, Little, Guira and Striped Cuckoos, Band-tailed and Nacunda Nighthawks, Swallow-tailed Hummingbird, Gilded Sapphire, White-tailed Goldenthroat, Blue-crowned Trogon, American Pygmy Kingfisher, Rufous-tailed Jacamar, Black-fronted Nunbird, Chestnut-eared Aracari, Toco Toucan, White, Little and Golden-green Woodpeckers, Narrow-billed Woodcreeper, Red-billed Scythebill, Chotoy Spinetail, Greater Thornbird, Great and Barred Antshrikes, Rusty-backed Antwren, Pearly-vented Tody-Tyrant, Rusty-fronted Tody-Flycatcher, Bran-coloured and Fuscous Flycatchers, Grey Monjita, Rufous Casiornis, Ashy-headed Greenlet, Black-capped Donacobius,

THE PANTANAL

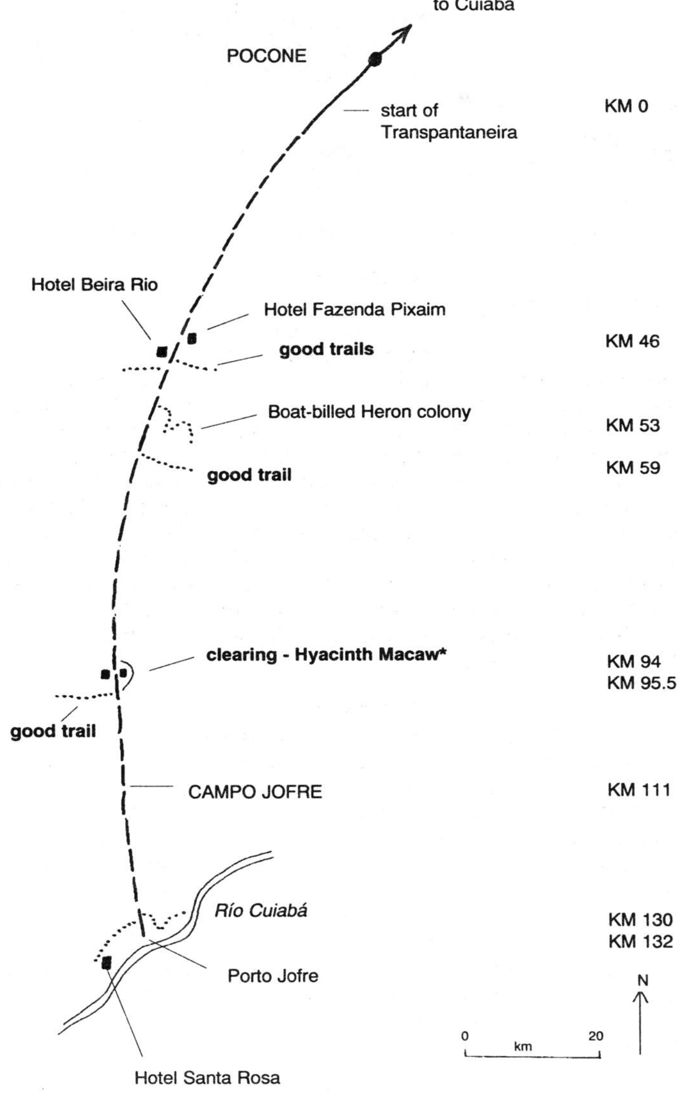

Thrush-like Wren, Masked Gnatcatcher, Yellowish Pipit, Flavescent Warbler, Saffron-billed Sparrow, Purple-throated Euphonia, Greyish Saltator, Crested Oropendola, Solitary Cacique, Epaulet Oriole, Troupial, Unicoloured, White-browed and Chopi Blackbirds, Bay-winged and Giant Cowbirds.

Other Wildlife

Yellow Caiman, Capybara, Black Howler, Silvery Marmoset, Black-capped Capuchin, Coati-mundi, Marsh Deer, Giant Otter, Giant Anteater, Ocelot, Jaguar, Maned Wolf.

Access

Heading south from Poconé the 'Transpantaneira' officially begins at km 109 from Cuiabá, 16 km south of Poconé. Directions from here on are based on the start of the 'Transpantaneira', which is signposted, being km 0.

Chestnut-bellied Guan* occurs in the area of Pousada Araras shortly after the start. Try opposite the entrance. Long-tailed Ground-Dove and Band-tailed Antbird occur in the scrub around the Hotel Beira Rio at km 46, Ashy-headed Greenlet occurs along the trail to the east at km 53 (which also leads to a Boat-billed Heron colony), and Mato Grosso Antbird and Helmeted Manakin occur along the trail to the east at km 59. Hyacinth Macaw* roosts around the hut on the east side of the road at km 94, and between the Hotel Santa Rosa and the camp site at Porto Jofre (km 132). Waterbirds may be particularly numerous around km 111, an area known as Campo Jofre, and Blue-throated Piping-Guan and Bare-faced Curassow occur along the trail to the east at km 130.

Accommodation: Poconé: Hotel Santa Cruz (B). Km 46: Pousada Pixiam (A). Km 132: Hotel Santa Rosa (A) (camping too; excellent birding).

The Pantanal is also accessible, albeit with more difficulty, from **Corumbá**, at the southern end. There is less forest and water here but Greater Rhea*, Ringed Teal, Hyacinth* and Golden-collared Macaws, Blue-crowned, Nanday, Blaze-winged and Green-cheeked Parakeets, Blue-fronted Parrot, White-fronted Woodpecker, Great Rufous Woodcreeper, Rufous Cachalote, Mato Grosso Antbird, Bearded Tachuri*, Cinereous Tyrant, Rufous Casiornis, White-rumped Monjita, Fawn-breasted Wren, Yellow-billed Cardinal, Black-and-tawny*, Marsh* and Grey-and-Chestnut* Seedeaters, and Black-backed Grosbeak all occur between Campo Grande and Corumbá, a distance of 403 km, as well as many of the species listed under 'The Pantanal'.

Accommodation: Fazenda Santa Clara, 98 km east of Corumbá (Hyacinth Macaw* occurs here).

RONDONIA

The rapidly disappearing forests of Rondonia, in southern Amazonia, west Brazil, support a number of rare endemics and specialities hard to see elsewhere in South America, notably White-breasted Antbird*.

Endemics

Hoffman's and Dusky-billed Woodcreepers, Glossy Antshrike, Rondonia Bushbird*, White-breasted Antbird*.

Specialities

Starred Wood-Quail, Crimson-bellied, Painted and Yellow-chevroned Parakeets, Black-banded Owl, Pale-rumped Swift, Fiery-tailed Awlbill*, Festive Coquette, Blue-tufted Starthroat, Pavonine Quetzal, Blue-cheeked Jacamar, Rufous-necked Puffbird, Rufous-capped Nunlet, Black-girdled Barbet, Red-necked Aracari, Yellow-ridged Toucan, Chestnut-throated Spinetail*, Bamboo and White-shouldered Antshrikes, White-eyed and Chestnut-shouldered Antwrens, Band-tailed Antbird, Black-spotted Bare-eye, Black-bellied Gnateater, White-browed Purpletuft, Pompadour Cotinga, Snow-capped Manakin, Plain Tyrannulet, Lawrence's Thrush, Tooth-billed Wren, White-winged Shrike-Tanager.

Others

Capped Heron, Harpy Eagle*, Black and Red-throated Caracaras, Spix's Guan, Pied Lapwing, Blue-and-yellow and Scarlet Macaws, Mealy Parrot, Crested Owl, Black-tailed and White-tailed Trogons, Striolated Puffbird, Lettered Aracari, Golden-green Woodpecker, Pygmy and Dot-winged Antwrens, Striated and Chestnut-tailed Antbirds, Screaming Piha, Spangled Cotinga, Southern Nightingale-Wren, Black-collared Swallow, Yellowish Pipit, Buff-rumped Warbler, Black-faced Tanager, Red-billed Pied Tanager, Thick-billed Euphonia, Paradise Tanager, Yellow-shouldered Grosbeak.

Access

From Pôrto Velho head south to Fazenda Rancho Grande, a farm and forest reserve with cabin accommodation and trails. Alternatively, this is road- and trackside-birding country. Accessibility changes constantly, and any logging tracks may be worth exploring. Previously productive spots include the 93-km stretch of road between Ariquemes (202 km southeast of Pôrto Velho) and Jaru. Harpy Eagle* and Black-girdled Barbet have been recorded along the logging track, leading south, 55 km from Ariquemes.

Guajará-Mirim is another good birding locality. Head southwest out of Pôrto Velho to Abuna, 215 km away, then south 130 km to the town of Guajará-Mirim on the Bolivian border. Turn left at the roundabout on entering the town. After 6 km follow the track straight on, ignoring the road to the left, then take the right fork after 2 km. After 1 km cross the bridge into Amazonian rainforest, where Blue-tufted Starthroat, Pavonine Quetzal, Black-girdled Barbet, White-eyed Antwren and White-winged Shrike-Tanager have been recorded.

Accommodation: Guajará-Mirim; Lima Palace Hotel.

The state of **Acre** in far west Brazil has been visited by very few birders, but the following species have been recorded around Cruzeiro do Sul near the Peruvian border and in the roadside forest 20 km west of Placido de Castro, east of the remote but thriving town of **Rio Branco**: Grey Tinamou, Pale-winged Trumpeter, Blue-headed Macaw, Black-capped and Cobalt-winged Parakeets, White-bellied Parrot, Pheasant

Cuckoo, Spot-tailed Nightjar, Chestnut and Bluish-fronted Jacamars, Chestnut-capped and Striolated Puffbirds, Yellow-billed Nunbird, Lemon-throated Barbet, Curl-crested Aracari, Rufous-breasted Piculet, Peruvian Recurvebill*, Bamboo and Bluish-slate Antshrikes, Sclater's and Rio Suno Antwrens, Black, Goeldi's and Sooty Antbirds, Ash-throated Gnateater, White-browed Purpletuft, Cinnamon Tyrant-Manakin and Johannes' Tody-Tyrant.

THE AMAZON

There are some superb birding areas along the world's longest and mightiest river.

Tabatinga, in the far northwest of Brazil, is just 4 km from Leticia in Colombia and only a few km from Peru. The birds are similar to those found near Leticia (see p. 200) and the specialities, most of which occur on the south side of the river near Benjamin-Constant, include Black-banded Crake, Red-billed Ground-Cuckoo, Needle-billed Hermit, Scarlet-crowned Barbet, Ocellated Woodcreeper, Dark-breasted Spinetail, Bluish-slate Antshrike, Plain-throated, Rufous-tailed and Chestnut-shouldered Antwrens, White-throated Antbird, Black-spotted Bare-eye, Plum-throated Cotinga, Johannes' Tody-Tyrant, Double-banded Pygmy-Tyrant, Little Ground-Tyrant, the endemic Grey Wren and Yellow-crested Tanager.

Access

From Tabatinga cross the Amazon to Benjamin-Constant on the south side. Then it is necessary to search for suitable habitat, and this may take some time and effort. To the west of the town has been productive in the past.

From Tabatinga and Benjamin-Constant it is possible to to visit the Leticia area in Colombia without too much border-control hassle, and to travel by boat (a three-day journey) to the Iquitos site in Peru (see p. 316).

Accommodation: Tabatinga: Solimoes Hotel (at army camp).

Tefé, approximately halfway between Tabatinga and Manaus, is a place for pioneers, although previous expeditions have recorded such goodies as Bartlett's Tinamou, Slaty-backed Forest-Falcon, Pale-winged Trumpeter, Black-banded Crake, Red-winged Wood-Rail, Short-tailed and Festive Parrots, Ladder-tailed Nightjar, White-bearded and Needle-billed Hermits, Fiery Topaz, White-chinned Jacamar, Brown-banded and Rufous-necked Puffbirds, Bar-bellied Woodcreeper, Curve-billed Scythebill, White-bellied and Red-and-white Spinetails, Orange-fronted Plushcrown, Short-billed and Ihering's Antwrens, Black-and-white, Slate-coloured, Plumbeous and Hairy-crested Antbirds, Plum-throated Cotinga, Brownish Elaenia, Riverside Tyrant, Citron-bellied Attila, Guianan Gnatcatcher, Pearly-breasted Conebill* and Blue-backed Tanager.

It is possible to hire a canoe from the 'canoe-rank' in town for the day and to explore river islands down the Amazon. However, to reach good forest around Tefé you will need to organise an expedition, or hire a helicopter!

MANAUS

Map p. 145

Manaus is a major Amazon port at the confluence of the Río Negro, over 2,250 km from the Atlantic. Thanks to the fluctuating water levels (up to 12 m) of the Amazon, the spread of Manaus has been checked, and large tracts of lowland rainforest still remain. The avifaunas of the north and south banks of the river, as well as the Manacapurú peninsula between the Río Negro and the Río Solimões (western Amazon), and the islands in the Río Negro, known as the Anavilhanas archipelago, are all different. Hence the Manaus area is one of the richest birding sites on earth, with over 600 species recorded.

The localised endemics include White-winged Potoo, Chestnut-headed Nunlet* and Klage's Antwren*, one of over 30 antbirds in the area. There are many star birds amongst the specialities, not least Zigzag Heron*, Rufous Potoo, Racket-tailed Coquette, Black-necked and Guianan Red-Cotingas, Guianan Cock-of-the-Rock and Dotted Tanager*.

Permission is required to visit the best sites, including two towers which breach the rainforest canopy and allow views of many rarely seen mega-birds, including Crimson Fruitcrow.

Endemics

White-winged Potoo, Chestnut-headed Nunlet*, Zimmer's Woodcreeper, Tail-banded Hornero, Glossy Antshrike, Klage's Antwren*.

Specialities

Zigzag Heron*, Little Chachalaca, Marail Guan, Crestless, Black and Wattled* Curassows, Grey-winged and Pale-winged Trumpeters, Maroon-tailed and Golden-winged Parakeets, Sapphire-rumped Parrotlet, Caica, Short-tailed, Dusky and Red-fan Parrots, Black-bellied and Pavonine Cuckoos, Rufous Potoo, Chapman's Swift, Needle-billed Hermit, Racket-tailed Coquette, Olive-spotted Hummingbird, Crimson Topaz, Pavonine Quetzal, Green Aracari, Guianan Toucanet, Yellow-ridged Toucan, Golden-collared Woodpecker, Chestnut-rumped Woodcreeper, Lesser Hornero, White-bellied Spinetail, Plain Softtail, Point-tailed Palmcreeper, Black-throated and Band-tailed Antshrikes, Cherrie's, Brown-bellied, Leaden, Spot-backed and Ash-winged Antwrens, Black-and-white, Black-headed, Ferruginous-backed, Rufous-throated and Dot-backed Antbirds, Reddish-winged Bare-eye, Spotted Antpitta, Chestnut-belted Gnateater, Black-necked and Guianan Red-Cotingas, Dusky Purpletuft, Purple-breasted and Pompadour Cotingas, Crimson Fruitcrow, Capuchinbird, Guianan Cock-of-the-Rock, Red-headed, White-throated, Yellow-crested and Flame-crested Manakins, Saffron-crested and Tiny Tyrant-Manakins, McConnell's Flycatcher, Snethlage's Tody-Tyrant, Painted Tody-Flycatcher, Ringed Antpipit, Brownish Elaenia, River Tyrannulet, Lesser Wagtail-Tyrant, Olive-green Tyrannulet, Double-banded Pygmy-Tyrant, Yellow-throated Flycatcher, Greater Schiffornis, Glossy-backed and Pink-throated Becards, Grey-chested Greenlet, Wing-banded Wren, Guianan Gnatcatcher, Fulvous Shrike-Tanager, Masked Crimson Tanager, Golden-sided Euphonia, Dotted Tanager*, Velvet-fronted Grackle.

From the towers overlooking the canopy near Manaus it is possible to see the incredible Crimson Fruitcrow; all crimson, except for the flight feathers and tail

Others

Great Tinamou, Horned Screamer, Capped Heron, King Vulture, Double-toothed and Plumbeous Kites, Tiny Hawk, Crested* and Harpy* Eagles, Ornate Hawk-Eagle, Red-throated Caracara, Lined Forest-Falcon, Spix's Guan, Azure Gallinule, Sungrebe, Sunbittern, Limpkin, Wattled Jacana, Large-billed Tern, Blue-and-yellow, Scarlet, Red-and-green and Red-bellied Macaws, White-eyed Parakeet, Red-lored and Festive Parrots, Dark-billed Cuckoo, Least Pygmy-Owl, Spectacled Owl, Grey Potoo, Band-rumped Swift, Long-tailed, Straight-billed and Dusky-throated Hermits, White-necked Jacobin, Versicoloured Emerald, Black-eared Fairy, Black-tailed, White-tailed and Black-throated Trogons, Blue-crowned Motmot, Yellow-billed, Paradise and Great Jacamars, White-necked Puffbird, Channel-billed Toucan, Yellow-throated, Golden-green, Chestnut, Lineated and Red-necked Woodpeckers, White-chinned, Long-tailed, Spot-throated, Cinnamon-throated, Red-billed, Barred and Lineated Woodcreepers, Speckled and Red-and-white Spinetails, Rufous-rumped and Olive-backed Foliage-gleaners, Short-billed Leaftosser, Fasciated, Blackish-grey, Mouse-coloured, Eastern Slaty, Amazonian, Dusky-throated and Cinereous Antshrikes, Pygmy, Streaked, Stipple-throated, White-flanked, Long-winged and Grey Antwrens, Grey, Warbling, White-plumed, Bicoloured and Scale-backed Antbirds, Rufous-capped and Black-faced Antthrushes, Thrush-like Antpitta, Screaming Piha, Spangled Cotinga, Amazonian Umbrellabird, White-crowned and Blue-backed Manakins, Dwarf Tyrant-Manakin, Slender-footed and White-lored Tyrannulets, Forest, Grey and Large Elaenias, Yellow-margined and Sulphur-rumped Flycatchers, Amazonian Black-Tyrant, Bright-rumped Attila, Greyish Mourner, Sirystes, Piratic Flycatcher, Slaty-capped Shrike-Vireo, Buff-cheeked and Tawny-crowned Greenlets, Pale-breasted and Black-billed Thrushes, Coraya and Musician Wrens, Collared Gnatwren, White-thigh ed Swallow, Red-billed Pied Tanager, Orange-headed, Yellow-backed, Fulvous-crested, Paradise, Spotted, Bay-headed and Opal-rumped Tanagers, Black-faced Dacnis, Chestnut-bellied Seedeater, Yellow-green, Slate-coloured and Blue-black Grosbeaks, Green Oropendola, Yellow-rumped Cacique, Oriole Blackbird.

Other Wildlife

Two species of river dolphin, Red Howler, Bearded Saki.

MANAUS

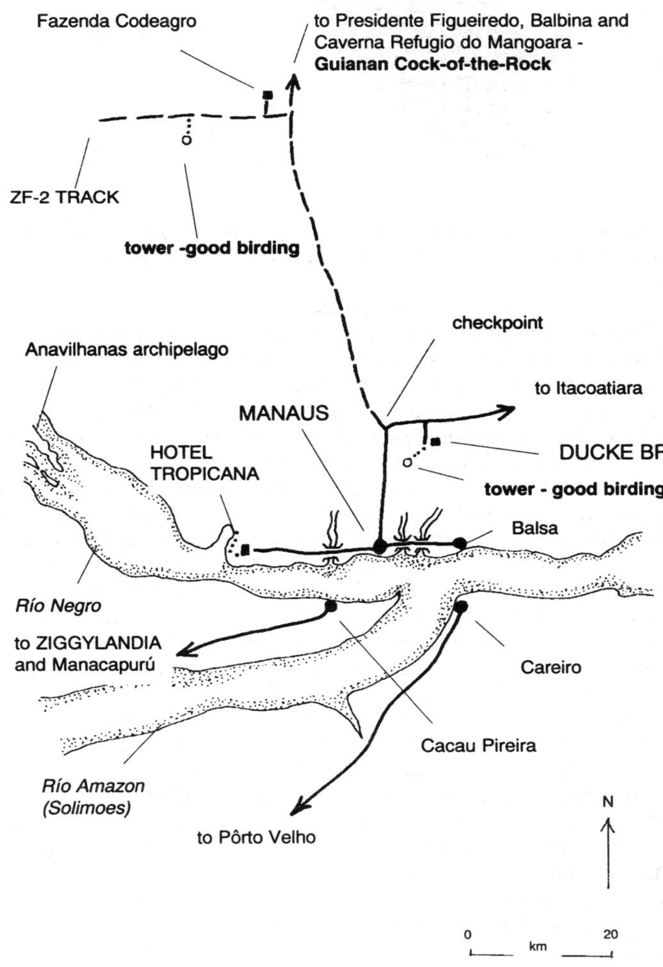

Access

Near Manaus, the grounds of the Hotel Tropical, on Ponta Negra west of the city, and the Japanese colony near Parque Dez, north of the city, are worth birding.

Head north out of the city on the road to Itacoatiara to reach the Ducke Biological Reserve, the entrance to which is on the south side of the road just east of the police checkpoint. Over 275 species have been recorded here, in what is excellent Amazonian rainforest. Birding on the ground is tough, but the 42-m high meterological tower allows rare views over the canopy; at dawn it is the best place to look for Harpy Eagle* and Crimson Fruitcrow as well as other raptors, parrots, Racket-tailed Coquette, Guianan Toucanet, Ash-winged Antwren, Purple-breasted and Pompadour Cotingas, Olive-green Tyrannulet, Buff-cheeked Greenlet, and Red-billed Pied Tanager. To visit and stay at Ducke, contact INPA.

Pale-bellied Mourner has been recorded further east along the road to Itacoatiara, where viewing the forest is somewhat easier.

Continuing north past the police checkpoint on the BR 174, turn west after 50 km on to the ZF-2 track, the entrance to the Smithsonian Institution Reserve. 15.5 km along this track look for a trail, on the south side, to another tower which is even better than the one at Ducke. Olive-green Tyrannulet has been recorded here. To visit and stay at this reserve contact BDFF.

Guianan Cock-of-the-Rock occurs along the trail to the caves of Caverna Refugio do Mangoara. To reach here, continue north along the BR 174 and turn east just before Presidente Figueiredo towards Balbina. Look for the caves' entrance 6.5 km along this road. The forest around Balbina is also worth birding, if time allows.

To bird the south side of the Amazon, take a ferry from Balsa east of the city, to Careiro. Explore any suitable habitat around here and along the BR 319 southwest to Pôrto Velho.

To visit the Manacapurú peninsula, take a ferry from Manaus to Cacau Pireira. 50 km southwest of here, on the road to Manacapurú, turn north to Ziggylandia, an ecological reserve run by IBAMA. Here it is possible to hire a boat to explore the south side of the Río Negro. IBAMA also run a reserve on the Anavilhanas archipelago, a series of islands in the Río Negro, northwest of Manaus. It may be possible to hire a boat to explore this area which consists of permanently flooded forest (igapo) where the birds are different again. Zigzag Heron*, Klages' Antwren* and Amazonian Umbrellabird occur here.

Accommodation: Hotel Tropical; Da Vinci Hotel.

The **Amazonia (Tapajós) NP**, 53 km southwest of Itaituba, is one of the few sites for the spectacular endemic, Golden Parakeet*. It occurs at Urua, 12.5 km beyond the Tracoa entrance. 16.6 km further southwest, on the road that traverses the NP, is the Ramal da Capelinha trail, on the north side of the road. As well as many birds, Tamandua Anteater, and Silvery and Tassel-ear Marmosets occur along here. Other endemics include White-crested Guan*, Dark-winged Trumpeter, Vulturine Parrot, Brown-chested Barbet, Hoffmann's Woodcreeper, and Harlequin and Pale-faced Antbirds, whilst specialities include Harpy Eagle*, Red-throated Piping-Guan, Razor-billed Curassow, Gould's Toucanet, Bar-breasted Piculet, Spix's Woodcreeper, Chestnut-winged

Foliage-gleaner, Saturnine Antshrike, Ornate Antwren, Dot-backed Antbird, Snow-capped and Flame-crested Manakins, Black-chested Tyrant, Short-tailed Pygmy-Tyrant and Rose-breasted Chat. The best way to get to Tapajós NP is by flying from Santarem to Itaituba. Both towns have travel agencies willing to organise visits and, possibly, meetings with the Park Director.

In between flights to Tapajós, it may be worth exploring the **Santarém** area. The dry cerrado-like habitat on the isolated peninsula near the village of Alter do Chão, 38 km west of Santarém and reached by canoe, supports Narrow-billed Woodcreeper and Rusty-backed Antwren. The recently described endemic, Kawall's Parrot, occurs in the Tapajós National Forest, run by IBAMA, south of remote Belterra and 50 km south of Santarém. Rubber-tappers trails allow rare access to the excellent forest here. Specialities of the area include Dark-winged Trumpeter, Vulturine Parrot, Scaled Ground-Cuckoo*, Silky-tailed Nightjar, Zimmer's Woodcreeper, Glossy Antshrike, Sclater's, Klage's* and White-eyed Antwrens, Bare-eyed Antbird, Black-bellied Gnateater, Black-necked and Guianan Red-Cotingas, White-tailed Cotinga, Opal-crowned Manakin, Zimmer's Tody-Tyrant*, Pale-bellied Mourner and Tooth-billed Wren.

The major port of **Belém** lies near the mouth of the Amazon in far north Brazil. The endemics, Buff-browed Chachalaca, Long-tailed Woodnymph and White-tailed Cotinga, occur near here as well as Golden-winged Parakeet, Spix's Woodcreeper, Chestnut-crowned Foliage-Gleaner, Spangled Cotinga, Black-chested Tyrant and Spotted Tody-Flycatcher. Permission is required from EMBRAPA (an agricultural organisation) to visit the **Guama Ecological Research Area**, just east of town. Some of the best forest near Belém is on **Mosqueiro Island**, 65 km north of Belem, 7 km into the island from the bridge. This and other forest patches, which need tracking down, support such specialities as Brazilian Tinamou, Scarlet-shouldered Parrotlet, Vulturine Parrot and Hooded Gnateater*. Scarlet Ibis, Red-bellied Macaw and Point-tailed Palmcreeper occur east of Atalaia beach, 12 km from **Salinópolis**, a seaside resort 228 km east of Belém.

CARAJÁS

At this little-known site, situated between the Xingu and Tocantins rivers, some 500 km southwest of Belém, near Marabá, Amazonian rainforest meets caatinga. Six species of macaw include Hyacinth*, and notable goodies recorded in the area include Pearly Parakeet, Scaled Ground-Cuckoo*, Long-tailed Potoo and White-tailed Cotinga.

Endemics

White-crested Guan, Dark-winged Trumpeter, Golden*, Jandaya and Pearly Parakeets, Vulturine Parrot, Scaled Ground-Cuckoo*, Dusky-billed Woodcreeper, Glossy Antshrike, Hooded Gnateater*, White-tailed Cotinga, Opal-crowned Manakin, Ash-throated Casiornis.

Specialities

Grey-bellied Goshawk*, Hyacinth Macaw*, Red-fan Parrot, Pavonine Cuckoo, Long-tailed Potoo, Fiery-tailed Awlbill*, Dot-eared and Racket-tailed Coquettes, Rufous-necked Puffbird, Black-girdled Barbet, Red-necked Aracari, Black-bellied Gnateater, White Bellbird, Snethlage's Tody-Tyrant, Pale-bellied Mourner, Guianan Gnatcatcher.

Others

Brazilian Tinamou, Harpy Eagle*, Amazonian Pygmy-Owl, Scissor-tailed Nightjar, Yellow-billed Jacamar, Blackish-grey Antshrike, Grey-chested Greenlet.

Access

A permit is required to visit Carajás, available from Companhia Vale Rio Doce (CVRD) in Marabá or Rio. It is a long trip from Belém, but flights are available.

Alta Floresta, in west-central Brazil, approximately 1,000 km north of Cuiabá, is one of the least explored areas of the Amazon basin. A number of specialities can be found here, in an island of rainforest which has been isolated by the Tapajós and Xingu rivers, as well as many species of wider southern Amazonia distribution which do not occur in the Manaus area to the north, but are possible in east Ecuador and Peru. Brazilian endemics include Dark-winged Trumpeter, Vulturine Parrot, Glossy Antshrike, Bare-eyed Antbird* and Golden-crowned Manakin*, whilst specialities include White-browed Hawk, Red-throated Piping-Guan, Blue-cheeked Jacamar, Striolated and Rufous-necked Puffbirds, Chestnut-throated Spinetail*, Peruvian Recurvebill*, Pearly and Saturnine Antshrikes, Sclater's and White-eyed Antwrens, Manu Antbird, Purple-throated Cotinga*, Amazonian Umbrellabird, Slender-footed and White-lored Tyrannulets, Snethlage's and Zimmer's* Tody-Tyrants, Black-backed Water-Tyrant, Thrush-like Schiffornis, Tooth-billed Wren and Amazonian Oropendola.

Bird the grounds of Hotel Floresta Amazonica in Alta Floresta, and the primary forest alongside the Río Cristalino on the boat trip to the Christalino River Lodge. The area around this lodge supports Red-throated Piping-Guan, Dark-winged Trumpeter, Chestnut-throated Spinetail*, and many more great birds.

NORTH OF THE AMAZON

The remote **Pico da Neblina NP** (2,200,000 ha) in far northwest Brazil, northeast of São Gabriel da Cachoeira, on the Río Negro, is a pioneer's paradise. To visit it requires considerable initiative, but help may be obtained from IBAMA. Specialities recorded here in the far past include Grey-legged and Barred Tinamous, Tawny-tufted Toucanet, Undulated Antshrike, Cherrie's Antwren, Grey-bellied and Chestnut-crested Antbirds, the endemic Pelzeln's Tody-Tyrant and Scaled Flower-piercer.

Northern **Roraima** is more accessible. It is possible to fly to Boa Vista from Manaus. Rio Branco Antbird occurs on the Ilha São José, 10 km up river from Boa Vista. Rio Branco Spinetail* is more tricky and requires a mini-expedition in the 'outback'. It occurs on the south side of the Río Tacutu at Conceiçao do Mau, near the Guyana border, northeast of Boa Vista. Both the species were formerly considered to be endemic to this tiny area, but have recently been discovered in nearby Guyana (see p. 278). The Río Anaua, Mucajai and Ilha de Maraca Biological Reserves, south, west and northwest of Boa Vista respectively, are virtually inaccessible although roads skirt their edges. Specialities of this area include many species similar to those present on the 'Escalera' in Venezuela (see p. 397).

Brazil's northeasternmost state, **Amapa**, is also fairly accessible now. From Macapá at the mouth of the Amazon, head 108 km north to Pôrto Grande. The road west from here to Serra do Naivo runs through excellent forest where Spot-tailed Antwren and Crimson Fruitcrow occur.

ADDITIONAL INFORMATION

Addresses

BDFF: Biological Dynamics of Forest Fragments, Rua Rio Purus 28, Manaus-AM-69011.
IBAMA: Director de Parques Nacionais, IBAMA, Av. L4 Norte - Setor Areas Isoladas Norte, Brasilia-DF-70620.
Pca. XV No. 2, 6 andar, sala 409 - Centro, Rio de Janeiro-RJ- 20010.
Av. Marechal Mascarenhas de Morais, 2487, Vitoria, Espirito Santo.
IEF (Instituto Estadual de Florestas): Director de Parques e Reserves, IEF, Rua Paracatu 304, 10 andar, Belo Horizonte-MG-30180.
INPA (Instituto Nacional de Pesquisas da Amazonia): Estrada V8 do Aleixo, Caixa Postal 478, Manaus-AM-69011.

Books

Birds in Brazil: A Natural History, Sick, H. 1993. Princeton UP.
Birding Brazil, Forrester, B. 1993. Privately published. Available from: Bruce C. Forrester, Knockshinnoch Bungalow, Rankinston, Ayrshire KA6 7HL, UK.
Birds of Southwestern Brazil, Dubs, B. 1992. Privately published.
Birds of Argentina and Uruguay: A Field Guide, Narosky, T. and Yzurieta, D. 1989. (This book covers over 90% of the birds found in Rio Grande do Sul, south Brazil.)

BRAZIL ENDEMICS (179)	BEST SITES
Yellow-legged Tinamou*	East: Fazenda Monte Claros
Lesser Nothura*	Central: Serra do Cipo, Brasilia NP and Emas NP
White-collared Kite*	Northeast: Pedra Talhada and Bananeira
White-necked Hawk*	East: Sooretama and Monte Pascoal
Buff-browed Chachalaca	East Amazonia: São Luís and Belém
White-crested Guan*	East Amazonia: Amazonia NP and Carajás
Chestnut-bellied Guan*	Southwest and Amazonia: Araguaia NP and Pantanal
White-browed Guan*	Northeast: Canudos
Alagoas Curassow*	Northeast: near Maçeio, but possibly extinct

Red-billed Curassow*	East: Sooretama and Monte Pascoal
Little Wood-Rail	East Coast and east Amazonia: Salavador and São Luís
Dark-winged Trumpeter	East Amazonia
Blue-eyed Ground-Dove*	Southwest: Serra das Araras near Cuiabá
Lear's Macaw*	Northeast: Canudos
Spix's Macaw*	Northeast: only one bird remains in the wild
Golden Parakeet*	East Amazonia: Amazonia NP and Carajás
Jandaya Parakeet	Northeast: Pedra Talhada, Fortaleza and Carajás
Golden-capped Parakeet*	East: Serra da Canastra, Rio Doce and Monte Claros
Caatinga Parakeet	East and northeast: Boa Nova and Canudos
Blue-throated Parakeet*	East: Monte Claros, Sooretama and Monte Pascoal
Pearly Parakeet	East: Amazonia: Carajás
Plain Parakeet	Southeast to northeast: many sites
Brown-backed Parrotlet*	Southeast: rare, Orgãos, Itatiaia and Serra do Mar
Golden-tailed Parrotlet*	East: rare
Vulturine Parrot	East Amazonia: most sites
Red-browed Parrot*	East: Sooretama, Monte Pascoal and Porto Seguro
Kawall's Parrot	East Amazonia: Santarém
Red-tailed Parrot*	Southeast: rare, Ilha do Cardoso
Blue-bellied Parrot*	Southeast: rare, Boraceia near São Paulo
Scaled Ground-Cuckoo*	East Amazonia: Santarém and Carajás
White-winged Potoo	Amazonia: rare, Manaus
Pygmy Nightjar*	Northeast: Pedra Talhada
Broad-tipped Hermit	East and northeast: Boa Nova and Canudos
Minute Hermit	East: Rio Doce and Sooretama
Saw-billed Hermit*	Southeast and East: Serra do Mar and Santa Theresa
Hook-billed Hermit*	East: Sooretama
Frilled Coquette	East and northeast: a few sites
Long-tailed Woodnymph	East and northeast: Monte Pascoal and Pedra Talhada
Flame-rumped Sapphire	East: known only from five specimens collected around 1881, and possibly a hybrid
Sombre Hummingbird	Southeast to northeast: a few sites
Brazilian Ruby	Southeast and east: a few sites
Hooded Visorbearer*	Northeast: Salvador and Morro do Chapeu
Hyacinth Visorbearer*	Central: Caraca NP and Serra do Cipo

Stripe-breasted Starthroat	East and northeast: Serra da Canastra and Canudos
Three-toed Jacamar*	Southeast and east: rare, near Orgãos
Crescent-chested Puffbird	Southeast and east: most sites
Chestnut-headed Nunlet*	Amazonia: Manaus
Brown-chested Barbet	East Amazonia: Amazonia NP
Spotted Piculet	East and northeast: Boa Nova
Varzea Piculet	Amazonia: rare from Río Madeira to west Para
Tawny Piculet	Northeast: Pedra Talhada and Forteleza
Ochraceous Piculet	East: lowlands of Ceará to Para
Yellow-eared Woodpecker	Southeast and east: most sites
Snethlage's Woodcreeper	East: rare in Minas Gerais
Moustached Woodcreeper*	Northeast: north of Canudos
Hoffmann's Woodcreeper	Amazonia: Rondonia and Amazonia NP
Zimmer's Woodcreeper	Amazonia: Manaus and Santarém
Dusky-billed Woodcreeper	Amazonia: Rondonia and Carajás
Long-tailed Cinclodes	South: Aparados NP
Tail-banded Hornero	East to Amazonia: a few sites
Striolated Tit-Spinetail	South: Aparados NP
Itatiaia Thistletail	Southeast: Orgãos and Itatiaia
Pinto's Spinetail*	Northeast: Pedra Talhada
Red-shouldered Spinetail	Northeast: Canudos
Pallid Spinetail	Southeast and east: most sites
Grey-headed Spinetail	East and northeast: Boa Nova and Pedra Talhada
Scaled Spinetail	East Amazonia: known only from Mexicana Island at the mouth of the Amazon.
Cipo Canastero*	Central: Serra do Cipo
Striated Softtail*	East: Sooretama and Boa Nova
Red-eyed Thornbird	Southeast: Orgãos and Serra do Mar
Pale-browed Treehunter	Southeast and east: most sites
Alagoas Foliage-gleaner*	Northeast: Pedra Talhada and Bananeira
White-collared Foliage-gleaner	Southeast and east: a few sites
Great Xenops*	Northeast: rare, Canudos
Silvery-cheeked Antshrike	East and northeast: Boa Nova and Canudos
Glossy Antshrike	Amazonia: most sites
Rondonia Bushbird*	Amazonia: Rondonia
Rufous-backed Antvireo	Southeast: most sites
Plumbeous Antshrike*	East: Monte Claros, Nova Lombardia and Sooretama
Klage's Antwren*	Amazonia: Manaus and Santarém
Star-throated Antwren	Southeast and east: most sites
Rio de Janeiro Antwren*	Southeast: near Orgãos, but not recorded since discovery in 1988.
Salvadori's Antwren*	Southeast and east: a few sites
Unicoloured Antwren*	Southeast to northeast: a few sites

Band-tailed Antwren*	East: Sooretama, Monte Pascoal and Porto Seguro
Pileated Antwren*	East and northeast: Boa Nova
Pectoral Antwren*	Northeast: Canudos
Narrow-billed Antwren*	East: Boa Nova
Serra Antwren*	East: Caraca NP, Ambiental de Peti and Monte Claros
Black-hooded Antwren*	Southeast: Serra do Mar
Ferruginous Antbird	Southeast and east: a few sites
Rufous-tailed Antbird*	Southeast: Orgãos and Itatiaia
Ochre-rumped Antbird*	Southeast and east: a few sites
Scaled Antbird	Southeast to northeast: most sites
Alagoas Antwren*	Northeast: Pedra Talhada
Rio de Janeiro Antbird*	Southeast and east: Orgãos, Itatiaia and Boa Nova
Bananal Antbird	South Amazonia: Araguaia NP
Fringe-backed Fire-eye*	Northeast: Salvador and Pedra Talhada
Slender Antbird*	East: Boa Nova
Scalloped Antbird*	East and northeast: Sooretama and Pedra Talhada
White-bibbed Antbird	Southeast and east: most sites
Squamate Antbird	Southeast: Serra do Mar
White-breasted Antbird*	Amazonia: Rondonia
Harlequin Antbird	East Amazonia: Amazonia NP
Bare-eyed Antbird	Amazonia: Santarém and Alta Floresta
Pale-faced Antbird	East Amazonia: Amazonia NP
Such's Antthrush	Southeast and east: most sites
White-browed Antpitta*	East and northeast: Boa Nova
Hooded Gnateater*	East Amazonia: Fortaleza, Belém and Carajás
Black-cheeked Gnateater	Southeast to northeast: most sites
Slaty Bristlefront*	Southeast: Itatiaia and Serra do Mar
Stresemann's Bristlefront*	East: known from only two speci mens, from coastal Bahia.
Brasilia Tapaculo*	Central: Serra da Canastra and Brasília NP
Bahia Tapaculo*	Northeast: Salvador area
White-breasted Tapaculo	Southeast and east: a few sites
Black-and-gold Cotinga*	Southeast: Orgãos and Itatiaia
Grey-winged Cotinga*	Southeast: rare, Orgãos
Hooded Berryeater*	Southeast and east: Orgãos, Serra do Mar and Nova Lombardia.
Black-headed Berryeater*	Southeast to northeast: Sooretama, Monte Pascoal, Porto Seguro and Pedra Talhada
Buff-throated Purpletuft*	Southeast and northeast: Serra do Mar and Pedra Talhada.
Kinglet Calyptura*	Southeast: although looked for, this colourful cotinga has not been seen this century. Formerly collected around Rio and Orgãos

Cinnamon-vented Piha*	Southeast and east: a few sites
Banded Cotinga*	East: Sooretama, Monte Pascoal and Porto Seguro
White-tailed Cotinga	East Amazonia: Santarém, Belém and Carajás
White-winged Cotinga*	East: Sooretama, Monte Pascoal and Porto Seguro
Opal-crowned Manakin	East Amazonia: Santarém and Carajás
Golden-crowned Manakin*	Amazonia: Alta Floresta
Pin-tailed Manakin	Southeast and east: most sites
Wied's Tyrant-Manakin	East and northeast: Nova Lombardia
Brown-breasted Bamboo-Tyrant	Southeast: Itatiaia, Serra do Mar and Aparados NP
Eye-ringed Tody-Tyrant*	Southeast: Orgãos and Serra do Mar
Hangnest Tody-Tyrant*	Southeast to northeast: a few sites
Pelzeln's Tody-Tyrant	North Amazonia: Pico da Neblina NP
Buff-breasted Tody-Tyrant*	Northeast: Pedra Talhada and Fortaleza
Kaempfer's Tody-Tyrant*	Southeast: Santa Catarina, formerly known from a 1929 specimen, this species was rediscovered in 1991 near the type-locality
Fork-tailed Tody-Tyrant*	Southeast: Serra do Mar
Buff-cheeked Tody-Flycatcher*	West Amazonia: known from a 1929 specimen
Yellow-lored Tody-Flycatcher	Southeast and east: most sites
Grey-capped Tyrannulet*	Southeast and east: Orgãos and Nova Lombardia
Noronha Elaenia	Noronha Islands off northeast coast
Grey-backed Tachuri*	Central: Serra da Canastra and Serra do Cipo
Minas Gerais Tyrannulet*	East: known only from a few specimens, from Minas Gerais
Oustalet's Tyrannulet*	Southeast and east: Serra do Mar and Nova Lombardia
Restinga Tyrannulet	Southeast: São Paulo and Santa Catarina (described in 1992)
Serra do Mar Tyrannulet*	Southeast: Orgãos, Itatiaia and Aparados NP
Alagoas Tyrannulet*	Northeast: Pedra Talhada
Velvety Black-Tyrant	Southeast and central: a few sites
Grey-hooded Attila	Southeast and east: most sites
Ash-throated Casiornis	Northeast: Canudos, Fortaleza and Carajás
White-naped Jay	Northeast: Canudos
Noronha Vireo	Noronha Islands off northeast coast
Grey-eyed Greenlet	Central: Serra da Canastra and Serra do Cipo

Long-billed Wren	Southeast and northeast: a few sites
Grey Wren	West Amazonia: Tabatinga
White-striped Warbler	Central: Brasília NP and Emas NP
Red-cowled Cardinal	East and northeast: Boa Nova, Canudos and Pedra Talhada
Crimson-fronted Cardinal	South Amazonia: Araguaia NP
Brown Tanager*	Southeast: Orgãos and Itatiaia
Cinnamon Tanager	Southeast and east: a few sites
Cone-billed Tanager*	Southwest: known from only a single specimen, taken in 1939 from west Mato Grosso
Scarlet-throated Tanager	Northeast: Canudos
Rufous-headed Tanager	Southeast and east: most sites
Cherry-throated Tanager*	East: known from only one specimen taken in 1870
Olive-green Tanager	Southeast: Itatiaia and Serra do Mar
Azure-shouldered Tanager*	Southeast and east: a few sites
Golden-chevroned Tanager	Southeast and east: most sites
Seven-coloured Tanager*	Northeast: Pedra Talhada
Brassy-breasted Tanager	Southeast: Orgãos, Itatiaia and Serra do Mar
Gilt-edged Tanager	Southeast and east: Itatiaia, Nova Lombardia and Boa Nova
Black-backed Tanager*	Southeast: rare, Ilha do Cardoso and Boraceia
Black-legged Dacnis*	Southeast: Serra do Mar
Bay-chested Warbling-Finch	Southeast: Orgãos and Itatiaia
Pale-throated Serra-Finch*	Central: Caraca NP and Serra do Cipo
Dubois' Seedeater	East: Nova Lombardia, Porto Seguro and Boa Nova
Hooded Seedeater*	Central: known from only one specimen taken in 1870 in Goias, possibly a hybrid
White-throated Seedeater	Northeast: Canudos and Pedra Talhada
Black-bellied Seedeater*	Central: Brasília NP
Forbes' Blackbird*	East and northeast: rare, Rio Doce and Pedra Talhada

North Amazonia = north of the Amazon.
West Amazonia = Tefé westwards.
Amazonia = Amazon basin, Manaus southwards.
East Amazonia = Santarém east to Belém and São Luís.
Northeast = North of Salvador.
East = Rio north to Salvador (not including Orgãos).
Southeast = Rio south to São Paulo.
South = Rio Grande do Sul.
Central = Belo Horizonte area and westwards to Cuiabá.
Southwest = Cuiabá westwards.

CHILE

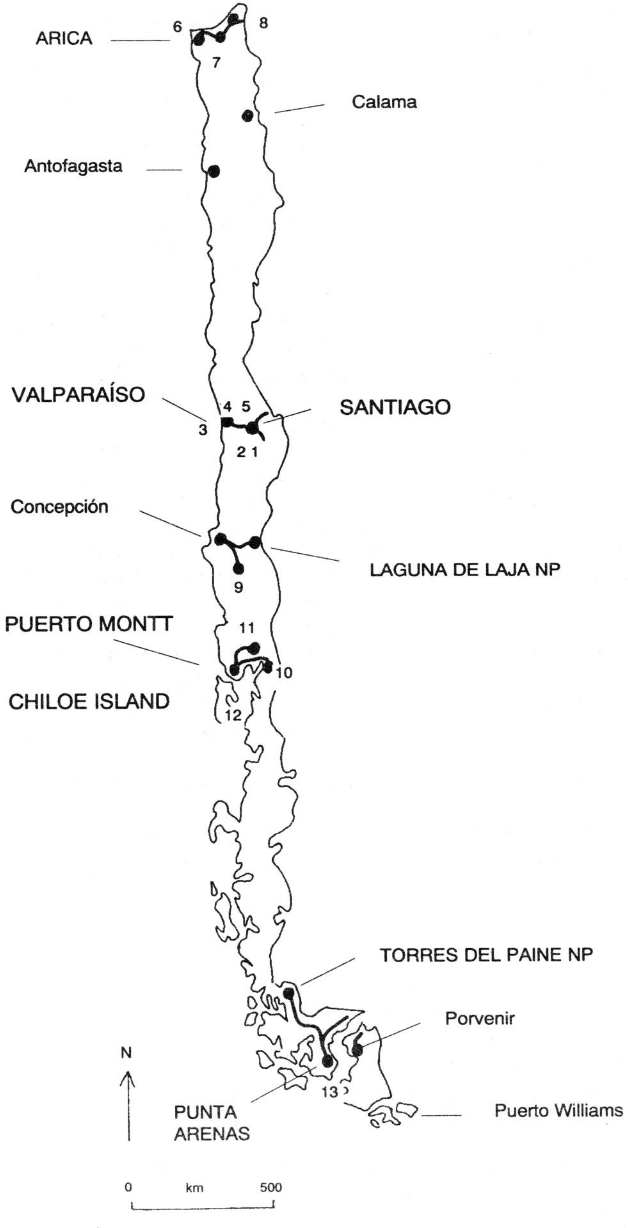

ARICA
6
8
7
Calama
Antofagasta
VALPARAÍSO
4 5
3
SANTIAGO
2 1
Concepción
LAGUNA DE LAJA NP
9
PUERTO MONTT
11
10
CHILOE ISLAND
12
TORRES DEL PAINE NP
Porvenir
N
13
PUNTA
ARENAS
Puerto Williams
0
km
500

1	Santiago Area	8	Lauca NP
2	Lake Peñuelas	9	Nahuelbuta NP
3	Valparaíso	10	Vicente Perez Rosales NP
4	Valparaíso-Zapallar Road	11	Puyehué NP
5	Cerro La Campana NP	12	Chiloé Island
6	Arica	13	Punta Arenas and T. del Fuego
7	Putre		

INTRODUCTION

Summary

Although seeing the best of Chile's birds involves covering some 4,000 km from north to south, the country's modern infrastructure, including an excellent internal air network, means it is possible to see most of the best birds on a short trip. Furthermore, the friendly atmosphere and many spectacular National Parks make travelling and birding here a delight. Owing to a complete lack of tropical forest, Chile's avifauna is short on quantity. However, never mind the list, think of the quality. There is no colour field guide, but most birders know what the beautiful Diademed Sandpiper-Plover* looks like, and this is the best country in which to look for it.

Size

Although only an average of 200 km wide, Chile is 4,329 km long so distances between the major sites are surprisingly large. For example, it is 2,100 km from Santiago in the centre to Arica in the north. At 756,945 km^2, it is nearly six times the size of England and a shade larger than Texas.

Getting Around

The best way to reach the main birding sites in north, central and south Chile is via the excellent internal air network. Although it is possible to travel by road from Santiago to Arica, there are few good roads away from the central area and other main towns. Driving to the sites from the major towns is not a major problem, but public transport is thin on the ground around Arica and Punta Arenas. Four-wheel drive high-clearance jeeps are recommended (but not absolutely essential) in these two areas, where cracked windscreens are commonplace owing to flying gravel.

Accommodation and Food

There is a wide range of accommodation, but Chile is not a cheap country, with prices resembling those in England and America. Even camping is relatively expensive. Those prepared to hunt around though may find cheaper accommodation, especially in Punta Arenas.

The food, especially seafood, if well cooked, is superb, but beware of cholera around Arica, where it is currently illegal to carry fruit.

Health and Safety

Apart from the risk of cholera around Arica, immunisation is recommended for hepatitis and typhoid.

The Chileans are very friendly, and the risk of crime is minimal.

Climate and Timing

The temperate climate means virtually any time of the year is good to visit the centre and the north. However, those in search of seabirds, Diademed Sandpiper-Plover* and tapaculos will do best between September and December. This is also the best time to visit the south since it is during the austral summer. Whilst the days are hot and the nights cool in the lowlands of the centre and the north, it is very cold at high-altitude Andean sites and around Punta Arenas, especially at night.

Beware of altitude sickness in the Andes, especially at Lauca NP (4,500 m/14,763 ft) in the far north. Drinking coca tea to combat the shortage of oxygen seems to be tolerated by the authorities, and is very effective.

Habitats

The Andes run north to south along the entire eastern border of Chile. This is rugged country with puna grasslands, bogs and lakes on the slopes of many mighty mountains and snow-capped volcanoes. The lower coastal range of hills to the west are separated by a central valley which is less defined in the north. These rocky hills support scrubby woodlands known as matorral.

The Atacama desert, one of the driest and most sparsely vegetated on earth, covers nearly a fifth of the low-lying land in the north. Only a few linear oases, formed by rivers spilling down from the Andes, interrupt what would otherwise be a huge, barren, almost lifeless habitat.

North and south of Santiago the lowland climate is similar to that of the Mediterranean, with dry summers and wet winters, and much of the land is used, very productively, for agriculture. South of here, between Concepción and Puerto Montt, rain falls throughout the year, sustaining a pictureseque landscape of lakes, rivers and forests of *Araucaria* (Monkey-puzzle) and *Nothofagus* (Southern Beech).

South of Puerto Montt, the southern fifth of Chile is a cold, remote wilderness of mountains, glaciers, fjords and forests, whilst in the far south, on and around Tierra del Fuego, the windswept Patagonian steppe dominates the landscape.

The cold Humboldt current runs along the western coastline, creating a nutrient-rich ribbon of water which supports a great variety of seabirds.

Conservation

Chile boasts an excellent and extensive conservation programme, with some of the best managed parks and reserves in South America. The 30 National Parks and 36 Forest Reserves managed by the Corporacion Nacional Forestal (CONAF) comprise some 70,000 km^2; over 10% of the land surface.

Ten threatened species, of which none are endemic, occur in Chile. One endemic bird is near-threatened.

Bird Families

Only 56 of the 92 families which regularly occur in South American are represented, the lowest total for a mainland country. These include seven of the 25 Neotropical endemic families and three of the nine South American endemic families. Thanks to the complete lack of tropical or subtropical forests there are no cracids, potoos, trogons, motmots, jacamars, toucans, woodcreepers, antbirds or manakins.

Well-represented families include rheas, waterfowl, seedsnipes, plovers, furnariids and tapaculos.

Bird Species

Only 440 species have been recorded in Chile, a result of being beyond the tropics. Although over 1,000 more species occur in Ecuador, a country less than half the size, Argentina supports only twice as many birds despite being 3.5 times larger.

Non-endemic specialities and spectacular species include Lesser Rhea*, Humboldt* and Magellanic Penguins, Peruvian* and Magellanic Diving-Petrels, Andean* and Puna* Flamingoes, Ruddy-headed Goose*, Torrent Duck, Andean Condor, Giant Coot, Magellanic Plover*, Grey-breasted and Least Seedsnipes, Magellanic Oystercatcher, Andean Avocet, Two-banded and Puna Plovers, Rufous-chested and Tawny-throated Dotterels, Diademed Sandpiper-Plover*, Dolphin Gull, Snowy-crowned and Inca Terns, Peruvian Sheartail, Chilean Woodstar*, Magellanic Woodpecker, Creamy-rumped Miner, White-throated Earthcreeper, Thorn-tailed Rayadito, Austral Canastero*, White-throated Treerunner, Black-throated Huet-huet, Chucao Tapaculo, Rufous-tailed Plantcutter, Black Siskin, Tamarugo Conebill* and Slender-billed Finch*.

Many of the specialities include species restricted to south Argentina and south Chile.

Endemics

Eight species occur only in Chile. They are a tinamou, a parakeet, a cinclodes, a unique furnariid known as Crag Chilia, a mockingbird, and three superb tapaculos: Chestnut-throated Huet-huet, Moustached Turca and White-throated Tapaculo.

An additional four endemics, a petrel, a hummingbird, a rayadito and a tit-tyrant, occur on the Juan Fernandez Islands.

Expectations

Do not expect to notch up a big list in Chile. However, it is possible to see over 50% of Chile's birds, between 230 and 250, in two or three weeks, providing the trip covers the whole country. Such a list could include all eight endemics which all occur from Santiago south to Puerto Montt.

SANTIAGO AREA

Map opposite

Chile's pleasant capital, situated in the Andean foothills in the middle of the country, is an ideal base from which to explore a number of excellent sites in the Andes, where the marvellous Diademed Sandpiper-Plover*, the highly localised Creamy-rumped Miner, and three endemics including the rare Crag Chilia, occur.

Endemics

Crag Chilia, Moustached Turca, Chilean Mockingbird.

Specialities

Diademed Sandpiper-Plover*, White-sided Hillstar, Creamy-rumped Miner, Grey-flanked and Dark-bellied Cinclodes, White-tailed* and

Great Shrike-Tyrants, Black-fronted Ground-Tyrant, Greater Yellow-Finch.

Others

Andean Condor, Red-backed Hawk, Grey-breasted Seedsnipe, Puna Plover, Black-winged Ground-Dove, Mountain Parakeet, Rufous-banded Miner, Scale-throated Earthcreeper, Lesser and Cordilleran Canasteros, White-browed and Ochre-naped Ground-Tyrants, Black-chinned Siskin, Grey-hooded, Plumbeous and Band-tailed Sierra-Finches.

Access

There are a number of sites worth birding around Santiago; Crag Chilia and Moustached Turca occur alongside the road up to **El Yeso reservoir** (2,286 m/7,500 ft), one of the few sites in South America, and the only one below 3,048 m (10,000 ft), where Diademed Sandpiper-Plover* occurs. Scour the gravelly plains and mass of streams beyond the reservoir for this bird, but do not despair if it remains elusive and you are also visiting Lauca NP (see p. 166). Grey-breasted Seedsnipe, Grey-flanked Cinclodes and Black-fronted Ground-Tyrant also occur here. Head southeast out of Santiago towards San Gabriel then turn east at the fork 24 km beyond San José de Maipó on to the rough road up to the reservoir.

Crag Chilia and Moustached Turca also occur near **Farellones** ski-resort (2,134 m/7, 000 ft), 51 km northeast of Santiago. Bird at various

SANTIAGO AND VALPARAÍSO

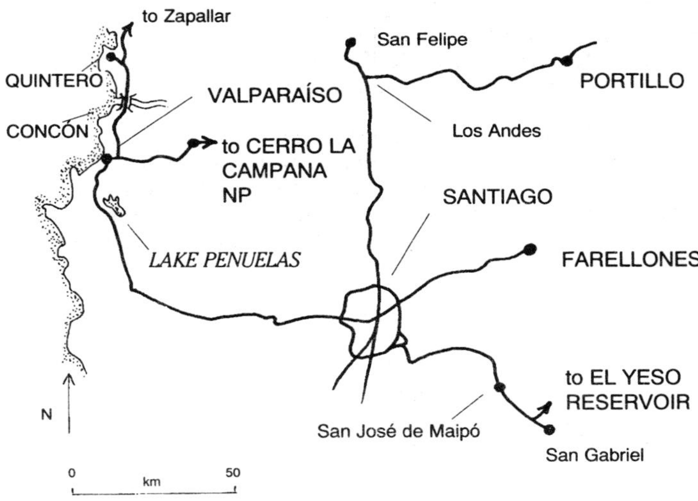

altitudes below the village. Creamy-rumped Miner occurs around **Portillo**, another ski-resort near the Argentine border, 145 km north-east of Santiago. Head north out of the city for 83 km to Los Andes, then turn east.

LAKE PEÑUELAS

Map p. 159

This large lake alongside the road between Santiago and Valparaíso, is a good site for the scarce Lake and Black-headed* Ducks.

Endemics

Chilean Mockingbird.

Specialities

Lake and Black-headed* Ducks, Dusky-tailed Canastero, Warbling Doradito, Patagonian Tyrant.

Others

White-tufted and Great Grebes, Chiloe Wigeon, Red Shoveler, Rosy-billed Pochard, Cinereous Harrier, Red-backed Hawk, White-winged, Red-gartered and Red-fronted Coots, Collared Plover, Black-winged Ground-Dove, Green-backed Firecrown, Striped Woodpecker, Chilean Flicker, Common Miner, Common Diuca-Finch, Long-tailed Meadowlark.

Access

The lake is opposite the junction to Quintay, on the north side of the Santiago–Valparaíso road. Bird from the narrow road which runs around the lake. The introduced California Quail also occurs here.

VALPARAÍSO

Map p. 159

The historic fishing port of Valparaíso and the seaside town of Viña del Mar, 9 km to the north, on the Pacific coast northwest of Santiago, both form good bases from which to visit some excellent coastal sites, with Cerro La Campana NP 50 km inland, and the Humboldt current a short distance offshore. This nutrient-rich current provides food for thousands of seabirds including Humboldt Penguin*, Royal Albatross, Pink-footed Shearwater*, Peruvian Diving-Petrel* and the unique Inca Tern. Whilst many may be seen by the ardent seawatcher in the right conditions, a pelagic trip between September and November is likely to produce the best views.

Endemics

Chilean Seaside Cinclodes.

Specialities

Humboldt Penguin*, Royal and Buller's Albatrosses, Pink-footed Shearwater*, Peruvian Diving-Petrel*, Inca Tern.

Others

Black-browed and Shy Albatrosses, Cape, Stejneger's and White-chinned Petrels, Buller's Shearwater, Wilson's and White-vented Storm-Petrels, Peruvian Booby, Guanay and Red-legged* Cormorants, Peruvian Pelican, Inca Tern, Chilean Skua.

Access

The best way to organise a boat trip is to ask at hotels or to approach local fishermen in Valparíaso. Such a trip is likely to be expensive.

VALPARAÍSO–ZAPALLAR ROAD

Map p. 159

There are a number of good birding sites along the road north from Valparaíso to Zapallar, including a couple of rocky headlands which are suitable for seawatching and good sites for the endemic Chilean Seaside Cinclodes. The road also crosses an estuary with adjacent reedbeds, where Stripe-backed Bittern occurs, before reaching the Zapallar valley which supports Giant Hummingbird and Green-backed Firecrown.

Endemics

Chilean Seaside Cinclodes.

Specialities

Humboldt Penguin*, Inca Tern, Green-backed Firecrown, Des Murs' Wiretail.

Others

White-tufted Grebe, Black-browed and Shy Albatrosses, Antarctic Giant Petrel, Cape and White-chinned Petrels, Guanay Cormorant, Peruvian Pelican, Stripe-backed Bittern, Plumbeous Rail, Spot-flanked Gallinule, Red-fronted Coot, Surfbird, Blackish Oystercatcher, Collared Plover, Rufous-chested Dotterel, Southern Lapwing, Grey Gull, Elegant and South American Terns, Chilean Skua, Giant Hummingbird, Dark-bellied Cinclodes, Thorn-tailed Rayadito, Andean Tapaculo, Rufous-tailed Plantcutter, White-crested Elaenia, Austral Negrito, Spectacled Tyrant, Andean Tapaculo, Yellow-winged Blackbird.

Other Wildlife

Sealion, Sea Otter.

Access

Chilean Seaside Cinclodes occurs at **Punta Concón**, just south of Concón, 6 km north of Viña del Mar. Seawatching can be good here. Stripe-backed Bittern occurs at the **Río Aconcagua estuary** just north of Concón. Explore the reedy pools and channels near the coast on the north side of the estuary. Seawatching can also be good from **Punta Liles** at **Quintero**, 17 km north of Concón. There is a Humboldt Penguin* colony at **Isla Cachagua**, just offshore, where Sea Otter also occurs. Giant Hummingbird and Green-backed Firecrown occur in the very pleasant **Zapallar valley**, a few km further north.

CERRO LA CAMPANA NP

Map p. 163

This excellent NP (1,000 m/3,281 ft), some 50 km inland from Valparaíso, protects scrubby woodland with *Ocoa* palms, where no less than five of Chile's eight endemics occur.

Endemics

Chilean Tinamou, Crag Chilia, Moustached Turca, White-throated Tapaculo, Chilean Mockingbird.

Specialities

Chilean Pigeon*, Rufous-legged Owl, White-sided Hillstar, Green-backed Firecrown, Des Murs' Wiretail, Dusky-tailed Canastero, Fire-eyed Diucon, Great Shrike-Tyrant, Patagonian Tyrant, Patagonian Sierra-Finch, Austral Blackbird.

Others

Red-backed Hawk, Picui Ground-Dove, Giant Hummingbird, Striped Woodpecker, Chilean Flicker, Thorn-tailed Rayadito, Plain-mantled Tit-Spinetail, White-throated Treerunner, Andean Tapaculo, Rufous-tailed Plantcutter, White-crested Elaenia, Tufted Tit-Tyrant, Austral Thrush, White-winged Diuca-Finch.

Access

The best way to reach this NP is from the coast. Head 40 km inland from Valparaíso to Limache, and continue to Olmue. The park entrance is 8 km beyond here just past Granizo. Bird the Sendero Andinista trail, which starts at the picnic area just inside the park and, more importantly, the Cerro Penitentes trail. Head up from the entrance for 6 km and take the left fork. The trail is on the right 1 km from the fork. Chilean Tinamou (high up), Crag Chilia (rocky ravines in the woodland), Moustached Turca (above the woodland) and White-throated Tapaculo all occur on this long, hot trail.

Cerro La Campana is probably the best site in Chile for the endemic Crag Chilia

CERRO LA CAMPANA NP

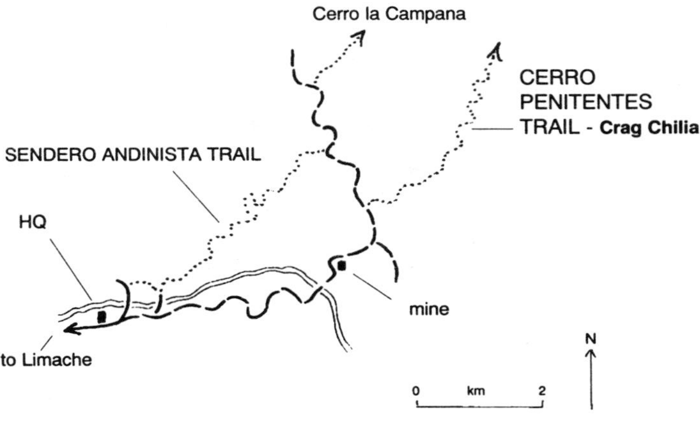

ARICA

Map p. 164

The pleasant town of Arica is over 2,000 km north of Santiago near the Peruvian border. It lies at the seaward edge of the Atacama desert alongside two meltwater rivers: the Azapa and Lluta. These small rivers form two linear oases, which support Peruvian Thick-knee, Peruvian Sheartail, the highly localised Chilean Woodstar* and the smart Slender-billed Finch*. Lying next to the Humboldt current, Arica is also an excellent place for seabirds, including Pink-footed Shearwater*, Peruvian Diving-Petrel* and Inca Tern.

This town is also the gateway to Lauca NP, one of the most spectacular NPs in South America.

Specialities

Humboldt Penguin*, Pink-footed Shearwater*, Peruvian Diving-Petrel*, Peruvian Thick-knee, Band-tailed and Grey Gulls, Peruvian and Inca Terns, Oasis Hummingbird, Peruvian Sheartail, Chilean Woodstar*, Slender-billed Finch*.

Others

Buller's Shearwater, Wilson's Storm-Petrel, Peruvian Booby, Guanay and Red-legged* Cormorants, Peruvian Pelican, Surfbird, Blackish Oystercatcher, Chilean Skua, Elegant and South American Terns, Croaking Ground-Dove, Southern Martin, Cinereous Conebill, Chestnut-throated Seedeater, Peruvian Meadowlark.

Access

Oasis Hummingbird occurs in Arica itself and may be seen visiting the town's fountains to drink and bathe. Peruvian Sheartail and Chilean

One of the best places to see the superb Peruvian Sheartail is in the Azapa valley, just outside Arica

Woodstar* occur in the **Azapa valley** which runs inland from Arica. The Faculty of Andean Studies garden just north of the road, past the garden centre in San Miguel, is usually worth a look, but any patches of flowering shrubs along the roadside here are worth checking.

Peruvian Thick-knee and Slender-billed Finch* occur in the Lluta valley.

ARICA

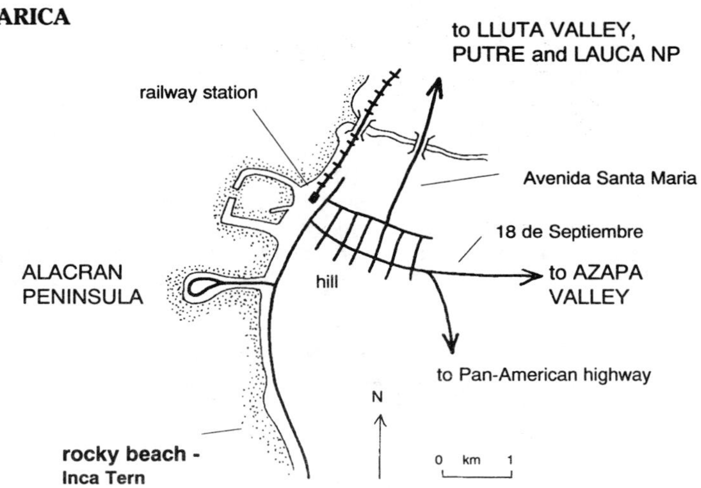

Head north out of Arica, turn east after 9 km and bird the fields and scrub inland from here, alongside the road to Putre and Lauca NP.

Inca Tern occurs on the Alacran Peninsula, a rocky causeway just south of Arica town. This is also an excellent seawatching point.

Accommodation: Hotel San Marcos (B).

Giant Conebill* has recently been discovered in Chile, alongside the road between the Lluta valley and Putre. The brilliant Andean Hillstar also occurs along here, especially near Putre.

PUTRE

Map below

The small, sleepy town of Putre lies at 3,048 m (10,000 ft), 120 km east of Arica. It is surrounded by surprisingly bird-filled hillsides which support a number of scarce Andean species including Tamarugo Conebill*. Those birders *en route* to Lauca NP would be wise to stop here for at least one day's birding and at least one night. It is a great way to acclimatise before ascending another 1,500 m (4,921 ft).

Specialities

Puna Miner, White-throated and Plain-breasted Earthcreepers, Streaked Tit-Spinetail, Tamarugo Conebill*, Black-throated Flower-piercer.

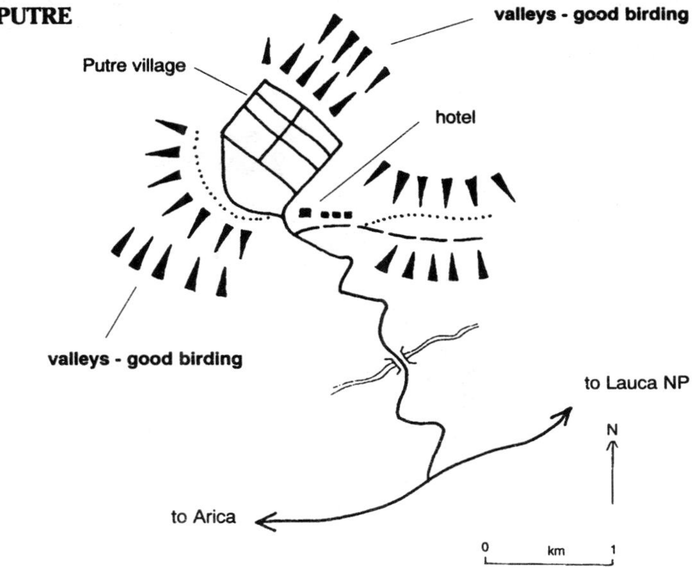

Others

Puna Hawk, Mountain Caracara, Bare-faced, Black-winged and Golden-spotted Ground-Doves, Mountain Parakeet, Band-winged Nightjar, Andean Swift, Sparkling Violet-ear, Andean Hillstar, Giant

Hummingbird, Straight-billed and Scale-throated Earthcreepers, White-winged Cinclodes, Cordilleran and Creamy-breasted Canasteros, Yellow-billed Tit-Tyrant, White-browed Chat-Tyrant, Black-billed Shrike-Tyrant, Spot-billed Ground-Tyrant, Chiguanco Thrush, Andean Swallow, Thick-billed Siskin, Cinereous Conebill, Blue-and-Yellow Tanager, Black-hooded, Mourning, Ash-breasted and Band-tailed Sierra-Finches, Bright-rumped and Greenish Yellow-Finches, Band-tailed Seedeater, Golden-billed Saltator.

Other Wildlife

Vicuña, Huemel Deer.

Access

Putre is reached by travelling east from the Lluta valley *en route* to Lauca NP and Bolivia. Bird the fields on the hillsides around the town and check any flowering shrubs in the town for Tamarugo Conebill*.

Accommodation: Restuarant El Oasis (C); Hosteria Las Vicuñas (A).

LAUCA NP

Map opposite

Situated at an altitude of 4,500 m (14,763 ft) in the Andes, Lauca NP presents a stunning setting in which to look for some very special birds, not least one of the rarest and most beautiful shorebirds in the world, Diademed Sandpiper-Plover*. Here, numerous bogs, pools and saline lakes punctuate the puna grasslands which lie below the barren scree slopes of two mighty snow-capped volcanoes, Pomerape and Parinacota.

In Lauca NP it is possible to see the extraordinary Diademed Sandpiper-Plover, in one of the world's wildest places. This would have to be one of the highlights of any birder's life*

LAUCA NP

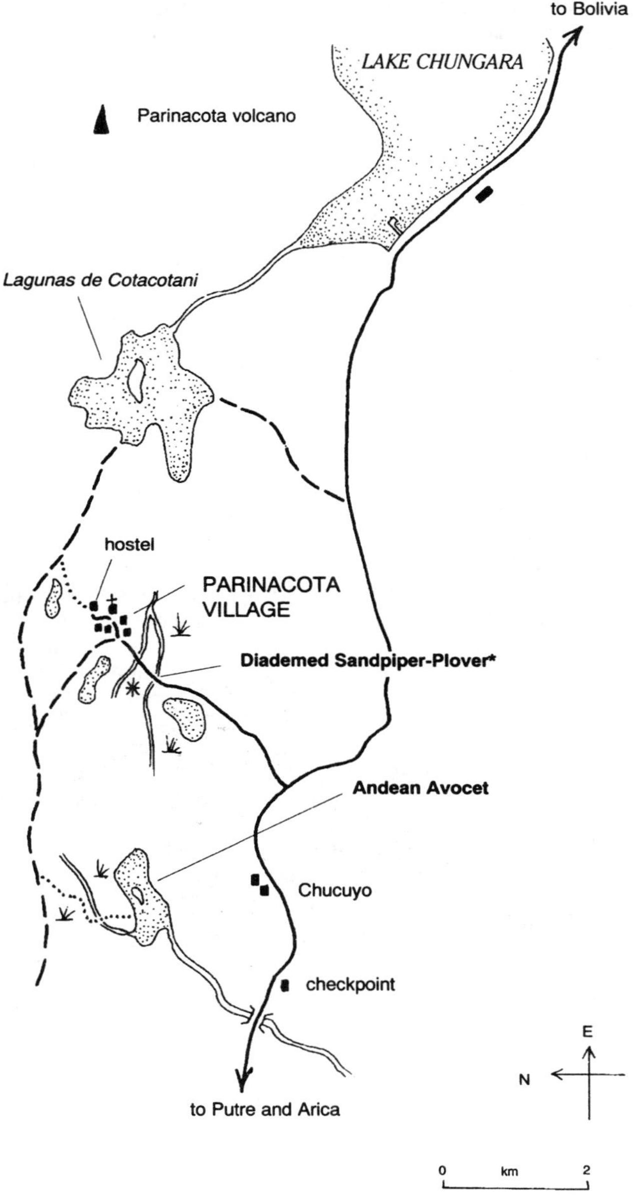
to Bolivia
LAKE CHUNGARA
Parinacota volcano
Lagunas de Cotacotani
hostel
PARINACOTA
VILLAGE
Diademed Sandpiper-Plover*
Andean Avocet
Chucuyo
checkpoint
to Putre and Arica
E
N
0
km
2

Specialities

Puna Tinamou, Andean* and Puna* Flamingos, Giant Coot, Diademed Sandpiper-Plover*, Rufous-bellied Seedsnipe, Andean Avocet, Puna Miner, Plain-breasted Earthcreeper, Streaked Tit-Spinetail, White-tailed Shrike-Tyrant*, Cinereous and White-fronted Ground-Tyrants, Red-backed and White-throated Sierra-Finches, Puna Yellow-Finch.

Others

Ornate Tinamou, Lesser Rhea*, Silvery Grebe, Andean Goose, Crested Duck, Puna Teal, Chilean Flamingo, Puna Ibis, Andean Condor, Puna Hawk, Mountain Caracara, Grey-breasted Seedsnipe, Puna Plover, Andean Lapwing, Andean Gull, Golden-spotted Ground-Dove, Andean Hillstar, Andean Flicker, Bar-winged and White-winged Cinclodes, Cordilleran Canastero, Rufous-naped and Puna Ground-Tyrants, Andean Negrito, Andean Swallow, Black Siskin, Black-hooded and Plumbeous Sierra-Finches, White-winged Diuca-Finch, Bright-rumped and Greenish Yellow-Finches.

Other Wildlife

Puma, Vicuña, Viscacha, Alpaca.

Access

The NP officially begins at Putre, but the best area is 50 km beyond here, towards Bolivia. Before the checkpoint near Chucuyo, scan the roadside grasslands for Puna Tinamou and Lesser Rhea*. Once past the checkpoint turn north just east of Chucuyo village. It is 4 km from this turning to the delightful **Parinacota** village, where there is a (very cold) hostel for visitors, complete with beds and cooking facilities (take food). Before heading in to the village check the flat area, below the small hillock west of the road, just before Parinacota, and the small streams to the east of the road opposite, for Diademed Sandpiper-Plover*.

Andean* and Puna* Flamingos, Andean Avocet and Puna Plover occur on the the large saline lake west of Parinacota. Take the track beyond the village to a point where it is closest to the lake, and walk out across the wet puna to the lakeshore. Alternatively, walk to the other side of the lake from the main road. Both ways seem never-ending at this altitude.

Lake Chungara, where the views are dramatic to say the least, is 12 km straight on from Chucuyo. Andean* and Puna* Flamingos also occur here.

The whole area surrounding Parinacota is excellent birding country and well worth exploring. This is high country though, so beware of altitude sickness, very cold nights, sunburn and windburn.

Accommodation: Hostel (B).

Puna Flamingo* and Horned Coot* occur on lakes some 400 km inland from **Antofagasta**, 700 km south of Arica. From Calama, 215 km northeast of Antofagasta, head 143 km southeast to the village of Tocanao, beyond the Pedro de Atacama customs post. Check the lakes just west of the road, south of Tocanao, the lake 10 km east on the old road, south of Toconao, and further south, 20 km beyond Socaire.

Humboldt Penguin* (colony), Peruvian Diving-Petrel*, Red-legged Cormorant* and Chilean Seaside Cinclodes occur at the coastal **Pan de Azucar** NP, 375 km south of Antofagasta. Turn west off the Pan-American highway 45 km north of Chanaral. The NP is 20 km from this turn-off.

NAHUELBUTA NP

Maps below and p. 170

This NP, approximately 225 km south of Concepción, in central Chile, protects pristine *Araucaria* (Monkey-puzzle tree) and *Nothofagus* (Southern Beech) forests with bamboo thickets. Although this site is off the beaten track, and remote from the other top quality sites, it is well worth visiting. The 'tapaculo-tracker' may see Black-throated Huet-huet, Chucao Tapaculo and the very scarce Ochre-flanked Tapaculo here.

Endemics

Chilean Tinamou, Slender-billed Parakeet*, Chilean Mockingbird.

Specialities

White-throated Hawk, Austral Parakeet, Rufous-legged Owl, Green-backed Firecrown, Magellanic Woodpecker, Dark-bellied Cinclodes, Des Murs' Wiretail, Black-throated Huet-huet, Chucao and Ochre-flanked Tapaculos, Fire-eyed Diucon, Patagonian Tyrant, Patagonian Sierra-Finch, Austral Blackbird.

Others

Bicoloured Hawk, Southern Lapwing, Striped Woodpecker, Chilean Flicker, Plain-mantled Tit-Spinetail, Thorn-tailed Rayadito, White-throated Treerunner, Andean Tapaculo, White-crested Elaenia, Tufted Tit-Tyrant, Austral Thrush, Chilean Swallow, Correndera Pipit, Black-chinned Siskin, Long-tailed Meadowlark.

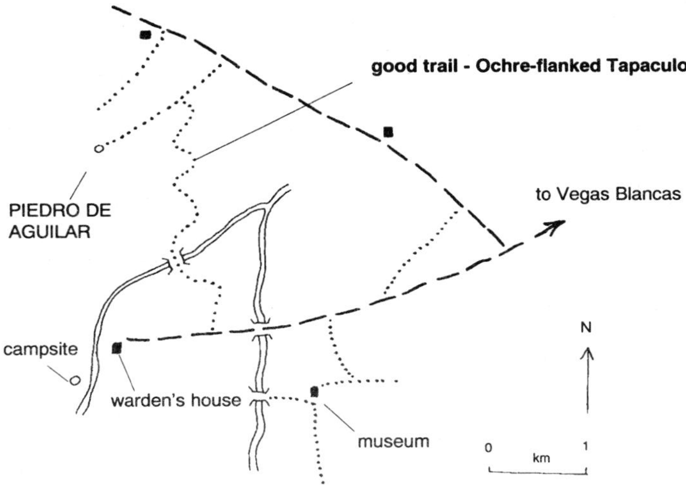

NAHUELBUTA NP

Access

The NP entrance is 37 km west of Angol. Turn north at Vegas Blancas, 30 km west of Angol. The riverine scrub around **Vegas Blancas** supports all three tapaculos (Map below). From here it is 7 km from the park entrance and 5 km from there to the warden's house. The 4-km trail from near here to to Piedro de Aguilar is best for the tapaculos.

NAHUELBUTA NP APPROACH

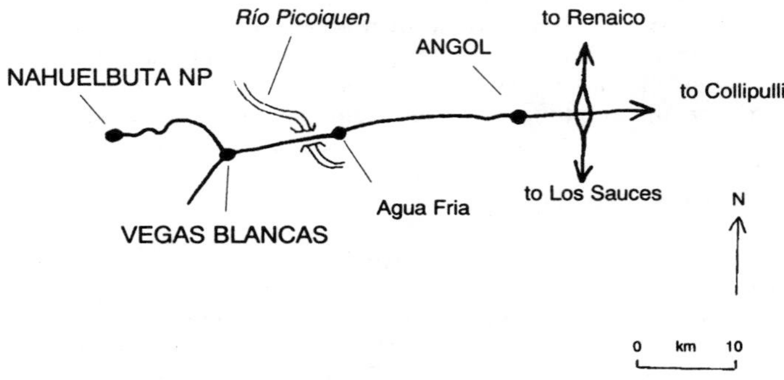

VEGAS BLANCAS

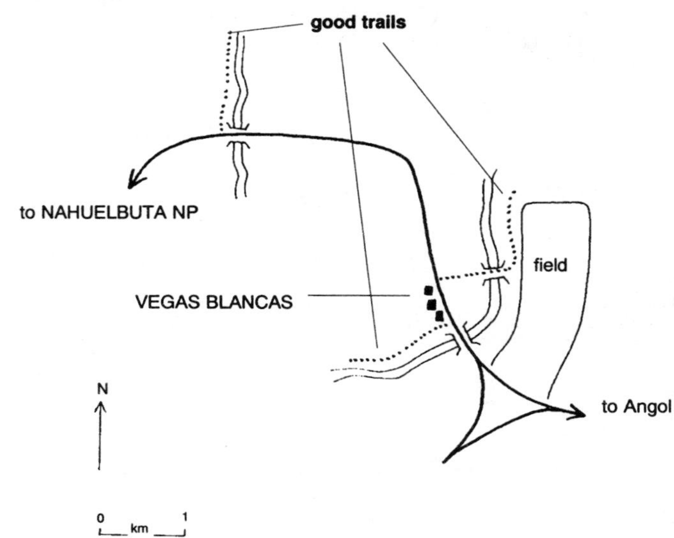

Accommodation: Camping only (take own food).

The endemic Chestnut-throated Huet-huet occurs at **Laguna de Laja NP**, approximately 300 km east of Concepción. Head northeast to Antuco from Los Angeles, approximately 200 km southeast of Concepción, and look for this rare bird along the trail to the south 0.5 km before the NP entrance, 24 km east of Antuco.

PUERTO MONTT AREA

Map below

This large town, 1,080 km south of Santiago, is the gateway to Chiloé Island and the excellent Vicente Perez Rosales and Puyehué NPs, where the specialities include Magellanic Woodpecker and three tapaculos.

PUERTO MONTT AREA

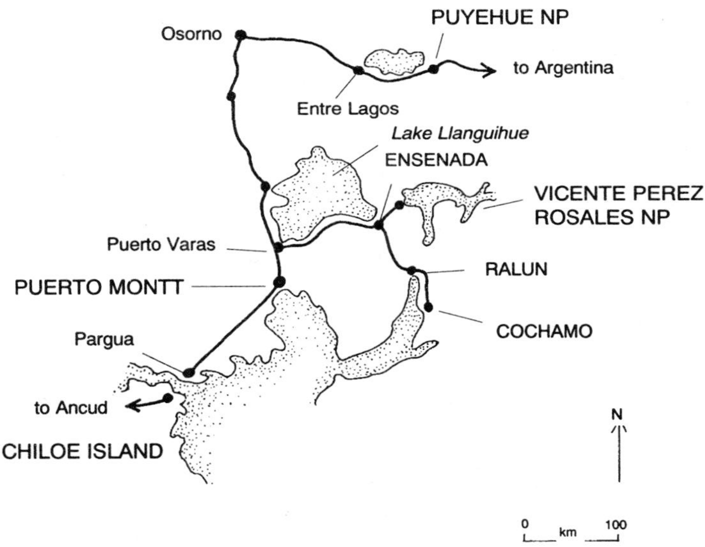

VICENTE PEREZ ROSALES NP

Map above

This NP and its surrounds, 100 km east of Puerto Montt, protects extensive forest where Magellanic Woodpecker and Black-throated Huet-huet occur.

Specialities

Ashy-headed Goose, Austral Parakeet, Magellanic Woodpecker, Grey-flanked and Dark-bellied Cinclodes, Black-throated Huet-huet, Chucao Tapaculo, Patagonian Tyrant.

Others

Chilean Pigeon*, Striped Woodpecker, Chilean Flicker, Thorn-tailed Rayadito, White-throated Treerunner, Tufted Tit-Tyrant.

Access

Head north out of Puerto Montt for 24 km and turn east at Puerto Varas on to the 50-km road to Ensenada. Turn south here and bird the forest alongside the Río Petrohue for 31 km to Ralun. Much of the forest south of Ralun, alongside the 16 km road to Cochamó, has been felled but the remaining scrub just south of town still supports Black-throated Huet-huet and Chucao Tapaculo.

PUYEHUÉ NP

Map p. 171

This NP, northeast of Puerto Montt, protects excellent *Nothofagus* forest which supports the endemic Slender-billed Parakeet* and a number of austral specialities.

Endemics

Chilean Tinamou, Slender-billed Parakeet*.

Specialities

Flying Steamerduck, Spectacled Duck*, Rufous-tailed Hawk*, Austral Parakeet, Green-backed Firecrown, Magellanic Woodpecker, Des Murs' Wiretail, Black-throated Huet-huet, Chucao and Ochre-flanked Tapaculos, Fire-eyed Diucon, Patagonian Tyrant.

Others

Torrent Duck, Chilean Pigeon*, White-throated Treerunner, Rufous-tailed Plantcutter, Black-chinned Siskin.

Other Wildlife

Pudu.

Access

Head north out of Puerto Montt 105 km to Osorno, then turn east past Lake Puyehué to Aguas Calientes where the park HQ is situated. Flying Steamerduck and Slender-billed Parakeet* occur *en route*. The Refugio Antillanca, a ski-resort in the NP, is the best birding area (and base). Many of the birds listed above including Slender-billed Parakeet*, Magellanic Woodpecker and Ochre-flanked Tapaculo occur in the forest around this resort.

CHILOÉ ISLAND

Map p. 171

This big island, 250 km long and 50 km wide, south of Puerto Montt, is often misty and wet. However, its tidal creeks and marshes, forest and bamboo, support some excellent birds including the endemic Slender-billed Parakeet*.

Endemics

Slender-billed Parakeet*.

Specialities

Magellanic Penguin, Magellanic Diving-Petrel, Ashy-headed Goose, Flightless Steamerduck, Dolphin Gull, Snowy-crowned Tern, Green-backed Firecrown, Short-billed Miner, Des Murs' Wiretail, Chucao Tapaculo, Dark-faced Ground-Tyrant, Patagonian Sierra-Finch.

Others

White-tufted Grebe, Black-browed Albatross, Antarctic Giant Petrel, Southern Fulmar, Imperial and Rock Shags, Red-legged Cormorant*, Chiloe Wigeon, Chilean Flamingo, Black-faced Ibis, South American Snipe, Two-banded Plover, Rufous-chested Dotterel, South American Tern, Southern Skua, Chilean Pigeon*, Rufous-tailed Plantcutter, Correndera Pipit, Mourning Sierra-Finch.

Other Wildlife

South American Sealion, Bottle-nosed Dolphin.

Access

Head south out of Puerto Montt to Pargua where the ferry crosses the Chacao Channel to Chiloé Island. Seawatching just west of Pargua and from the 15-minute ferry may produce Magellanic Penguin and Magellanic Diving-Petrel. The inlet north of the road between Chacao, where the ferry docks, and Ancud, is good for shorebirds. Slender-billed Parakeet* occurs around Quellón, 180 km south of Ancud.

It is possible to take a ferry from Puerto Montt south through the Moraleda Channel to Chacabuco, a 22-hour trip via remote fjords where seabirds abound. From Chacabuco the rarely visited Queulat, Walkin and Coyhaique NPs may be worth exploring.

PUNTA ARENAS AND TIERRA DEL FUEGO

Map p. 174

The town of Punta Arenas, 2,140 km south of Santiago, is the gateway to the Chilean half of Tierra del Fuego, and the spectacular Torres del Paine NP. Together with the area around Punta Arenas itself, the seas, lake-dotted steppe grasslands and *Nothofagus* forests here support a superb selection of seabirds, including Magellanic Diving-Petrel, water-

The dove-like Magellanic Plover is one of the 'megabirds' to be found around Punta Arenas and on Tierra del Fuego at the end of the world!*

fowl, including the rare Ruddy-headed Goose*, and shorebirds, including the unique Magellanic Plover*, which is in a family of its own.

Specialities

Elegant Crested-Tinamou, Magellanic Penguin, Magellanic Diving-Petrel, Upland, Kelp, Ashy-headed and Ruddy-headed* Geese, Flightless and Flying Steamerducks, Spectacled Duck*, Rufous-tailed Hawk*, White-throated Caracara, Magellanic Plover*, Magellanic Oystercatcher, Dolphin Gull, Austral Parakeet, Magellanic Woodpecker, Short-billed Miner, Grey-flanked Cinclodes, Thorn-tailed Rayadito, Austral Canastero*, Fire-eyed Diucon, Chocolate-vented Tyrant, Dark-faced and Cinnamon-bellied Ground-Tyrants, Patagonian Sierra-Finch, Canary-winged and Yellow-bridled Finches, Patagonian Yellow-Finch, Austral Blackbird.

PUNTA ARENAS AND TIERRA DEL FUEGO

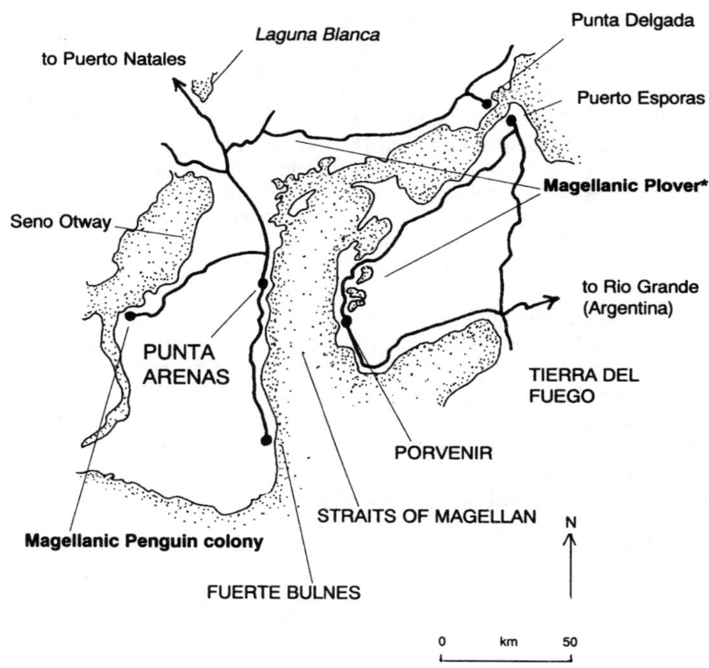

Others

Lesser Rhea*, Great Grebe, Black-browed Albatross, White-tufted, Great and Silvery Grebes, Antarctic Giant Petrel, Southern Fulmar, Wilson's Storm-Petrel, Imperial and Rock Cormorants, Black-necked and Coscoroba Swans, Torrent Duck, Chiloe Wigeon, Red Shoveler, Chilean Flamingo, Black-faced Ibis, Andean Condor, Cinereous Harrier, Red-backed Hawk, Chimango Caracara, Peregrine Falcon, Least Seedsnipe, Two-banded Plover, Rufous-chested and Tawny-throated Dotterels, Brown-hooded Gull, South American Tern, White-crested Elaenia, Tufted Tit-Tyrant, Austral Negrito, Austral Thrush, Patagonian Mockingbird, Chilean Swallow.

Other Wildlife

Guanaco, Commerson's Dolphin.

Access

Lesser Rhea*, Magellanic Oystercatcher, Two-banded Plover, and Rufous-chested and Tawny-throated Dotterels occur at the Magellanic Penguin colony on Seno Otway, northwest of Punta Arenas. Thorn-tailed Rayadito, Fire-eyed Diucon and Austral Blackbird occur in the woods around town and just north of Fuerte Bulnes, 55 km south of town. The magnificent Magellanic Woodpecker also occurs here, in the best patch of *Nothofagus* forest near Punta Arenas.

A car ferry connects Punta Arenas with Porvenir, on **Tierra del Fuego**. This ferry can be very busy so it is wise to book a 'return' in advance. Magellanic Penguin, Black-browed Albatross, Antarctic Giant Petrel, Magellanic Diving-Petrel and Commerson's Dolphin all occur in the Straits of Magellan, which this ferry crosses, and may be seen from the boat. Ruddy-headed Goose*, Magellanic Plover* and Chocolate-vented Tyrant occur alongside the road between Porvenir and Puerto Esporas, which runs along the northwest coast of Tierra Del Fuego. It is possible to return to the mainland at the end of this road, on another car ferry which connects Puerto Esporas with Punta Delgada. Magellanic Plover* and Least Seedsnipe occur alongside the road between Punta Delgada and the Punta Arenas–Puerto Natales road.

Accommodation: Punta Arenas: Nena's Guesthouse, 366 Boliviana (C). It is also possible to stay in Porvenir, on Tierra del Fuego.

In November 1992 the very rare Hooded Grebe*, previously considered to be endemic to Argentina, was reported from roadside lagoons between Punta Arenas and Puerto Natales, 250 km to the north.

The lakes, meadows and rushing rivers below the spectacular glaciers and granite towers of **Torres del Paine NP**, 392 km north of Punta Arenas (145 km northwest of Puerto Natales) support Torrent and Spectacled* Ducks, Andean Condor, White-throated Caracara, Tawny-throated Dotterel, Chocolate-vented Tyrant and Yellow-bridled Finch. It is possible to stay here.

Some of the species listed above including Magellanic Woodpecker and White-throated Treerunner, occur as far south as **Puerto Williams**.

JUAN FERNANDEZ ISLANDS NP

These islands are a National Park. They lie 650 km west of Chile and support a few interesting birds including four endemics: Defilippe's Petrel*, on Isla Santa Clara, Juan Fernandez Firecrown*, on Isla Robinson Crusoe, Mas Afuera Rayadito* on Isla Alejandro Selkirk and Isla Mas Afuera, and Juan Fernandez Tit-Tyrant. The islands are accessible by boat (every three weeks from Valparaíso) which takes 36 hours or plane (flights from Santiago and Valparaíso, subject to demand).

EASTER ISLAND

Although this island, 3,790 km west of Chile and famous for its rock carvings, is not strictly 'South American' it is accessible from Chile so a brief list of the best birds occurring there follows: Christmas Island Shearwater, Red-tailed and White-tailed Tropicbirds, Great Frigatebird, Grey-backed and Sooty Terns, Blue-grey and Brown Noddies, and Common White-Tern.

ADDITIONAL INFORMATION

Books

Guia de Campo de Las Aves de Chile (A Field Guide to the Birds of Chile), Modinger, B. and Holman, G. 1989. (Spanish)
Birds of Argentina and Uruguay: A Field Guide, Narosky, T and Yzurieta, D. 1989. (This book covers 85% of the birds found in Chile.)

CHILE ENDEMICS (8)	**BEST SITES**
Chilean Tinamou	Central and central-south: Campana NP, Nahuelbuta NP and Puyehué NP
Slender-billed Parakeet*	Central-south: Nahuelbuta NP, Chiloé Island and Puyehué NP
Chilean Seaside Cinclodes	North to central-south: Viña del Mar, Zapallar road and Arica
Crag Chilia	Central: rare, Farellones and Campana NP
Chestnut-throated Huet-huet	Central-south: rare, Laguna de Laja NP
Moustached Turca	Central: Farellones and Campana NP
White-throated Tapaculo	Central: Campana NP
Chilean Mockingbird	Central and central-south: Lake Peñuelas , Campana NP and Nahuelbuta NP

North = Arica area.
Central = Santiago area to coast and south to Concepción.
Central-south = Concepción south to Puerto Montt and Chiloé Island.

COLOMBIA

1	San Lorenzo Ridge	11	Munchique NP
2	Tayrona NP	12	Purace NP
3	Ríohaca	13	San Agustín
4	Santa Marta–Barranquilla Road	14	Cuevo de los Guacharos NP
5	Salamanca NP	15	Garzón–Florencia Road
6	Los Katios NP	16	Pasto–Mocoa Road
7	Bogotá	17	La Planada NR
8	Laguna de Sonso Res.	18	Río Nambi NR
9	Cali–Buenaventura Road	19	Río Blanco
10	Pichinde Valley	20	Amacayacu NP

INTRODUCTION

Summary

Ironically, Colombia, the country that supports more birds than any other in the world, and which has South America's best field guide, is not a popular birding destination. This is because of the rather poor infrastructure, which can mean lengthy travel times and, more importantly, the long-standing drug and guerrilla wars, which can make travelling very dangerous. However, hyper-careful birders with plenty of time have made successful trips to Colombia in recent years and been rewarded with some of the continent's rarest and most spectacular birds, notably tanagers.

Size

Colombia is the fourth biggest country in South America (1,138,914 km^2), nearly nine times the size of England, nearly twice the size of Texas and four times larger than Ecuador.

Getting Around

Although only 10% of the roads are paved, buses are the best mode of transport. They are ubiquitous and reach almost every village, but the best ones are expensive for South America. Car-hire is also expensive, which is probably a blessing in disguise because most roads are in poor condition and frequented by crazy drivers. The internal air network is well developed and extensive. River transport is quite expensive as well as time-consuming, but the only way into Amazonia apart from by air. Forget the trains.

Accommodation and Food

A wide variety of accommodation exists but most would suit only the budget birder who is prepared to go without home comforts. The cheapest residencias and hospedajes are usually near the bus terminals, whereas the more comfortable places are near the town plazas. Make sure the place looks safe before booking a room.

Rice, soup and stew dishes are ubiquitous. Western food can be hard to find away from the main towns and is usually expensive.

Health and Safety

Immunisations against malaria and hepatitis at least is recommended, and obtain a yellow fever vaccination if arriving from an infected area. Do not drink the water without at least boiling it.

Unfortunately Colombia is not for the fainthearted birder. It hosts the largest guerrilla rebellion in South America and is the world's biggest producer of cocaine. Violence is rife, especially in Bogotá, Medellín and Cali. Even the Amazon basin and Santa Marta region are troublesome. However, the intrepid birder loaded with common sense could travel to Colombia and probably return home safely with a bag full of goodies few others have seen. Check with the Colombian embassy and the Foreign Office for the latest news before planning a trip.

Climate and Timing

Broadly speaking, the two dry seasons in the Andean region fall between December and March, and July and August. However, the topography is very complicated and wide variations occur within short distances, especially in the south. The Llanos is driest between December and March whilst Amazonia is wet throughout the year. The best time to visit the Santa Marta region is in May.

Habitats

Colombia is a country of two halves. The west is mountainous, the east is low-lying Llanos and rainforest.

In west Colombia the Andes splits into three fingers or cordilleras which run north to south: the low Occidental (west), the high Central and the medium Oriental (east), separated by two big valleys, the Cauca (west) and Magdalena (east). The forests on the Pacific slope of the Occidental range are the wettest in South America and support a distinct avifauna. A fourth mountain range rises to 5,775 m (18,947 ft) from the northern coastline: the Sierra Nevada de Santa Marta, which also boasts a distinct avifauna.

The Llanos lie east and north of the Andes and Amazonian rainforest and still cover much of the southeast of the country.

The variety of habitat is completed by desert scrub, woodland and dry forest in the far north, mainly east of Santa Marta.

Conservation

Colombia has over 30 National Parks and reserves (covering nearly five million ha), most of which are run by INDERENA, a branch of the Ministry of Agriculture. However, habitat destruction outside and inside these parks is still rife, and many of the country's fine endemic birds are restricted to remote remnant forest.

55 threatened species, of which no less than 28 are endemic, occur in Colombia. A further ten endemics are near-threatened.

Bird Families

79 of the 92 families which regularly occur in South American are represented, three fewer than Ecuador and seven fewer than Peru. These include 20 of the 25 Neotropical endemic families and five of the nine South American endemic families. Rheas, seriemas, Magellanic Plover and seedsnipes are missing, as well as Sharpbill.

Well-represented families include cracids, wood-quails, hummingbirds (over 140), jacamars, puffbirds, toucans, antbirds, cotingas, manakins and tanagers.

Bird Species

Colombia supports more birds than any other country in the world. Its staggering list of at least 1,700 species (potentially over 1,800) is slightly larger than Peru's, a country of equal size, and it has some 40 more species than Brazil, a country 7.5 times larger. However, there are only 150 more birds here than in Ecuador which is just a quarter the size.

Non-endemic specialities and spectacular species include Northern Screamer*, Plumbeous Forest-Falcon*, Nocturnal Curassow, Bearded Helmetcrest, White-tipped Quetzal, White-eared Jacamar, Lanceolated Monklet*, White-faced Nunbird, Toucan Barbet*, Black-necked Red-Cotinga, Orange-breasted Fruiteater, Blue and Black-tipped* Cotingas, Long-wattled Umbrellabird*, Black Solitaire, White-capped, Scarlet-and-white, Golden-chested and Moss-backed Tanagers, Masked Mountain-Tanager*, and Tanager Finch*.

Endemics

59 bird species occur only in Colombia. This is the third highest country endemic list in South America, behind Brazil (179) and Peru (104), but above Venezuela (41). No less than 16 (27%) are restricted to the Santa Marta mountain range in the north of the country. The 11 hummingbirds, four antpittas and 11 'tanagers' include such spectacular species as Colourful Puffleg*, Multicoloured Tanager* and Torquoise Dacnis-Tanager*.

Expectations

Few birders have been to Colombia, but with its list over 1,700 one might expect to see a few hundred species with some ease, especially in the south.

SANTA MARTA AREA

Map opposite

The coastal resort of Santa Marta on Colombia's north coast lies near an impressive range of habitats, ranging from mangrove-lined lagoons, freshwater marshes and arid woodland along the coast to subtropical and temperate forests on the slopes of the Sierra Nevada de Santa Marta which rise in dramatic fashion just inland. Guerrilla and drug wars sometimes affect this region of Colombia so check before planning a trip here.

Red-footed Booby may be seen from the seafront at Santa Marta.

SAN LORENZO RIDGE - SANTA MARTA MOUNTAINS

Map opposite

The forested slopes of the snow-capped Sierra Nevada de Santa Marta mountains (rising to 5,800 m/19,000 ft), just 40 km southeast of Santa Marta, support no less than 16 of Colombia's 59 endemic species, an incredibly high degree of endemism.

Endemics

Chestnut-winged Chachalaca, Santa Marta Parakeet*, Blossomcrown*, White-tailed Starfrontlet, Black-backed Thornbill, Santa Marta Woodstar, Rusty-headed and Streak-capped Spinetails, Santa Marta Antpitta, Santa Marta Tapaculo, Santa Marta Bush-Tyrant*, Santa Marta Wren, Yellow-crowned Redstart, Santa Marta and White-lored Warblers, Santa Marta Brush-Finch, Santa Marta Mountain-Tanager.

Specialities

Band-tailed Guan, Bearded Helmetcrest, White-tipped Quetzal, Black-hooded Thrush, Rufous-browed Conebill, Orange-crowned Oriole.

Others

Lesser Yellow-headed Vulture, Andean Condor, Semicollared* and White-rumped Hawks, Black-and-chestnut Eagle*, Sickle-winged Guan, Scaly-naped Parrot, Green and Sparkling Violet-ears, Mountain Velvetbreast, Tyrian Metaltail, Masked Trogon, Blue-crowned Motmot, Emerald Toucanet, Black-banded Woodcreeper, Spotted Barbtail, Montane Foliage-gleaner, Grey-throated Leaftosser, Slaty Antwren,

SANTA MARTA AREA

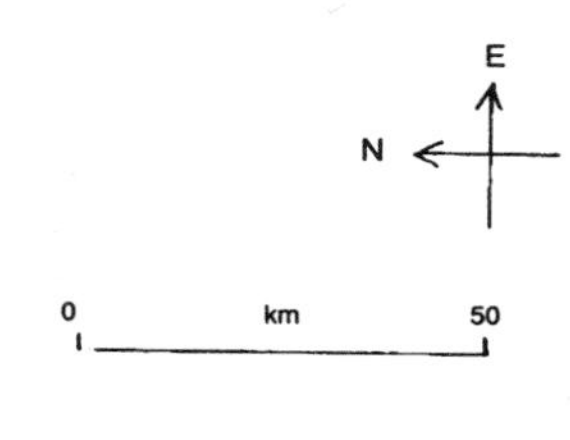

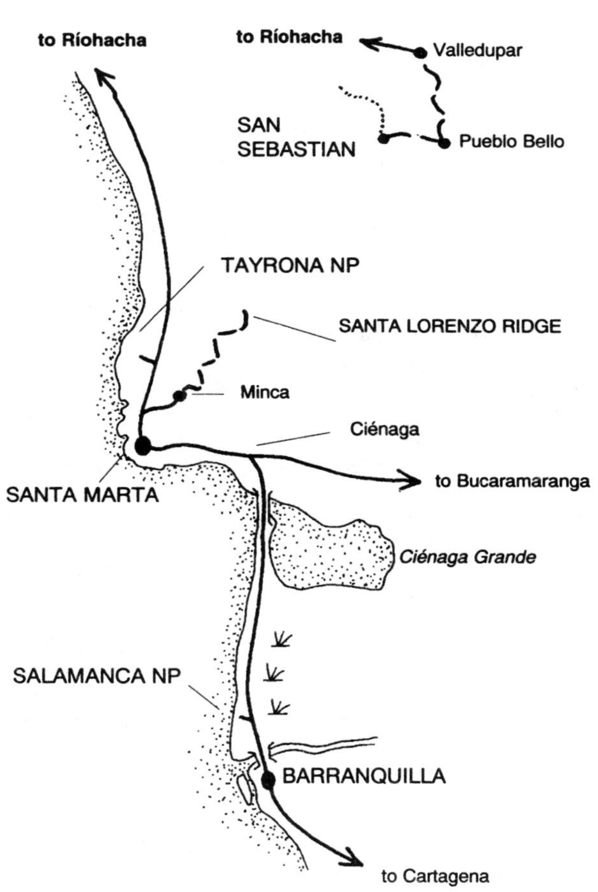

Long-tailed Antbird, Rusty-breasted Antpitta, Rufous-vented Tapaculo, Red-crested Cotinga, Golden-breasted Fruiteater, Black-throated Tody-Tyrant, Black-capped, Golden-faced and White-throated Tyrannulets, Mountain Elaenia, White-throated Spadebill, Cinnamon Flycatcher, Yellow-bellied and Rufous-breasted Chat-Tyrants, Streak-throated Bush-Tyrant, Black-chested Jay, Orange-billed Nightingale-Thrush, Mountain Wren, Grey-breasted Wood-Wren, Brown-bellied Swallow, Andean Siskin, Buff-breasted Mountain-Tanager, Thick-billed Euphonia, Blue-naped Chlorophonia, Black-capped and Black-headed Tanagers, Plush-capped Finch, Slaty Finch, Stripe-tailed Yellow-Finch, Slaty Finch, White-sided Flower-piercer, Crested Oropendola, Yellow-backed Oriole.

Access

Only one rough road penetrates the mountains from the northwest side, leading up to the San Lorenzo ridge, and this is not always open. A four-wheel-drive is recommended, especially in the wet season (September to December). Turn south off the Ríohacha road a few km east of Santa Marta (just past the police checkpoint) towards Minca. Bird the road-side forest along the 13-km stretch to Minca, then continue up and up, keeping left at all forks after 26 km, to km 32, and bird the roadside for the next ten km to km 42. Also bird above and below the INDERENA experimental station to the end of the road, where the second telecommunication tower and the treeline are reached.

It is also possible to reach the southeast side of the mountains. Head south from Ríohacha to Valledupar and then west to Pueblo Bello, where jeeps may be hired to get to **San Sebastian**, an Indian village, where there are a number of trails worth exploring. The sixteenth Santa Marta endemic, Santa Marta Sabrewing*, has been recorded near here (and on the north slopes in the upper Macotama valley). It does not occur on the northwest side of the mountains.

TAYRONA NP

Map p. 181

This small NP, 35 km east of Santa Marta, supports remnant dry, semi-deciduous woodland, and wet, humid tropical forest, where the goodies include Shining-green Hummingbird and Black Antshrike.

Specialities

Pale-bellied Hermit, Shining-green Hummingbird, Black Antshrike, Golden-winged Sparrow, Rosy Thrush-Tanager, Crimson-backed Tanager, Black-faced Grassquit, Orange-crowned Oriole.

Others

Green Ibis, King Vulture, Pearl and Double-toothed Kites, White Hawk, Black Hawk-Eagle, Blue Ground-Dove, Blue-and-yellow and Military Macaws, Ruby-topaz and Violet-bellied Hummingbirds, Long-billed Starthroat, Black-tailed Trogon, Rufous-tailed Jacamar, White-necked Puffbird, Keel-billed Toucan, Golden-green Woodpecker, Buff-throated Woodcreeper, Red-billed Scythebill, Pale-legged Hornero, Plain Xenops, Great Antshrike, Jet, Bare-crowned and White-bellied Antbirds, Lance-tailed, White-bearded and Striped Manakins, Slate-headed Tody-

Flycatcher, Sooty-headed, Yellow and Pale-tipped Tyrannulets, Black-tailed Flycatcher, Thrush-like Schiffornis, Cinnamon and One-coloured Becards, Golden-fronted Greenlet, Band-backed, Black-bellied, Rufous-breasted and Rufous-and-White Wrens, Long-billed Gnatwren, Rufous-capped Warbler, White-eared Conebill, Trinidad Euphonia, Red-legged Honeycreeper, Yellow-tailed Oriole.

Other Wildlife

Tamandua.

Access

The best wet forest lies just inside the Canaveral entrance, north of the Santa Marta–Ríohacha road, 35 km east of Santa Marta. Maicao and Ríohacha buses stop at this entrance. Bird the trails around the cabins, restaurant and car park.

Accommodation: Canaveral; cabins (C).

RÍOHACHA **Map p. 181**

The desert scrub west of Ríohacha, 135 km east of Santa Marta, supports a number of specialities, many of which are restricted to northeast Colombia and northwest Venezuela. This can be a dangerous area, however, and it may be more prudent to look for these birds in Venezuela.

Specialities

Rufous-necked Wood-Rail, Bare-eyed Pigeon, Blue-crowned Parakeet, Buffy Hummingbird, White-whiskered Spinetail, Black-backed Antshrike, Slender-billed Tyrannulet, Venezuelan Flycatcher, Stripe-backed Wren, Tocuyo Sparrow, Glaucous Tanager, Black-faced Grassquit, Vermilion Cardinal, Orinocan Saltator.

Others

Crested Bobwhite, Russet-throated Puffbird, Black-crested Antshrike, Pearly-vented Tody-Tyrant, Pale-eyed Pygmy-Tyrant, Cattle Tyrant, Bicoloured Wren, Pileated Finch, Dull-coloured Grassquit, Yellow Oriole.

Access

Good patches of desert scrub can be found at km 130 and km 140 from Santa Marta, 20 km west of Ríohacha.

SANTA MARTA–BARRANQUILLA ROAD **Map p. 181**

The 95–km coastal road west from Santa Marta to Barranquilla runs alongside marshes and woodland where the rare Northern Screamer* and localised endemic Sapphire-bellied Hummingbird* occur.

Endemics

Sapphire-bellied Hummingbird*.

Specialities

Northern Screamer*.

Others

White-cheeked Pintail, Capped Heron, Rufescent Tiger-Heron, Stripe-backed Bittern, Bare-faced Ibis, Snail Kite, Savanna Hawk, Limpkin, Wattled Jacana, Blue Ground-Dove, Red-and-green Macaw, Scarlet-fronted Parakeet, Blue-winged Parrotlet, Yellow-crowned and Orange-winged Parrots, Dwarf, Grey-capped and Little Cuckoos, Red-billed Emerald, Blue-crowned Motmot, Groove-billed Toucanet, Keel-billed Toucan, Golden-olive and Crimson-crested Woodpeckers, Pale-legged Hornero, Pale-breasted and Yellow-chinned Spinetails, Pied Water-Tyrant, White-headed Marsh-Tyrant, Rusty-margined Flycatcher, Black-chested Jay, Black-capped Donacobius, Swallow-Tanager, Crested Oropendola, Yellow-hooded Blackbird.

Access

Check the huge Ciénaga Grande (lake), where Sapphire-bellied Hummingbird* occurs in mangroves, and the marshes south of the road, especially opposite the Los Cocos entrance to Salamanca NP.

More productive marshes where Northern Screamer* occurs may be found south from Ciénaga on a rough road, especially between Fundación and Pivijay.

SALAMANCA NP **Map p. 181**

This small NP, 85 km west of Santa Marta, supports mangroves, tidal pools and desert scrub on a sand spit which separates the Caribbean from Ciénaga Grande. The rare endemic Sapphire-bellied Hummingbird* occurs here.

Endemics

Chestnut-winged Chachalaca, Sapphire-bellied Hummingbird*.

Specialities

Bare-eyed Pigeon, Russet-throated Puffbird, Chestnut Piculet.

Others

Magnificent Frigatebird, Reddish Egret, Lesser Yellow-headed Vulture, Common Black-Hawk, Black-collared Hawk, Yellow-headed Caracara, Collared Plover, Southern Lapwing, Brown-throated Parakeet, Pied Puffbird, Red-rumped Woodpecker, Straight-billed Woodcreeper, Northern Scrub-Flycatcher, Panama Flycatcher, Bicoloured Conebill, Blue-black Grassquit, Bronzed Cowbird.

Access

This NP is one hour's drive (85 km) west of Santa Marta on the road to Barranquilla. Head south out of Santa Marta to Ciénaga, then turn west. The park entrance is on the north side of the road 10 km east of Barranquilla. Bird the pools at the west end of the NP by the Los Cocos HQ, the desert scrub at the east end of the NP and, most importantly, the 1-km boardwalk through the mangroves; also at the east end of the

NP, the best spot for Sapphire-bellied Hummingbird*.

LOS KATIOS NP

Map below

This remote NP near the Panama border in northwest Colombia supports a wilderness of swamps and forest where the rare endemic Sooty-capped Puffbird*, which was only discovered in 1975, and a number of restricted-range specialities, including Tacarcuna Wood-Quail* and Sooty-headed Wren occur. No vehicles are allowed in the park (all travelling is done by river) and permission from INDERENA (office in Turbo) must be obtained before visiting.

Endemics

Sooty-capped Puffbird*.

Specialities

Northern Screamer*, Plumbeous Hawk*, Tacarcuna Wood-Quail*, Great Green Macaw, Grey-capped Cuckoo, Sapphire-throated Hummingbird, Dusky-backed Jacamar, Black Antshrike, Speckled Mourner, Golden-collared Manakin, White-headed and Sooty-headed Wrens, Black Oropendola.

LOS KATIOS NP

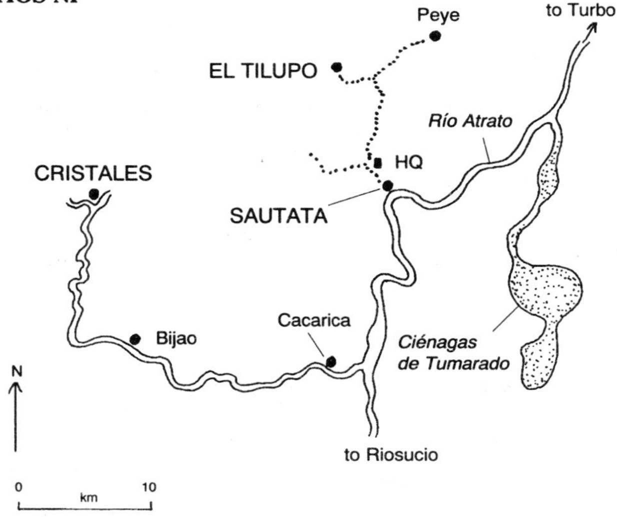

Others

Bare-throated Tiger-Heron, King Vulture, Grey-headed Chachalaca, Crested Guan, Grey-breasted Crake, Grey-chested Dove, Blue-and-yellow and Red-and-green Macaws, Red-lored Parrot, Violet-headed and Violet-bellied Hummingbirds, Green-crowned Brilliant, Purple-crowned Fairy, Collared Aracari, Keel-billed Toucan, Black-cheeked and

Cinnamon Woodpeckers, Checker-throated Antwren, Chestnut-backed Antbird, Purple-throated Fruitcrow, Black-tailed, Panama and Grey-capped Flycatchers, Black-chested Jay, Bay Wren, White-eared Conebill, Dusky-faced Tanager, Chestnut-headed Oropendola, Yellow-backed Oriole.

Access

Take one of the regular boats from Turbo up the Río Atrato to Sautata where the NP HQ is situated. Sooty-capped Puffbird and Black Antshrike occur around here. From here it is possible to hike to the 125-m high El Tilupo waterfall, where Crested Guan occurs, and Peye. It is also possible to continue up the Río Atrato to Cacarica (motorised canoes for Bijao and Cristales), where Speckled Mourner, Golden-collared Manakin and Black Oropendola occur.

BOGOTÁ

Around Colombia's often troubled, whacky capital city, which lies at 2,650 m (8,694 ft), there are a few birding sites worth visiting.

PARQUE LA FLORIDA

This small marsh next to the airport supports two rare endemics. However, it is under threat from development.

Endemics

Bogotá Rail*, Apolinar's Wren*.

Specialities

Noble Snipe.

Others

Masked Duck, Least Bittern, Spot-flanked Gallinule, Band-tailed Seedeater.

Access

The park lies at the end of the El Dorado airport runway. Head out of town towards the airport and take the last turn on the right.

CHOACHI

The remnant temperate forest and shrubs near the pleasant village of Choachi, approximately 50 km southeast of Bogotá, support some excellent hummingbirds including Bearded Helmetcrest.

Specialities

Coppery-bellied Puffleg, Bronze-tailed Thornbill, Bearded Helmetcrest, Rufous-browed Conebill.

Others

Tawny-breasted Tinamou, Glowing Puffleg, Black-tailed Trainbearer, White-throated Tyrannulet, Brown-backed Chat-Tyrant, Great Thrush, Brown-bellied Swallow, Andean Siskin, Pale-naped Brush-Finch, Scarlet-bellied and Buff-breasted Mountain-Tanagers, Grassland Yellow-Finch, Band-tailed and Plain-coloured Seedeaters.

Access

There is a good trail before the top of the road and the village, on the right opposite the first house on the left. Hummingbirds favour the shrubs at the top on the right. It is also possible to take the cable-car to Monserrat and walk to the Bogotá-Choachi road from there.

The recently discovered endemic Cundinamarca Antpitta occurs at 2,250 m (7,382 ft) along the new road between Monterrendondo and El Guaitiquia, southeast of Bogotá. Bearded Helmetcrest has been recorded from the Paramo de la Laguna Verde, some 50 km north of Bogotá, and in Chingaza NP, near La Calera, east of Bogotá.

The endemic Apolinar's Wren* occurs at **Lago de Tota**, over 250km northeast of Bogotá, via Tunja and south of Sogamoso. The endemic Gorgeted Wood-Quail* and Black Inca* have been recorded in subtropical oak forest near Virolín approximately 100 km north of here.

There are a few excellent birding sites *en route* west and south from Bogotá to Cali; the endemic Velvet-fronted Euphonia has been recorded around **Melgar**, approximately 100 km southwest of Bogotá on the Simon Bolivar highway to Girardot. Shining-green Hummingbird, Bar-crested Antshrike, White-eared Conebill and Scrub Tanager also occur here. Further west still, the endemic Greyish Piculet, Moustached Antpitta* (recently discovered in Ecuador), and endemic Apical Flycatcher occur at Finca La Esmeralda near Montenegro, just west of **Armenia**, approximately halfway between Bogotá and Cali. Bar-crested Antshrike and Scrub Tanager also occur here. From Armenia it is possible to travel north to Pereira (30 km) and Manizales (86 km).Cauca Guan*, Chestnut Wood-Quail*, Rufous-fronted Parakeet*, Bicoloured Antpitta*., Multicolored Tanager* and Red-bellied Grackle* all occurin **Los Nevados NP**, east of Periera. The little-known spectacular endemics Black-and-gold Tanager* (on the west slope of Cerro Tatama above Todo La Fierra), and Gold-ringed Tanager* (at Alto de Pisones), have recently been rediscovered near here, northwest of **Manizales**.

CALI AREA

The large, dangerous city of Cali lies in the Cauca valley, in southwest Colombia. To the north, near Buga, is the Laguna de Sonso Reserve, to the west, along the road to Buenaventura, there is remnant Pacific slope forest on the flanks of the west Andes, and to the southwest is the Pichinde valley. These three sites support some spectacular endemics and restricted-range species, not least the amazing Multicoloured Tanager*.

CALI AREA

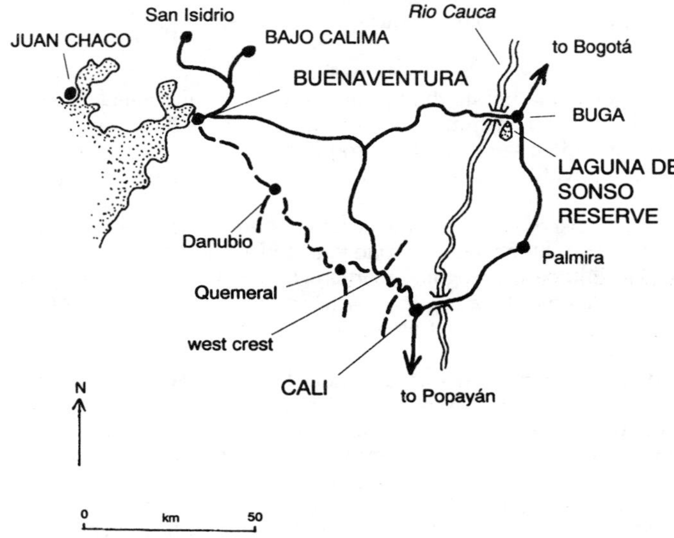

LAGUNA DE SONSO RESERVE

Map above

Three rare endemics have been recorded around this old oxbow of the Río Cauca, and in nearby secondary forest, set amongst farmland, near Buga, on the road between Bogotá and Cali. However, these great rarities may no longer be present owing to the continued spread of agriculture.

Endemics

Cauca Guan*, Apical Flycatcher, Turquoise Dacnis-Tanager*.

Specialities

Yellow-breasted Crake, Spectacled Parrotlet, Crested Ant-Tanager, Orange-crowned Euphonia.

Others

Horned Screamer, Pinnated Bittern, Snail Kite, Collared Plover, Grey-headed Dove, Dwarf, Dark-billed, Little and Striped Cuckoos, Spot-breasted Woodpecker, Pale-breasted Spinetail, Jet Antbird, Sooty-headed, Golden-faced and Mouse-coloured Tyrannulets, Pied Water-Tyrant, Cinereous Becard, Crimson-backed Tanager.

Access

Buga is 45 km north of Cali. From here head west towards the Río Cauca bridge and Buenaventura, for 4 km, on a road which runs alongside the north shore of the Laguna de Sonso, allowing good views. Just before the Río Cauca bridge turn south onto a track through several haciendas, then take the first left to a private ranch. Ask permission here to proceed to the Laguna.

To bird the south end of the Laguna cross the Río Cauca and turn first left to the village of **Bosque Yotoco**. Cauca Guan* and Torquoise Dacnis-Tanager* have been recorded around here. Take the narrow road on the far side of village which bears left to the Río Cauca. Cross the river and continue by foot along the dykes.

CALI–BUENAVENTURA ROAD

Map p. 188

The road northwest from Cali to the coast climbs out of Cali over the Andes and down through a rainshadow valley to the very wet lowlands at Buenaventura. Many spectacular species restricted to west Colombia and northwest Ecuador, especially tanagers, including the rare endemic Multicoloured Tanager*, have been recorded along this road in the past, mainly up to the mid 1980s. However, deforestation may have led to its demise since.

Endemics

Multicoloured Tanager*.

Although still regularly seen in Panama, the Black-tipped Cotinga is a rare sight in South America. The Buenaventura area is one of the most likely sites for this shy species

Specialities

Saffron-headed Parrot, White-whiskered Hermit, Empress Brilliant, White-eyed Trogon, Five-coloured Barbet*, Choco Toucan, Narino Tapaculo, Blue and Black-tipped* Cotingas, Long-wattled Umbrellabird*, Rufous-naped Greenlet, White-headed Wren, Slate-throated Gnatcatcher, Scarlet-and-white, Ochre-breasted and Tawny-crested Tanagers, Crested Ant-Tanager, Golden-chested and Purplish-mantled Tanagers, Yellow-collared Chlorophonia, Glistening-green, Grey-and-gold, Blue-whiskered*, Emerald, Rufous-throated and Metallic-green Tanagers, Scarlet-thighed Dacnis.

Others

Blue-footed Booby, Barred Hawk, Scaled Pigeon, Blue-headed and Mealy Parrots, Buff-tailed Coronet, Greenish Puffleg, Wedge-billed Hummingbird, Blue-crowned Motmot, Pied Puffbird, Chestnut-mandibled Toucan, Brown-billed Scythebill, Streak-capped Treehunter, Uniform Antshrike, Streaked Antwren, Jet, Chestnut-backed and Immaculate Antbirds, Undulated and Fulvous-bellied Antpittas, Green-and-black Fruiteater, Handsome and White-ringed Flycatchers, Slaty-capped Shrike-Vireo, Dusky-faced and Golden-naped Tanagers.

Access

Head northwest out of Cali on the road to Buenaventura and bird the following:

(i) the San Antonio road, which runs south 16 km west of Cali. 2 km after the 'holiday village' bird the roadside forest along the left fork and the trail on top of the steep bank a few metres up the right fork. Uniform Antshrike and Undulated Antpitta occur here;

(ii) the track north of the 'West Crest' at 18 km, where there is a low pass (1,800 m/5,906 ft); turn north between the two restaurants to a small patch of woodland beyond 0.5 km, where Multicoloured Tanager* has been recorded;

(iii) the 'Old Road' that runs southwest from the fork 21 km west of Cali. The 'Old Road' is best below Danubio where cliffs and waterfalls punctuate the otherwise luxuriant cover of very wet forest. However, this area is about three hours drive from Cali, and Buenaventura would be a better base. From there turn right at the checkpoint outside town and bird between km 60 and km 76 (where Scarlet-and-white and Golden-chested Tanagers occur), and between km 80 and km 115. At Danubio, a bridge across the Río Anchicaya allows access to a HEP Plant far up the valley where it is possible to stay and look for Long-wattled Umbrellabird*. Permission is needed from Corporacion Autonoma del Cauca (CVC) in Cali.

The best birding method to employ here would be to bird the 'West Crest' *en route* from Cali to Buenaventura on the new road, then bird the 'Old Road' from Buenaventura, thereby missing out the deforested section of the old road between the new road turn-off, Queremal and Danubio;

(iv) extra-keen birders may wish to attempt birding the Tokio forest. At Queremal (1,460 m/5,790 ft) take the track south from the plaza up to Tokio towers and the fine trail through cloud forest in the Farallones NP, where Narino Tapaculo and Purplish-mantled Tanager occur.

The best birding near Buenaventura is along the Bajo Calima road which runs northeast 100 km west of Cali, a few miles east of where the old road meets the new. Blue and Black-tipped* Cotingas occur here. Black-tipped Cotinga* also occurs around **Juan Chaco**, west of Buenaventura.

PICHINDE VALLEY

Multicoloured Tanager* and Andean Cock-of-the-Rock have been recorded in this valley, an hour's drive southwest of Cali.

Endemics

Greyish Piculet, Multicoloured Tanager*.

Specialities

Black-billed Peppershrike, Rufous-naped Greenlet, Crested Ant-Tanager.

Others

Torrent Duck, Bronzy Inca, Booted Racket-tail, Crested and Golden-headed Quetzals, Red-headed Barbet, Spotted Barbtail, Red-faced Spinetail, Slaty Antwren, Andean Cock-of-the-Rock, Sooty-headed and Golden-faced Tyrannulets, Glossy-black Thrush, Ashy-throated Bush-Tanager, Blue-naped Chlorophonia, Golden-naped and Metallic-green Tanagers.

Access

Head up from the entrance to the Cali Water Treatment Plant on the road to the statue of Christ (El Christo Rey). After passing through the 'holiday village', stay left just after the river and stay right at all forks until reaching a house called 'Casa Blanca'. Bird the road from here, especially the forest in the valley below which is in the Reserva de los Gallos de Monte (Cock-of-the-Rock Reserve). Better still, bird the trails to the river; two trails enter the forest just below the first house on the left past 'Casa Blanca' and one starts further up the road before the store, where the river is close to the road. This latter trail, and where the previous trails cross the river, are the best places to see Andean Cock-of-the-Rock and Crested Ant-Tanager (formerly endemic to Colombia but recently discovered in Ecuador).

POPAYÁN AREA

Map p. 192

This pleasant colonial town in southwest Colombia is the gateway to a number of superb birding sites.

MUNCHIQUE NP

The subtropical Pacific slope forest in this NP, two hours drive from Popayán, supports the very rare Colourful Puffleg*, which was recorded here in 1987, 20 years after its original discovery.

Endemics

Colourful Puffleg*

Specialities

Yellow-eared Parrot*, Dusky Bush-Tanager, Tanager Finch*.

POPAYÁN AREA

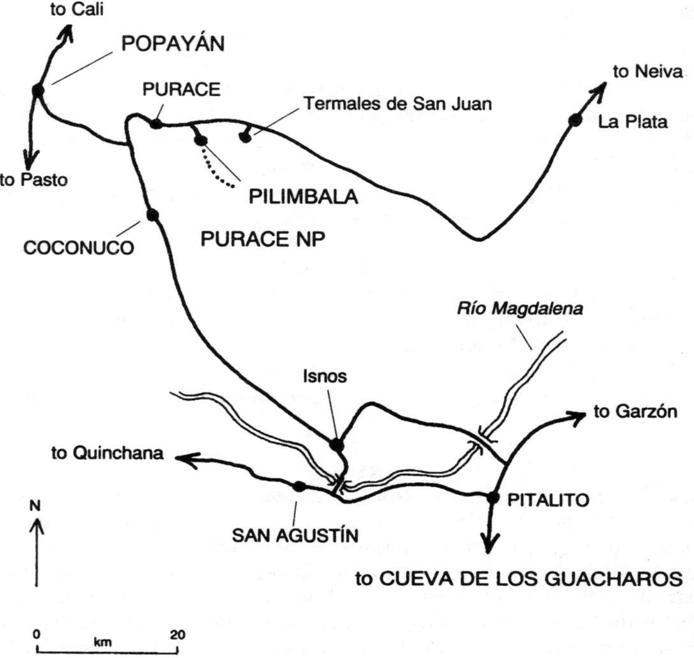

Others

Collared Inca, Sword-billed Hummingbird, Tourmaline Sunangel, Rufous Spinetail, Pearled Treerunner, Long-tailed Antbird, Chestnut-crowned Antpitta, Red-crested Cotinga, Barred Fruiteater, Rufous-headed Pygmy-Tyrant, White-banded Tyrannulet, Yellow-bellied and Rufous-breasted Chat-Tyrants, Sepia-brown Wren, Citrine Warbler, Rufous-naped Brush-Finch, Capped Conebill, Grey-hooded Bush-Tanager, Grass-green Tanager, Hooded Mountain-Tanager.

Access

Head west out of Popayán, then (i) bear left at the first fork and head towards Tambo and Cerro Munchique. Near the base of the mountain, at km 52, take the track on the right which leads to the telecommunications tower at the top of the mountain. Bird the area around the tower and just below. The little-known endemic Colourful Puffleg* has been recorded at Charquayaco 12 km from here. Back at the main road, continue to the pass and bird down the road for 10 km, checking mixed flocks for the rare Tanager Finch*, and (ii) bear right at the first and second forks, passing Uribe (km 18) and Romelia *en route* to the NP (km 41). Bird the HQ at the entrance, a site for Colourful Puffleg*, and one to two km beyond, where Tanager Finch* occurs.

This site is best at dawn before the fog and rain develops.

PURACE NP

Map p. 192

Purace (830 km^2) is considered by many to be Colombia's most beautiful NP. It is situated in the central Andes and protects temperate forest, shrub-covered slopes and paramo on the lofty slopes of seven active snow-capped volcanoes, from 2,500 m (8,203 ft) to 4,700 m (15,420 ft). A few very rare birds have been recorded here including Mountain Avocetbill, Crescent-faced Antpitta* and Masked Mountain-Tanager*, all of which may also be seen in Ecuador.

Specialities

Noble Snipe, Oilbird, Golden-breasted Puffleg, Mountain Avocetbill, Crescent-faced Antpitta*, Agile Tit-Tyrant, White-rimmed Brush-Finch*, Black-backed Bush-Tanager, Masked Mountain-Tanager*.

Others

Torrent Duck, Andean Guan, Swallow-tailed Nightjar, Shining Sunbeam, Glowing Puffleg, Purple-backed Thornbill, Black-billed Mountain-Toucan*, Andean Tit-Spinetail, Tawny Antpitta, Tufted Tit-Tyrant, Pale-naped and Slaty Brush-Finches, Black-headed Hemispingus, Rufous-crested Tanager, Black-chested, Scarlet-bellied and Buff-breasted Mountain-Tanagers, Plush-capped Finch, Plumbeous Sierra-Finch, Plain-coloured Seedeater.

Access

The most spectacular part of the park is the southwest. Head east out of Popayán on the road to Neiva, stopping 7 km east of the town to look for Crescent-faced Antpitta*, and continue to 21 km from Popayán (km post 172). Turn south here to Coconuco. The road enters the park beyond here, at 3,000 m (9,843 ft) and winds through the almost magical misty woodland of bamboo and stunted trees strewn with bromeliads, where the rare Masked Mountain-Tanager* occurs. This road connects with the Isnos–San Agustín road (see that site below) in the Magdalena valley, where Black-billed Mountain-Toucan* occurs.

Ignoring the Coconuco turning continue north and east past the village of Purace to km 153, where Condors roost on the cliffs to the left. At km post 152 take the right hairpin south to Pilimbala, the park HQ. Bird around the HQ and along the trail beyond.

Back on the road to Neiva, continue east to the bamboo forests between km posts 141 and 144, and the trail-head to Termales de San Juan, both worth birding. 10 km further on there is an Oilbird cave, and 1.5 hours' drive beyond there is Finca Merenberg, a private farm with superb subtropical forest, where Rufous-crested Tanager occurs.

Masked Mountain-Tanager* has also been recorded at San Rafael in Purace NP.

SAN AGUSTÍN

Map p. 192

The remnant forest and scrub surrounding this popular Archaelogical Park, one of the most important historical sites in South America, supports the endemic Dusky-headed Brush-Finch*.

Endemics

Dusky-headed Brush-Finch*.

Specialities

Bar-crested Antshrike, Crimson-backed and Scrub Tanagers, Grey Seedeater.

Others

Olivaceous Piculet, Booted Racket-tail, White-backed Fire-eye, Yellow-faced Grassquit, Rusty Flower-piercer.

Access

The park, 2 km west of San Agustín village, 35 km west of Pitalito, is open from 08.00 to 18.00.

CUEVA DE LOS GUACHAROS NP

Colombia's first NP (2,040 m/6,693 ft) was opened in 1960 to protect a chain of caves formed by the Río Suaza, south of Pitalito. Up to 2,000 Oilbirds have been known to breed in these caves (from December to June). However, seeing them and the rare endemic Red-bellied Grackle*, which has also been recorded here, involves a six-hour trek there and back. A permit is also required from INDERENA in Pitalito.

Endemics

Dusky-headed Brush-Finch*, Red-bellied Grackle*.

Specialities

Yellow-eared Parrot*, Spectacled Parrotlet, Oilbird, Tawny-bellied Hermit, Scrub Tanager.

Others

Bronzy Inca, Olive-backed Woodcreeper, Striped Woodhaunter, Unicoloured Tapaculo, Andean Cock-of-the-Rock, Rufous-breasted Flycatcher, Golden-faced Tyrannulet, Rufous-tailed Flycatcher, Yellow-bellied Siskin, Yellow-throated and Pale-naped Brush-Finches, Saffron-crowned Tanager.

Access

It is a long walk to the caves from Palestina. Head along the 2-km track to the Guarapas bridge, and once across, climb up the hill on a rough path for 25 km. A guide is recommended.

Accommodation: Cabanas are available and camping is possible in the park, or there is a basic hospedaje in Palestina on the road from Pitalito.

GARZÓN–FLORENCIA ROAD

The road over the eastern 'Oriental' cordillera, from Garzón, over 100 km northeast of San Agustín on the road to Neiva, south to Florencia, allows rare access to east Andean slope forest in Colombia. Rarely seen birds such as White-capped Tanager and the endemic Red-bellied Grackle* occur along this road, and a few Amazonian species reach their western limits at the bottom of the road, near Florencia.

Endemics

Red-bellied Grackle*.

Specialities

Amazonian Umbrellabird, White-capped Tanager, Golden-collared Honeycreeper.

Others

Solitary* and Black-and-chestnut* Eagles, Speckled Chachalaca, Andean Guan, Grey-chinned Hermit, Lettered Aracari, Black-billed Mountain-Toucan*, Ash-browed Spinetail, Stripe-chested and Yellow-breasted Antwrens, Andean Cock-of-the-Rock, Yellow-browed Tody-Flycatcher, Barred Becard, Grey-mantled Wren, Short-billed Bush-Tanager, Golden-crowned, Paradise and Golden-eared Tanagers.

Access

Bird high and low. Highland specialities include the beautiful White-capped Tanager. Lowland specialities such as Red-bellied Grackle* occur between km posts 52 and 56, where the road is close to the Río Hacha. Lower down, the forest is badly fragmented but still worth birding.

PASTO AREA

Map p. 197

This area of southwest Colombia, near the Ecuador border, is one of the richest places on earth for birds. Some 1,300 species may exist in the tiny department of Narino, and yet it is only a quarter of the size of England and one twentieth the size of Texas. Some 40 species, many of which are spectacular hummingbirds and tanagers, occur only in west Colombia and northwest Ecuador, an endemic centre known as the 'Choco'. Many of these species, some of which are rare and difficult to access in Ecuador, occur at sites around Pasto. Complete with on-site accommodation and trails, two of the sites are without doubt amongst the best birding sites in Colombia and South America. The good news is that this is probably the safest region of Colombia to visit. The bad news is that permission to visit some specific sites is needed, at least to confirm accomodation

PASTO–MOCOA ROAD

This rough road east of Pasto is one the best sites in South America for the stunning White-capped Tanager.

Specialities

Dusky Piha, White-rimmed Brush-Finch*, White-capped Tanager.

Others

Sulphur-bellied Tyrannulet, Chestnut-bellied Thrush, Turquoise Jay, Orange-eared, Golden-eared, Flame-faced and Blue-browed Tanagers, Deep-blue Flower-piercer.

Despite recent sightings in Ecuador of this 'tanager-with-snow-on-its-head', the road between Pasto and Mocoa is still one of the most reliable sites in the Andes for seeing a group of the noisy, flighty White-capped Tanager

Access

Two hours drive either way from Laguna La Cocha or Mocoa (the best bases), there is a 25-km stretch of excellent forest (between 1,800 m/5,906 ft and 2,300 m/7,546 ft), well worth prolonged birding. At the Nariño/Putumayo border the road is a one-way nightmare, so make sure the most experienced driver is at the wheel, or take the bus!.

LA PLANADA NR

Map opposite

This superb 3,600-ha reserve, situated at 1,800 m (5,906 ft), was set up in 1983 to protect remnant cloud forest. Complete with accommodation and trails, this is one of the best sites in Colombia. Nearly 250 species have been recorded including many restricted-range goodies such as White-faced Nunbird, Toucan Barbet*, Orange-breasted Fruiteater, Beautiful Jay*, and Scarlet-and-white Tanager. Arrangements to visit should be made in advance (see Additional Information, p. 202).

Specialities

Dark-backed Wood-Quail*, Empress Brilliant, Brown Inca, Buff-tailed and Velvet-purple Coronets, Hoary Puffleg*, Violet-tailed Sylph, Purple-throated Woodstar, White-faced Nunbird, Toucan Barbet*, Plate-billed Mountain-Toucan*, Fulvous Treerunner, Uniform Treehunter, Narino Tapaculo, Orange-breasted Fruiteater, Club-winged Manakin, Beautiful Jay*, Black-billed Peppershrike, Black Solitaire, Dusky Bush-Tanager, Scarlet-and-white and Purplish-mantled Tanagers, Yellow-collared Chlorophonia, Glistening-green and Rufous-throated Tanagers, Indigo Flower-piercer, Black-winged Saltator.

PASTO AREA

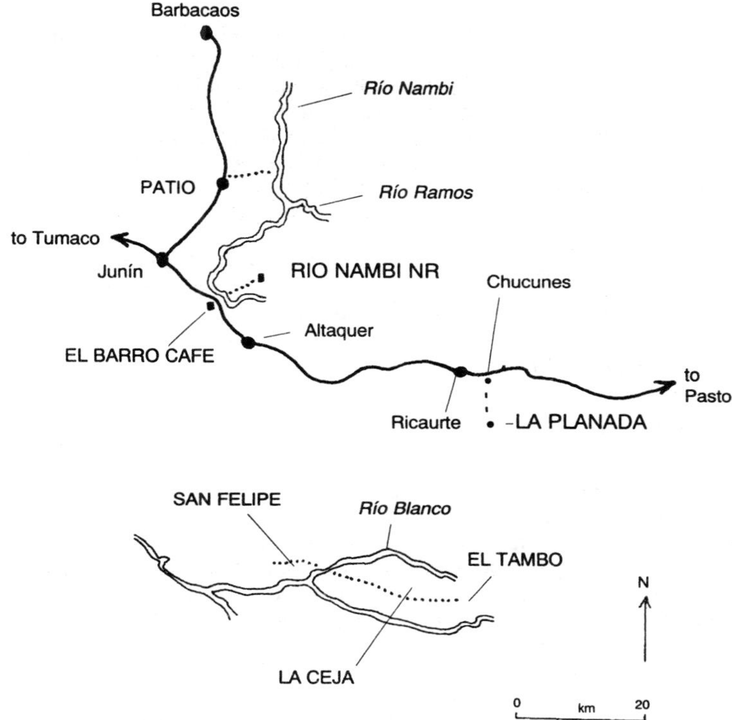

Others

Barred, White-rumped and Short-tailed Hawks, Black-and-chestnut Eagle*, Barred Forest-Falcon, Sickle-winged Guan, Plumbeous and Ruddy Pigeons, White-throated Quail-Dove, Barred Parakeet, Bronze-winged Parrot, Rufescent Screech-Owl, Mottled Owl, Grey Potoo, Chestnut-collared Swift, Tawny-bellied Hermit, Andean Emerald, Rufous-tailed Hummingbird, White-tailed Hillstar, Booted Racket-tail, Golden-headed Quetzal, Masked Trogon, Crimson-rumped Toucanet, Golden-olive and Powerful Woodpeckers, Tyrannine, Strong-billed and Spot-crowned Woodcreepers, Azara's, Slaty and Red-faced Spinetails, Rusty-winged and Spotted Barbtails, Lineated and Scaly-throated Foliage-gleaners, Streak-capped Treehunter, Tawny-throated Leaftosser, Uniform Antshrike, Long-tailed Antbird, Rufous-rumped Antwren, Rufous-breasted Antthrush, Chestnut-crowned and Ochre-breasted Antpittas, Unicoloured Tapaculo, Green-and-black and Scaled Fruiteaters, Olivaceous Piha, Golden-winged Manakin, Streak-necked Flycatcher, Bronze-olive and Rufous-headed Pygmy-Tyrants, Rufous-crowned Tody-Tyrant, Golden-faced Tyrannulet, Sierran Elaenia, Fulvous-breasted Flatbill, White-throated Spadebill, Ornate, Handsome and Acadian Flycatchers, Rufous-tailed Tyrant, Dusky-capped Flycatcher, Barred Becard, Pale-eyed and Glossy-black Thrushes, Mountain Wren, Grey-breasted Wood-Wren, Olive-crowned Yellowthroat, Buff-rumped Warbler, Tricoloured and Chestnut-capped Brush-Finches, Rufous-crested, Blue-capped, White-winged and Flame-rumped Tanagers, Blue-winged and Black-chinned Mountain-Tanagers, Fawn-breasted, Silver-throated, Saffron-crowned, Flame-faced, Golden-naped, Metallic-green, Blue-and-black and Black-capped Tanagers, Yellow-bellied and Blue Seedeaters, Yellow-faced Grassquit, Rusty and White-sided Flower-piercers.

Access

Head south from Pasto then west towards Ricaurte, a 3–6 hour journey. The reserve is 15 km southeast of Ricaurte. Turn south in the village of Chucunes. There are plenty of trails to explore.

Accommodation: (A).

RIO NAMBI NR Map p. 197

The forest protected by the Río Nambi NR (2,000 ha) is at a lower altitude (1,400 m/4,593 ft) than La Planada NR, and supports even more restricted-range species, making it one of the best birding sites in Colombia and South America. Nearly 150 species have been recorded here including Choco Vireo, which was only discovered, here, in 1991. Another species, a woodpecker similar to Yellow-throated, may also be new to science. The many goodies include Plumbeous Hawk*, Plumbeous Forest-Falcon*, Baudo Guan*, Lanceolated Monklet*, Moss-backed Tanager, and Scarlet-thighed Dacnis.

This reserve is run by FELCA, who prefer visitors to book in advance (see Additional Information, p. 202).

Endemics

Choco Vireo.

Specialities

Fasciated Tiger-Heron*, Plumbeous Hawk*, Plumbeous Forest-Falcon*, Dark-backed Wood-Quail*, Baudo Guan*, Blue-fronted Parrotlet, White-whiskered Hermit, Tooth-billed* and Purple-chested Hummingbirds, Empress Brilliant, Brown Inca, Velvet-purple Coronet, Purple-bibbed Whitetip, Violet-tailed Sylph, White-eyed Trogon, Lanceolated Monklet*, Toucan Barbet*, Stripe-billed Aracari, Choco Toucan, Uniform Treehunter, Orange-breasted Fruiteater, Club-winged Manakin, Tufted Flycatcher, Beautiful Jay*, Rufous-brown* and Black Solitaires, Slate-throated Gnatcatcher, Scarlet-and-white, Ochre-breasted, and Moss-backed Tanagers, Yellow-collared Chlorophonia, Glistening-green, Grey-and-gold, Blue-whiskered* and Rufous-throated Tanagers, Golden-collared Honeycreeper, Scarlet-thighed Dacnis, Great-billed Seed-Finch*, Indigo Flower-piercer.

Others

Grey-headed Kite, Barred Hawk, Barred Forest-Falcon, Sickle-winged Guan, Pale-vented Pigeon, Maroon-tailed Parakeet, Bronze-winged and Mealy Parrots, Rufescent Screech-Owl, Mottled Owl, Grey Potoo, Chestnut-collared, White-collared and Band-rumped Swifts, Tawny-bellied and Little Hermits, White-tipped Sicklebill, Green-fronted Lancebill, White-necked Jacobin, Rufous-tailed Hummingbird, Green-crowned Brilliant, White-tailed Hillstar, Purple-crowned Fairy, Golden-headed Quetzal, Broad-billed and Rufous Motmots, Red-headed Barbet, Crimson-rumped Toucanet, Yellow-vented, Smoky-brown, Golden-olive and Cinnamon Woodpeckers, Plain-brown, Long-tailed, Wedge-billed and Spotted Woodcreepers, Slaty and Red-faced Spinetails, Spotted Barbtail, Buffy Tuftedcheek, Striped Woodhaunter, Scaly-throated Foliage-gleaner, Plain Xenops, Spot-crowned Antvireo, Slaty and Rufous-rumped Antwrens, Immaculate Antbird, Plain-backed Antpitta, Green-and-black and Scaled* Fruiteaters, Olivaceous Piha, Golden-winged and Green Manakins, Tawny-rumped Tyrannulet, Yellow-bellied and Lesser Elaenias, Scale-crested Pygmy-Tyrant, Eye-ringed and Fulvous-breasted Flatbills, Ornate, Tawny-breasted, Sulphur-rumped and Black-tailed Flycatchers, Slaty-backed Chat-Tyrant, White-ringed Flycatcher, Thrush-like Schiffornis, Cinnamon Becard, Masked Tityra, Black-billed Peppershrike, Slaty-capped Shrike-Vireo, Sepia-brown and Bay Wrens, White-breasted and Grey-breasted Wood-Wrens, Tawny-faced Gnatwren, White-thighed Swallow, Golden-bellied and Buff-rumped Warblers, Orange-billed Sparrow, Olive Finch, Yellow-throated Bush-Tanager, Dusky-faced, Emerald, Golden, Silver-throated, Flame-faced, Golden-hooded and Masked Tanagers, Variable Seedeater, Slate-coloured and Blue-black Grosbeaks.

Access

Head south from Pasto then west towards Ricaurte and Junín. Stop at the El Barro café, 5 km northwest of Altaquer and 6 km southeast of Junín, a 4–6 hour drive from Pasto. Ask at the cafe for a guide to escort you along the 4-km trail (excellent birding) to the reserve HQ. Choco Vireo occurs in mixed-feeding flocks around here. Also bird the degraded habitat near the road around the café. Club-winged Manakin and Great-billed Seed-Finch* occur here.

Accommodation: Basic lodge, including food (B).

Nearby **Patio** (Map p. 197) is another superb birding area. Head north out of Junín on the road to Barbacoas for 15 km. Good forest can be reached about three hours walk east from here through plantations and logged forest. The forest, 1 km up the Río Nambi from the confluence of the Río Nambi and Río Ramos, is particularly productive, with Plumbeous Forest-Falcon*, Baudo Guan*, Lanceolated Monklet*, Slate-throated Gnatcatcher and Yellow-green Bush-Tanager* present. Ask at Rio Nambi NR for a guide. Take supplies for camping.

RIO BLANCO

Map p. 197

This remote area above the Río Blanco valley on the Ecuador border supports a number of scarce high-altitude species in paramo grassland

and *Polylepis* woodland, including Andean Potoo and Chimborazo Hillstar, until recently considered to be endemic to Ecuador. A mini camping expedition and prior permission are needed to visit this site

Specialities

Carunculated Caracara, Noble Snipe, Rufous-banded and Buff-fronted* Owls, Andean Potoo, Swallow-tailed Nightjar, Chimborazo Hillstar, Gorgeted Sunangel, Golden-breasted Puffleg, Viridian Metaltail, Rainbow-bearded Thornbill, Mountain Avocetbill, Toucan Barbet*, Plate-billed Mountain-Toucan*, Stout-billed Cinclodes, White-chinned Thistletail, Many-striped Canastero, Plain-tailed Wren, Giant Conebill*, Dusky Bush-Tanager.

Others

Andean Condor, Red-backed Hawk, Black-and-chestnut Eagle*, Andean Guan, Scaly-naped Parrot, White-throated Screech-Owl, Speckled Hummingbird, Shining Sunbeam, Mountain Velvetbreast, Great Sapphirewing, Collared Inca, Sword-billed Hummingbird, Purple-backed Thornbill, Tyrian Metaltail, Crested Quetzal, Crimson-mantled and Powerful Woodpeckers, Tyrannine Woodcreeper, White-browed Spinetail, Rusty-winged and Spotted Barbtails, Pearled Treerunner, Tawny Antpitta, Unicoloured, Rufous-vented and Andean Tapaculos, Rufous-headed Pygmy-Tyrant, Black-capped, Tawny-rumped, Sulphur-bellied and White-banded Tyrannulets, Crowned, Yellow-bellied, Rufous-breasted and Brown-backed Chat-Tyrants, Red-rumped and Smoky Bush-Tyrants, Black-billed Shrike-Tyrant, Plain-capped Ground-Tyrant, Slaty-backed Nightingale-Thrush, Great Thrush, Paramo Pipit, Andean Siskin, Spectacled Redstart, Rufous-naped and Stripe-headed Brush-Finches, Blue-backed Conebill, Grass-green Tanager, Grey-hooded Bush-Tanager, Black-capped and Black-eared Hemispinguses, Hooded, Scarlet-bellied and Buff-breasted Mountain-Tanagers, Blue-and-black Tanager, Plush-capped Finch, Band-tailed, Plain-coloured and Paramo Seedeaters.

Access

Write to the La Planada NR (see Additonal Information) for directions to, and permission to stay at, **El Tambo** (3,350 m/10,991 ft), a five-hour trip southwest of Pasto. Carunculated Caracara, Chimborazo Hillstar and Giant Conebill* occur around here. From El Tambo you will need a tent and supplies to bird down the road in the very wet forest at **La Ceja** (2,700 m/8,858 ft) where Andean Potoo and Swallow-tailed Nightjar occur, and the Cloud forest at **San Felipe** (2,200 m/7,218 ft), where Buff-fronted Owl*, Gorgeted Sunangel and Black-eared Hemispingus occur.

LETICIA

Colombia's main Amazon town, 3,200 km from the Atlantic, is best reached by air. Although birding the tracks north of town, its side-trails, and the Incer road east out of Leticia into Brazil (particularly the north branch of this road), can be productive, nearby Monkey Island and Amacayacu NP are far better. Leticia is also only 4 km from Tabatinga in Brazil (see p. 142)

The whacky White-eared Jacamar may be seen hunting from exposed perches around Leticia or in the nearby Amacayacu NP

The 9-km long, 1-km wide **Monkey Island**, situated in the Amazon river, supports Undulated Tinamou, Sungrebe, five macaws, Hoatzin, five nightjars, four aracaris, Long-billed Woodcreeper, Castelnau's Antshrike, and Ash-breasted Antbird. There is an expensive lodge (A) on the island, which is a three-hour, 40-km boat trip up river from Leticia. There is a trail, and boat trips can be arranged to visit the Río Cayaru on the Peruvian bank, and Isla Corea, Quebrada Tucuchira and other trails on the north bank. Book through Turamazonas, Hotel Parador Ticuna, Apto. Aereo 074, Leticia, or, in USA, 855A, U.S., 19S, Tarpon Springs, FL 33581 (tel: 934-9713).

AMACAYACU NP

This big (3,000 km^2) NP north of Leticia supports many Amazonian specialities including such spectacular rarities as Orange-breasted Falcon*, Nocturnal Curassow and Black-necked Red-Cotinga. Some 500 species have been recorded.

Specialities

Zigzag Heron*, Orange-breasted Falcon*, Nocturnal Curassow, Maroon-tailed and Cobalt-winged Parakeets, Sapphire-rumped Parrotlet, Black-headed and Short-tailed Parrots, Tawny-bellied Screech-Owl, Straight-billed Hermit, White-eared, White-chinned, Bluish-fronted and Purplish Jacamars, White-chested Puffbird, Yellow-billed Nunbird, Many-banded Aracari, Plain-breasted Piculet, Chestnut-winged Hookbill, Black-crested and Plain-winged Antshrikes, Black Bushbird, Río Suno and Leaden Antwrens, White-shouldered Antbird, Reddish-winged Bare-eye, Black-necked Red-Cotinga, Black-and-white Tody-Tyrant, Brownish Elaenia, Royal Flycatcher, Little Ground-Tyrant, Band-tailed Oropendola.

Others

Horned Screamer, Capped and Agami Herons, Green Ibis, King Vulture, Slender-billed Kite, Tiny, Slate-coloured and Grey-lined Hawks, Crested Eagle*, Ornate Hawk-Eagle, Black and Red-throated Caracaras, Collared Forest-Falcon, Spix's Guan, Blue-throated Piping-Guan, Marbled Wood-Quail, Azure Gallinule, Sungrebe, South American Snipe, Ruddy Quail-Dove, Blue-and-yellow and Red-and-green Macaws, White-eyed and Painted Parakeets, Blue-winged Parrotlet, Orange-cheeked and Yellow-crowned Parrots, Hoatzin, Pheasant Cuckoo, Black-banded Owl, Great and Grey Potoos, Blackish and Ladder-tailed Nightjars, Little Hermit, White-chinned Sapphire, Black-tailed Trogon, Rufous and Blue-crowned Motmots, Brown and Yellow-billed Jacamars, Spotted and Collared Puffbirds, Lanceolated Monklet*, Lettered Aracari, Bar-breasted and Rufous-breasted Piculets, Cream-coloured and Red-necked Woodpeckers, Long-billed and Striped Woodcreepers, Curve-billed Scythebill, Ruddy and Red-and-white Spinetails, Chestnut-winged Foliage-gleaner, Rufous-tailed Xenops, White-shouldered and Spot-winged Antshrikes, Grey, Banded and Dot-winged Antwrens, Yellow-browed, Silvered, White-plumed, Bicoloured and Spot-backed Antbirds, Spotted Antpitta, Purple-throated, Plum-throated and Spangled Cotingas, Amazonian Umbrellabird, Dwarf Tyrant-Manakin, Sepia-capped Flycatcher, Yellow-browed Tody-Flycatcher, Ringed Antpipit, River Tyrannulet, Short-tailed Pygmy-Tyrant, Brownish Flycatcher, Rufous-tailed and Olivaceous Flatbills, Yellow-breasted Flycatcher, Amazonian Black-Tyrant, Riverside Tyrant, Cinnamon and Dull-capped Attilas, Greyish Mourner, Dusky-chested Flycatcher, Greater Schiffornis, Collared Gnatwren, Bicoloured Conebill, Black-faced Dacnis, Green Oropendola, Red-rumped Cacique.

Access

To arrange a visit to this excellent park it is best to contact the INDERENA office in Leticia, at AA006. The southern end of the park is accessible via Matamata, 60 km by boat upstream from Leticia. The northern end can be reached by flying to Ipíranga in Brazil, where a boat meets the plane and then travels up river into Colombia to Tarapacá. From here it is possible to hire a boat to travel up the Río Cotuhe to Lorena.

Accommodation: Matamata Hostel (C).

The remote town of **Mitú**, in east Colombia near the Brazilian border, is accessible by air from Bogotá. The lowland rainforest here, accessible via the Río Vaupes at Urania and Tucanare to the east, and the Río Querary north to Tapurucuara, supports such goodies as Fasciated Tiger-Heron*, Salvin's Curassow, Giant Snipe, Dusky-billed Parrotlet, Red-fan Parrot, Ladder-tailed Nightjar, Azure-naped Jay, and White-bellied Dacnis*.

ADDITIONAL INFORMATION

Addresses

INDERENA, the National Parks Office, is on Diagonal 34, No. 5-84 (tel: 285 1172), Bogotá, and AA 1384, Pireira (25 No. 23–47, Ground Floor).
RESERVA NATURAL LA PLANADA, AA 1562, Pasto, Narino.
FELCA (The Foundation for the Ecology of Hummingbirds in

Altaquer),The Secretary, FELCA, AA 384, Pasto, Nariño, Colombia.
CARDER, AA 3559, Pireira (25 No 23–47, Floor 11).

Books

A Guide to the Birds of Colombia, Hilty, S. and Brown, W. 1986. Princeton UP.

COLOMBIAN ENDEMICS (59)	BEST SITES
Chestnut-winged Chachalaca	North: Santa Marta
Cauca Guan*	West: Laguna de Sonso and Los Nevados NP
Blue-knobbed Curassow*	North: humid forest
Chestnut Wood-Quail*	North and West: Los Nevados NP
Gorgeted Wood-Quail*	Central: near Virolín
Bogotá Rail*	Central: rare, Bogotá
Tolima Dove*	Central: rare, Andes
Santa Marta Parakeet*	North: Santa Marta
Flame-winged Parakeet*	Central: Boyaca and Cundinamarca depts
Rufous-fronted Parakeet*	West: Nevado del Ruiz and Los Nevados NP
Indigo-winged Parrot*	Central: Andes
White-chested Swift*	South: rare, Andes
Santa Marta Sabrewing*	North: rare, Santa Marta
Sapphire-bellied Hummingbird*	North: rare, Salamanca NP
Indigo-capped Hummingbird	Central: foothills and lowlands
Chestnut-bellied Hummingbird*	Central: rare, Boyaca and Bolívar depts
Blossomcrown*	North: Santa Marta
Black Inca*	Central: near Virolín
White-tailed Starfrontlet	North: Santa Marta
Dusky Starfrontlet*	North: known only from one specimen taken in 1953, (possibly a dark morph of Golden-bellied Starfrontlet)
Colourful Puffleg*	South: Munchique NP
Black-backed Thornbill	North: Santa Marta
Santa Marta Woodstar	North: Santa Marta
Sooty-capped Puffbird*	Northwest: rare, Los Katios NP
White-mantled Barbet*	Central: rare, humid forest
Greyish Piculet	West: near Armenia and Pichinde valley
Silvery-throated Spinetail	Central: Boyaca and Cundinamarca depts
Rusty-headed Spinetail	North: Santa Marta
Streak-capped Spinetail	North: Santa Marta
Argus Bare-eye*	Southeast: known only from a single specimen taken in 1951, and thought to be a possible hybrid
Santa Marta Antpitta	North: Santa Marta and Los Nevados NP

Cundinamarca Antpitta	Central: near Bogotá
Bicoloured Antpitta*	West: Los Nevados NP
Brown-banded Antpitta*	West: possibly extinct, not seen since 1911
Santa Marta Tapaculo	North: Santa Marta
Antioquia Bristle-Tyrant*	North: lower Cauca valley
Santa Marta Bush-Tyrant*	North: Santa Marta
Apical Flycatcher	West: near Armenia and Laguna de Sonso
Choco Vireo	South: Río Nambi
Apolinar's Wren*	Central: Bogotá and Lago de Tota
Niceforo's Wren*	Central: seen in 1989, the first records since its discovery in 1945
Santa Marta Wren	North: Santa Marta
Yellow-crowned Redstart	North: Santa Marta
Santa Marta Warbler	North: Santa Marta
White-lored Warbler	North: Santa Marta
Santa Marta Brush-Finch	North: Santa Marta
Olive-headed Brush-Finch*	Central: seen in 1989, the first records since 1967
Dusky-headed Brush-Finch*	South: San Agustín and Guacharos NP
Sooty Ant-Tanager*	North and central: rare, west Andean foothills
Black-and-gold Tanager*	West: rare, Cerro Tatama
Gold-ringed Tanager*	West: rare, Alto de Pisones
Santa Marta Mountain-Tanager	North: Santa Marta
Velvet-fronted Euphonia	Central: Melgar
Multicoloured Tanager*	West: Cali area and Los Nevados NP
Torquoise Dacnis-Tanager*	Central and west: near Bogotá and Laguna de Sonso
Chestnut-bellied Flower-piercer*	West: high in Andes
Baudo Oropendola*	Northwest: known only from three specimens, taken around 1900
Red-bellied Grackle*	Central: rare, Guacharos NP, Garzón-Florencia road and Los Nevados NP
Mountain Grackle*	Central: rare, east Andes

North = Santa Marta area and south to Medellín.
Northwest = Los Katios NP.
West = Medellín south to Cali area.
South = Popayán southwards.
Central = Cucuta south to the Bogotá area.

ECUADOR

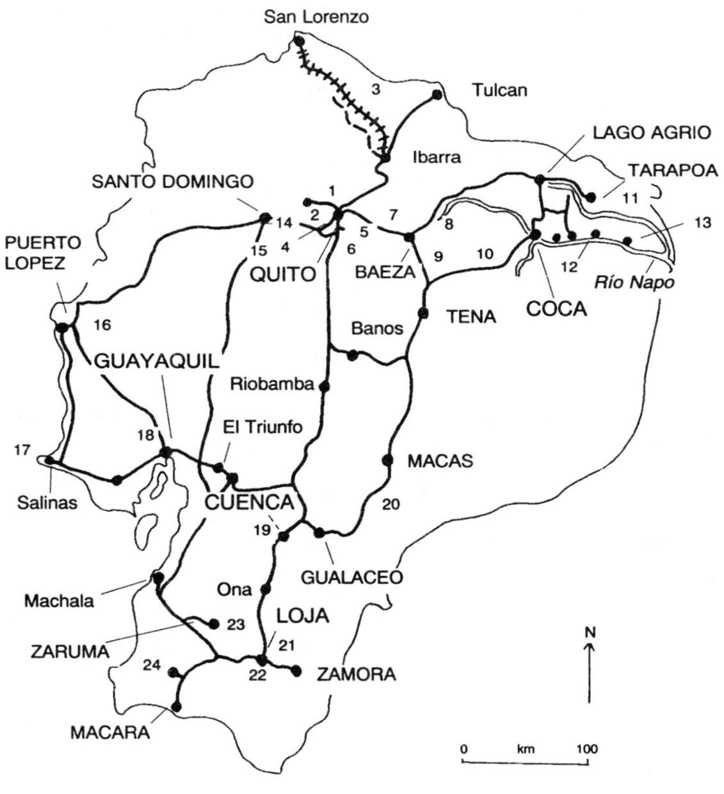

1	The Nono–Mindo Road and Mindo	8	Coca Falls
		9	Baeza–Tena Road and Tena
2	Yanacocha	10	Loreto Road
3	El Placer	11	Cuyabeno Reserve
4	Chiriboga Road	12	La Selva
5	Pasochoa Forest Reserve	13	Panacocha
6	Cotopaxi NP	14	Tinalandia
7	Papallacta Pass–Baeza Road	15	Río Palenque SC

16 Machalilla NP
17 Santa Elena Peninsula
18 Cerro Blanco
19 Las Cajas NRA
20 Gualaceo–Macas Road
21 Loja–Zamora Road
22 Podocarpus NP (north)
23 Buenaventura
24 Celica

INTRODUCTION

Summary

Ecuador's unique geographical position, where the equator crosses the Andes, small size, stable government, and good infrastructure makes it the ideal destination for birders interested in South America, especially first-timers. Roads and lodges offer the birder rare access to two of the richest avifaunal areas in the world: the east Andean slope temperate–subtropical forest belt and west Amazonian tropical lowland rainforest, a habitat hat-trick full of toucans, antbirds and tanagers. Although the proposed field guide has not reached the shelves yet, most species are illustrated in various books.

Size

Ecuador, named after the equator on which it lies, is one of the smallest South American countries (283,561 km^2). Only the archipelagos, the guianas and Uruguay are smaller. It is just twice the size of England, only half the size of Texas, and a third the size of Venezuela. Hence birding this country involves the minimum amount of travelling and the maximum amount of birding.

Getting Around

Although the roads between major cities and towns are paved, most of the others are poorly maintained or merely rough tracks which are often steep and treacherous, hence a four-wheel-drive with plenty of clearance and added traction, is recommended (but not absolutely essential). Beware of fast-moving trucks and buses, and a frustrating lack of roadsigns. If hiring a car always carry your driving license with you as there are constant checks. The bus network is cheap and extensive. There is also an excellent, reliable and cheap internal air network including air taxis to virtually anywhere, although boats are needed to reach some sites in the east, along the Río Napo and Río Aguarico.

Accommodation and Food

A wide range of accommodation is available from the relatively luxurious hotels to the basic, cheap and ubiquitous pensions and hosterias, although there are few higher class hotels outside the provincial capitals and resorts of Salinas and Playas.

When visiting remote sites, stock up in big city supermarkets since many small towns and villages have few shops and even fewer goods.

Health and Safety

Immunisation against amoebic dysentery, typhoid, hepatitis, malaria, yellow fever and rabies is recommended, and beware of altitude sickness in the high Andes. Boil and sterilise the water, or buy bottled mineral water, and avoid uncooked vegetables and salads.

Personal safety is not a major problem in Ecuador, but police searches are regular so always carry your passport. You may end up in prison without it.

Some roads are very steep and badly maintained in the Andes. Make sure the most experienced driver is at the wheel.

Climate and Timing

One of the many advantages of Ecuador is that its special birds can be seen virtually all year round, although the best times for the north and east are July to January, and for the south February to March (the time of the annual rains when many birds are singing).

The coastal region is hot and humid, and especially wet from January to April. In the central Andean region it is often chilly, misty and damp, and can rain most afternoons, especially from November to May. The eastern tropical lowlands are also hot and humid, and wet most of the year, but especially from June to August.

Habitats

Despite its small size, Ecuador has an extraordinary range of habitats. Just over 300 km separate the northern tip of Atacama desert in the southwest and the western edge of the Amazonian rainforest in the east.

The Andes form the country's backbone, running north to south through the middle of it, with two parallel cordilleras separated by a 40 km wide trough. Within the Andean zone there are over 30 volcanoes, including at least eight active ones, which rise to 6,310 m (20,702 ft) at Chimborazo. Paramo grasslands and lakes surround their snow-capped cones, whilst lower down, both the west and, particularly, the east slopes of the Andes, still retain extensive tracts of elfin, temperate and subtropical forests, full of great birds and famous for their mixed tanager flocks in particular.

The world's wettest forests, which run primarily along the west coast of Colombia, extend south to northwest Ecuador, whilst a few hundred km to the south lies the northern tip of the Atacama desert. The rest of the coastal plain has mainly been turned over to agriculture, although some remnant patches of deciduous forest remain, especially in Machalilla NP and near Guayaquil.

To the east of the Andes is the 'Oriente', the western extremity of Amazonia. The lowland rainforest here is the richest habitat in the world for birds with up to 300 species per sq km in some places.

Conservation

Some of Ecuador's major habitats are 'protected' in six National Parks, six Reserves and two National Recreation Areas, as well as a number of private nature reserves, which together encompass some 10% of the land surface. Unfortunately, logging, oil exploitation, mining, ranching and poaching all occur within most of these areas, and very few conservation-minded staff are employed. There is a standard charge (B) for entry to all NPs.

Ecotourism is alive and well in Ecuador and in many ways this country is showing the way to sustainable use of its resources. In some cases, indigenous people have been employed to build and maintain lodges, as well as to act as guides once they are up and running. These people, who are as much a part of the habitat as the trees and birds, seem happy about this, since their home, the forest, is preserved and not destroyed as happens with logging, mining, ranching and so on. This seems to be the way ahead, and many countries would do well to follow some of Ecuador's examples.

41 threatened species, of which nine are endemic, occur in Ecuador. A further two endemics are near-threatened, leaving just one endemic off the critical list.

Bird Families

82 of the 92 families which regularly occur in South America are represented in Ecuador, including 21 of the 25 Neotropical endemic families and six of the nine South American endemic families. There are no rheas, seriemas or Magellanic Plover.

Well-represented families include hummingbirds, motmots, jacamars, toucans, antbirds, cotingas and 'tanagers'.

Bird Species

Approximately 1,550 species have been recorded in Ecuador, a remarkable total for such a tiny country. Only Brazil, Colombia and Peru, all countries over four times larger, support a greater variety of species.

Non-endemic specialities and spectacular species include Zigzag Heron*, Andean Condor, Carunculated Caracara, Noble and Imperial* Snipes, Peruvian Thick-knee, Great Green Macaw, Banded Ground-Cuckoo*, Long-tailed Potoo, many hummingbirds, White-eared Jacamar, Toucan Barbet*, Grey-breasted*, Black-billed* and Plate-billed* Mountain-Toucans, many antbirds, Chestnut-belted Gnateater, Ocellated Tapaculo, Black-necked Red-Cotinga, Orange-breasted, Scarlet-breasted* and Fiery-throated* Fruiteaters, Purple-throated Cotinga*, Long-wattled Umbrellabird*, Andean Cock-of-the-Rock, and numerous multi-coloured tanagers including Grass-green and White-capped Tanagers, Masked Mountain-Tanager*, Golden-crowned Tanager, Yellow-collared Chlorophonia and Golden-collared Honeycreeper.

Many of the country's specialities occur only in northwest Ecuador and west Colombia (over 60 species are endemic to this area known as the 'Choco'), in southwest Ecuador and northwest Peru (some 50 species are endemic to this area, known as 'Tumbesia'), the Andes of Ecuador and Colombia, and west Amazonia.

Endemic Species

Only twelve species are endemic to Ecuador. These birds include two parakeets, four hummingbirds, and Orange-crested Manakin. Half of them are very rare and only half are likely to be seen on a trip that takes in both the east and the south of the country.

Expectations

Expect to see a lot of birds in Ecuador. In August–September 1992, a Danish Ornithological Society trip recorded an incredible 844 species in just 27 days, setting, in the process, a new world bird tour record (for trips of less than one month). This record would be hard to beat, and all but impossible for an individual or small team to repeat, but well prepared, experienced and energetic birders may see over 600 species on a three-week trip that takes in the north, east, and south. Even 700 is possible on such a trip, since Birdquest recorded 709 species in 22 days during July 1992.

Birders who prefer a gentler pace and who chose to make two separate short trips, may see over 300 species in the north and east, and over 400 species in the south.

QUITO

QUTIO AREA

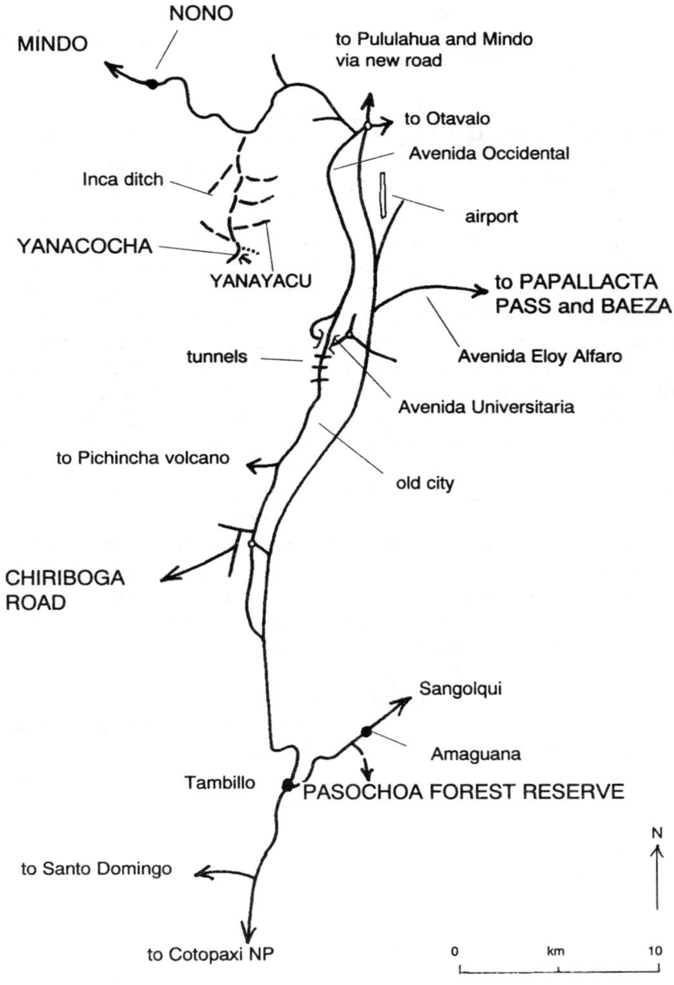

Crisp, spring-like Quito, the capital of Ecuador, lies just 25 km south of the equator in north Ecuador, at 2,850 m (9,350 ft). This pleasant city is the gateway to some exceptional birding with subtropical forest, temperate forest, paramo, and even *Polylepis* woodland all nearby.

The wooded gullies and parks of the captital, such as La Carolina, support Eared Dove, Grey-rumped Swift, Sparkling Violet-ear, Giant Hummingbird, Sapphire-vented Puffleg, Black-tailed and Green-tailed Trainbearers, Streak-throated Bush-Tyrant, Great Thrush, Hooded Siskin, Rufous-naped Brush-Finch, Blue-and-yellow Tanager, and Rusty Flower-piercer.

Accommodation: Hotel Colon (A); Hotel Inca (B); Hosteria Los Andes (C); Residencia Dapsilla (parking); Hotel Majestic (C).

THE NONO–MINDO ROAD AND MINDO

Maps p. 209 and p. 212

This road and the area around the town of Mindo, just 84 km northwest of Quito, is one of the best birding sites in South America. Some 430 species have been recorded, including 370 around Mindo alone.

The remaining west Andean slope temperate and subtropical forests here (1,524 m/5,000 ft to 2,286 m/7,500 ft) support the endemic Pale-mandibled Aracari*, and some 20 species restricted to west Colombia and northwest Ecuador (an area known as the 'Choco'), on a par with La Planada NR in south Colombia, but not as impressive as the Río Nambi area near there (see p. 196 and p. 198) or the El Placer area in northwest Ecuador (see p. 215). These specialities include Hoary Puffleg*, Toucan Barbet*, Esmeraldas Antbird, Long-wattled Umbrellabird*, Yellow-collared Chlorophonia and Black-chinned Mountain-Tanager. A number of exotic spectaculars also occur, not least Plate-billed Mountain-Toucan*, Scaled Fruiteater*, Andean Cock-of-the-Rock and many tanagers, including Grass-green Tanager.

The best time to visit is July, though any time of year can be good. Although accommodation in the town is quite basic, the trails are conveniently close and varied, making Mindo a very pleasant place to 'stay a while'. Permission from people in the town is required to walk some of the trails, but this is not usually a problem.

Many of the birds present also occur at sites elsewhere in northwest Ecuador and southwest Colombia, but none are so accessible as this one.

Endemics

Pale-mandibled Aracari*.

Specialities

Fasciated Tiger-Heron*, Baudo Guan*, Dark-backed Wood-Quail*, White-throated and Uniform Crakes, Blue-fronted Parrotlet, Violet-bellied Hummingbird, Fawn-breasted and Empress Brilliants, Brown Inca, Buff-tailed and Velvet-purple Coronets, Gorgeted Sunangel, Hoary Puffleg*, Violet-tailed Sylph, Toucan Barbet*, Scarlet-backed and Guayaquil Woodpeckers, Uniform Treehunter, Esmeraldas Antbird, Giant Antpitta*, Long-wattled Umbrellabird*, Club-winged Manakin,

Although a globally threatened species, the Toucan Barbet is still relatively common around Mindo*

Rufous-winged Tyrannulet, Snowy-throated Kingbird, Beautiful Jay*, Black-billed Peppershrike, Black Solitaire, Ecuadorian Thrush, Plain-tailed Wren, White-winged Brush-Finch, Dusky Bush-Tanager, Crested Ant-Tanager, Black-chinned Mountain-Tanager, Yellow-collared Chlorophonia, Glistening-green and Rufous-throated Tanagers, Black-winged Saltator.

Others

Torrent Duck, Semicollared*, Bicoloured and Barred Hawks, Ornate Hawk-Eagle, Black-and-chestnut Eagle*, Barred Forest-Falcon, Sickle-winged Guan, Sunbittern, White-throated Quail-Dove, Barred Parakeet, Little Cuckoo, Mottled and Rufous-banded Owls, Lyre-tailed Nightjar, Chestnut-collared Swift, Tawny-bellied Hermit, Booted Racket-tail, Purple-crowned Fairy, Crested and Golden-headed Quetzals, Broad-billed and Rufous Motmots, Plate-billed Mountain-Toucan*, Black-mandibled Toucan, Crimson-mantled and Powerful Woodpeckers, Rusty-winged and Spotted Barbtails, Pearled Treerunner, Streaked Tuftedcheek, Scaly-throated and Ruddy Foliage-gleaners, Tawny-throated Leaftosser, Uniform Antshrike, Long-tailed and Dusky Antbirds, Black-headed Antthrush, Plain-backed and Ochre-breasted Antpittas, Narino Tapaculo, Green-and-black and Scaled* Fruiteaters, Olivaceous Piha, Andean Cock-of-the-Rock, Golden-winged and Green Manakins, Torrent Tyrannulet, White-throated Spadebill, Yellow-bellied and Slaty-backed Chat-Tyrants, Thrush-like Schiffornis, Barred and Black-and-white Becards, Andean Solitaire, Slaty-backed Nightingale-Thrush, Sepia-brown Wren, Russet-crowned Warbler, Tricoloured Brush-Finch, Grass-green Tanager, Hooded Mountain-Tanager, Chestnut-breasted Chlorophonia, Flame-faced and Golden-naped Tanagers, Plush-capped and Slaty Finches, Blue Seedeater, Russet-backed Oropendola.

Other Wildlife

Spectacled Bear.

Access

To reach the Nono–Mindo road head north out of Quito on the Avenida Occidental. 11 km north of the junction with Avenida Universitaria turn west towards Nono on the San Miguel de Los Bancos road. Mindo is 63 km from the Avenida Occidental, 55 km beyond Nono. Alternatively, it is now possible to reach the lower part of the road more quickly, via the new road to Tandayapa (41 km from the Avenida Occidental). Continue north past the junction with the old San Miguel de Los Bancos Road for 2 km, then turn north.

MINDO

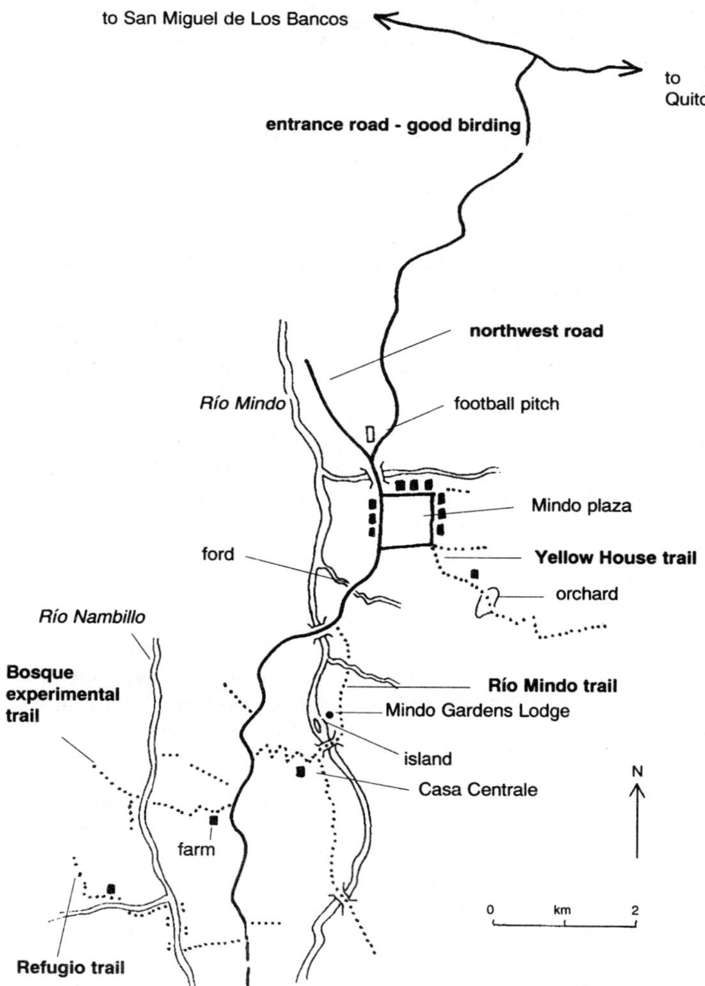

8.5 km along the San Miguel de Los Bancos road is the turning to Yanacocha (see p. 214) and 54 km further on, beyond Nono and Tandayapa, is the turning south to Mindo. The remnant forest either side of Tandayapa supports Toucan Barbet*, Plate-billed Mountain-Toucan* and Andean Cock-of-the-Rock. Roadside birding along the 8.5-km stretch from the main road to Mindo can be also productive. Broad-billed and Rufous Motmots, Black-mandibled Toucan, Narino Tapaculo, and Grass-green and Rufous-throated Tanagers occur along here.

There are a number of trails to bird around Mindo:

(i) Río Mindo trail: take the road southwest out of Mindo square, cross the ford at the Río Sangoonbi and turn left on to the trail just before the first bridge over the Río Mindo. Look for White-throated Crake at the first stream, and Fasciated Tiger-Heron*, Sunbittern and Andean Cock-of-the-Rock from the bridge over the river;

(ii) Bosque Experimental trail: take the steep trail to the right, 7 km from the Río Mindo bridge southwest of Mindo square. From here the trail crosses the Río Nambillo and heads up into good forest. Dark-backed Wood-Quail*, Uniform Crake, Barred Parakeet, Toucan Barbet*, Ochre-breasted Antpitta, Andean Cock-of-the-Rock (lek) and Club-winged Manakin (lek) occur along here. The local conservation organisation, Pacaso, control access to this trail. Ask for Milton Navarez in Mindo;

(iii) Refugio trail: continue beyond the Bosque Experimental trail turn-off to the end of the road. Take the right fork here to the Río Nambillo, then turn north, following the river where Plate-billed Mountain-Toucan* occurs, then west along a stream, then north again up to the refugio. Beyond here is an Andean Cock-of-the-Rock lek and a tunnel where Tawny-throated Leaftosser occurs. Black Solitaire also occurs along this trail. The local conservation organisation, Amigos de Natura, control access to this trail. Contact them for permission and a guide (useful);

(iv) Yellow House trail: head southeast out of Mindo square on the trail to a yellow house, owned by Señora Garzón (who charges a minimal entrance fee and keeps a logbook with the latest sightings in). From this house continue southeast to an orchard where Esmeraldas Antbird occurs near the watertank on side-trail to the left, and through fields to a valley on the left where Yellow-collared Chlorophonia occurs, then up in to forest where you may find Uniform Antshrike;

(v) Northwest road: head north out of Mindo square on the road out and take the road to the west opposite the Hosteria El Bijao. Pass a farm, then take the right fork, the left fork and continue north, looking for Grass-green Tanager and Blue Seedeater, through orchards, and over fords to another fork some 10 km from Mindo. The left fork goes to the Río Mindo, where Torrent Tyrannulet occurs, and halfway along there is a side-trail into scrub where Plate-billed Mountain-Toucan* and Slaty Finch occur.

Accommodation: Mindo: Salon Nor Occidental (C); Hosteria El Bijau (C). Mindo Gardens Lodge (A) (Giant Antpitta*).

For more details contact the Amigos de Natura, Correo Central de Mindo, Provincia de Pichincha, Ecuador. This is a local conservation organisation backed by the Worldwide Fund for Nature.

Over 250 species have been recorded in the **Maquipucuna Biological Reserve** (3,000 ha), run by Fundacion Natura. Turn north 30 km west of Nono. From here it is 27 km on an increasingly rough road through Las Delicias to the reserve entrance.

YANACOCHA

Map p. 209

Situated on the northern slopes of the Pichincha volcano, at an altitude of 3,000 m (9,843 ft) to 3,500 m (11,483 ft), Yanacocha can be a difficult site to reach without a four-wheel-drive vehicle. The temperate forest and paramo here support the very rare Imperial Snipe* and some scarce hummers.

Specialities

Tawny-breasted and Curve-billed Tinamous, Imperial Snipe*, Buff-winged Starfrontlet, Golden-breasted Puffleg, Rainbow-bearded Thornbill, Mountain Avocetbill, Ocellated Tapaculo.

Others

Andean Guan, Andean Pygmy-Owl, Swallow-tailed Nightjar, Sparkling Violet-ear, Giant Hummingbird, Shining Sunbeam, Mountain Velvetbreast, Great Sapphirewing, Sapphire-vented Puffleg, Black-tailed Trainbearer, Purple-backed Thornbill, White-browed Spinetail, Undulated and Rufous Antpittas, Unicoloured Tapaculo, Red-crested Cotinga, Barred Fruiteater, Rufous-headed Pygmy-Tyrant, Tufted Tit-Tyrant, Crowned and Rufous-breasted Chat-Tyrants, Streak-throated and Smoky Bush-Tyrants, Spot-billed and Plain-capped Ground-Tyrants, Rufous Wren, Paramo Pipit, Rufous-naped and Stripe-headed Brush-Finches, Blue-backed, Capped and Giant* Conebills, Hooded, Black-chested and Scarlet-bellied Mountain-Tanagers, Golden-crowned Tanager, Ash-breasted Sierra-Finch, Grassland Yellow-Finch, Paramo Seedeater, Glossy, Black and Masked Flower-piercers.

Access

Turn south 8 km along the Nono–Mindo road. 1.3 km along here take the left fork, then the right fork, then the right fork again. The rough track on the left after here (before the road takes a sharp left) goes to **Yanayacu**, where *Polylepis* woodland supports Rainbow-bearded Thornbill and Giant Conebill*. This is also one of the few known sites in South America for Imperial Snipe*. This bird displays at night between October and May so dawn and dusk during this period are the best times to look for it. Beyond the Yanayacu track the Yanacocha road takes a sharp left where another road joins from the right. Just beyond this junction there is an irrigation ditch on the left. Golden-breasted Puffleg, Mountain Avocetbill and Black-chested Mountain-Tanager occur along the path which runs alongside this ditch.

The very rare endemic Black-breasted Puffleg*, thought by some to be the most endangered hummingbird in the world, occurs only on the

Pichincha and Atacazo volcanoes. It has recently been recorded (1993) on Pichincha.

Approximately one hour's drive along the main road north from Quito, just west of the main road and 3 km south of Puellero, is the village of Alchipichi. Ask the locals here to point you in the direction of a small ravine beyond the football pitch, to the northwest of the village, where Oilbird roosts during the day. Scrub Tanager also occurs here.

There are some good remnant forest patches *en route* north from Quito to Otavalo, especially during the final 16 km to Otavalo (which is 93 km north of Quito). Maroon-chested Ground-Dove and Golden-breasted Puffleg occur here. Scrub Tanager occurs in the gardens and orchards surrounding the Hosteria Chorlavi, to the left of the Otavalo–Ibarra road, 3 km before Ibarra. Silvery Grebe occurs on Lago Yaguarcocha, which is encircled by a race-track, to the east of the main road, just north of Ibarra, 27 km north of Otavalo.

From Ibarra it is possible to travel by train to **El Placer**, one of a number of villages along the railway to San Lorenzo on the northwest coast of Ecuador, which lie close to good forest (see next major site). The train service is unreliable but a new road is planned to connect Ibarra with these villages. Check before planning your trip to Ecuador because this area could become a major birding site, although no facilities are available at present.

Another good site, north of Ibarra, is the **Paramo del Angel**, near the Colombian border. Head west out of Tulcan and bird the roadside between there and Maldonado (86 km), especially around Lagunas Verdes, approximately halfway between the two towns, and/or bird the track north from the village of El Angel, some 50 km southwest of Tulcan. There is a good trail to the left some 15 km along here. Andean Condor, Carunculated Caracara, Yellow-eared Parrot*, Chimborazo Hillstar, Golden-breasted Puffleg, White-tailed Shrike-Tyrant* and Turquoise Jay all occur here, and Bearded Helmetcrest, known otherwise only from much further north, has been reported.

EL PLACER

Map p. 217

This remote village is one of a number along the Ibarra–San Lorenzo railway which lie close to remnant west Andean slope subtropical forest where over 30 Choco endemics (birds which occur only in west Colombia and northwest Ecuador) have been recorded, the best selection for any site within the Choco area. These specialities include Banded Ground-Cuckoo*, Orange-fronted Barbet, Black-tipped Cotinga*, Orange-breasted Fruiteater, Long-wattled Umbrellabird*, Yellow-collared Chlorophonia, and some superb tanagers.

Unfortunately, this superb birding area is difficult to reach on the irregular trains and lacks any facilities. However, the new road planned to connect Ibarra with San Lorenzo should enable access to at least one of these villages, from where some exploration may be needed to find suitable habitat.

Birders intending to visit Ecuador, who have a particular interest in Choco endemics, may find it wise (and considerably more exciting) to visit this site instead of, or in addition to, the Nono-Mindo road and Mindo (see p. 210, especially if up-to-date enquiries reveal that the area has opened up. Otherwise, south Colombia may be the preferred destination (see p. 195 and p. 200).

Specialities

Plumbeous Hawk*, Plumbeous Forest-Falcon*, Baudo Guan*, Rufous-fronted Wood-Quail, White-throated Crake, Dusky Pigeon, Blue-fronted Parrotlet, Rose-faced Parrot, Banded Ground-Cuckoo*, Oilbird, White-whiskered Hermit, Velvet-purple Coronet, Purple-bibbed Whitetip, Violet-tailed Sylph, White-eyed Trogon, Barred Puffbird, Orange-fronted Barbet, Stripe-billed Aracari, Choco Toucan, Uniform Treehunter, Stub-tailed and Esmeraldas Antbirds, Rufous-crowned Antpitta, Narino Tapaculo, Orange-breasted Fruiteater, Black-tipped Cotinga*, Long-wattled Umbrellabird*, Club-winged Manakin, Orange-crested Flycatcher, Ochraceous Attila*, Snowy-throated Kingbird, Black-billed Peppershrike, Rufous-brown Solitaire*, Stripe-throated Wren, Slate-throated Gnatcatcher, Golden-bellied Warbler, Yellow-green Bush-Tanager*, Scarlet-and-white, Spectacled, Ochre-breasted, Scarlet-browed, Flame-rumped, Golden-chested and Moss-backed Tanagers, Yellow-collared Chlorophonia, Grey-and-gold, Blue-whiskered* and Rufous-throated Tanagers, Golden-collared Honeycreeper, Scarlet-breasted Dacnis*, Black-winged Saltator.

Others

Great Tinamou, Barred Hawk, Ornate Hawk-Eagle, Crested Guan, Tawny-faced Quail, Sapphire Quail-Dove, Great Green Macaw, Bronze-winged and Red-lored Parrots, Mottled, Black-and-white, Crested and Striped Owls, Crowned Woodnymph, Bronze-tailed Plumeleteer, Purple-crowned Fairy, Black-throated Trogon, Broad-billed and Rufous Motmots, White-whiskered Puffbird, Crimson-rumped Toucanet, Black-cheeked and Crimson-bellied Woodpeckers, Barred and Spotted Woodcreepers, Brown-billed Scythebill, Slaty and Red-faced Spinetails, Spotted Barbtail, Striped Woodhaunter, Ruddy Foliage-gleaner, Tawny-throated Leaftosser, Great, Western Slaty and Russet Antshrikes, Spot-crowned Antvireo, Checker-throated Antwren, Bicoloured and Ocellated Antbirds, Streak-chested Antpitta, Scaled Fruiteater*, Rufous Piha, Andean Cock-of-the-Rock, Green Manakin, Bronze-olive Pygmy-Tyrant, Grey Elaenia, Black-capped and Scale-crested Pygmy-Tyrants, White-throated Spadebill, Tawny-breasted and Sulphur-rumped Flycatchers, Rufous Mourner, Thrush-like Schiffornis, Band-backed Wren, Tawny-faced and Long-billed Gnatwrens, Tropical Gnatcatcher, Dusky-faced, Tawny-crested, Emerald, Silver-throated, Rufous-winged and Golden-hooded Tanagers, Scarlet-thighed Dacnis, Large-billed Seed-Finch*.

Access

To reach El Placer (650 m /2,133 ft) book the train from Ibarra well in advance. Accommodation and food are difficult to find and camping is probably the only way this site can be birded thoroughly. As well as El Placer, there are a number of other villages nearby, all of which lie close to various tracks and trails.

The village of **Lita** (570 m/1,870 ft) is 23 km east of El Placer, and even now, occasionally accessible by four-wheel-drive from Ibarra. Great Green Macaw, Banded Ground-Cuckoo*, and Scarlet-browed, Grey-and-Gold and Blue-whiskered* Tanagers have been recorded near here. Scarlet-and-white Tanager has been recorded near the villages of

EL PLACER

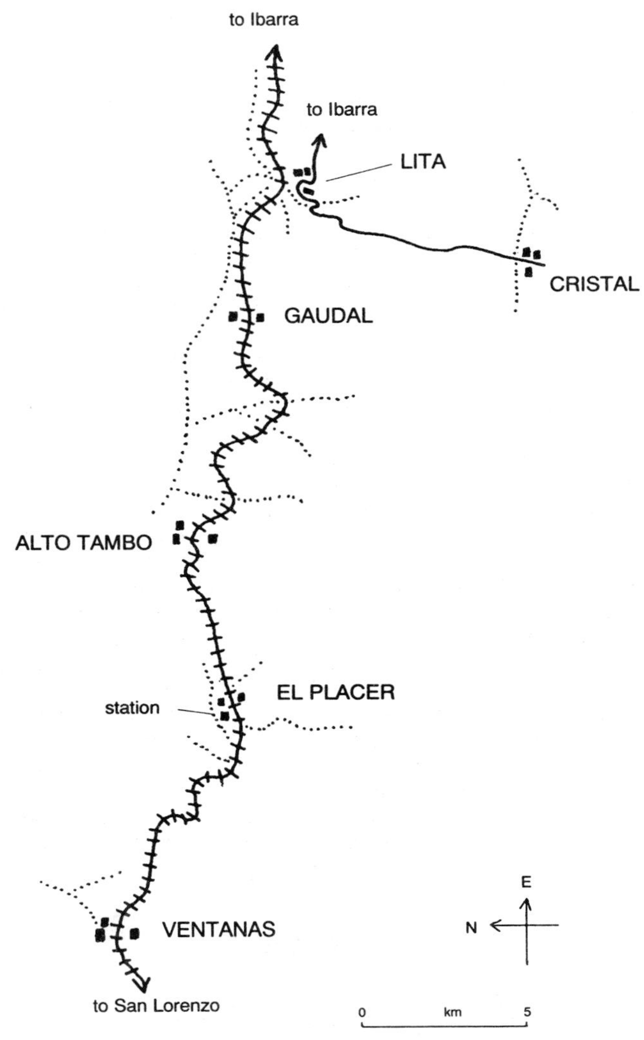

Guadual, 7 km west of Lita, and **Alto Tambo**, 15 km west of Lita, both of which lie at 600 m (1,969 ft). The remnant pockets of trees and scrub around the village of **Cristal** (1,100 m/3,609 ft), 12 km south of Lita, may still support Bronze-winged Parrot, Ochraceous Attila* and Rufous-throated Tanager, whilst good forest may still exist to the south. Great Green Macaw, White-eyed Trogon, Barred Puffbird, Black-tipped Cotinga*, Ochraecous Attila*, Slate-throated Gnatcatcher, Scarlet-and-white, Spectacled and Golden-chested Tanagers, and Scarlet-breasted Dacnis* have been recorded around the village of **Ventanas** (450 m/1,476 ft), west of El Placer (35 km west of Lita).

CHIRIBOGA (OLD SANTO DOMINGO) ROAD

Map p. 209

The quiet old road southwest from Quito to Santo Domingo traverses the west Andean slope, passing through some large tracts of temperate and subtropical forests which support similar birds to the Mindo area, including such goodies as Hoary Puffleg*, Toucan Barbet*, Plate-billed Mountain-Toucan*, Black-chinned Mountain-Tanager and Yellow-collared Chlorophonia. Species occuring more reliably here than at Mindo include Purple-throated Woodstar and Barred Puffbird. The road is also one of the best sites in Ecuador for Torrent Duck.

The higher part of the road is best worked from Quito, whilst the lower part of the road is closer to Tinalandia (see p. 234) and Santo Domingo.

Specialities

Tawny-bellied Hermit, Brown Inca, Buff-tailed Coronet, Gorgeted Sunangel, Hoary Puffleg*, Purple-bibbed White-tip, Purple-throated Woodstar, Barred Puffbird, Toucan Barbet*, Esmeraldas Antbird, Narino Tapaculo, Agile Tit-Tyrant, Turquoise Jay, Plain-tailed Wren, Rufous-chested and Ochre-breasted Tanagers, Black-chinned Mountain-Tanager, Yellow-collared Chlorophonia.

The extraordinary Torrent Duck negotiates rapids and slippery rocks with ease. They are often seen on the river alongside part of the Chiriboga road

Others

Torrent Duck, Andean Guan, Bronze-winged Parrot, Band-winged and Lyre-tailed Nightjars, Green-fronted Lancebill, Green Thorntail, Andean Emerald, Speckled and Sword-billed Hummingbirds, Sapphire-vented Puffleg, Booted Racket-tail, Purple-crowned Fairy, Crested and Golden-headed Quetzals, Broad-billed Motmot, Plate-billed Mountain-Toucan*, Crimson-mantled and Powerful Woodpeckers, Pale-legged Hornero, Slaty and Red-faced Spinetails, Streak-capped Treehunter, Ruddy Foliage-gleaner, Long-tailed Antbird, Rufous-breasted Antthrush, White-bellied and Ochre-breasted Antpittas, Unicoloured Tapaculo, Green-and-black and Scaled* Fruiteaters, Andean Cock-of-the-Rock, Golden-winged Manakin, Streak-necked Flycatcher, Rufous-crowned Tody-Tyrant, Ashy-headed and Golden-faced Tyrannulets, Ornate and Tawny-breasted Flycatchers, Yellow-bellied and Rufous-breasted Chat-Tyrants, Barred and Cinnamon Becards, Rufous, Sepia-brown and Bay Wrens, Tricoloured Brush-Finch, Grass-green Tanager, Black-capped Hemispingus, Hooded, Scarlet-bellied and Buff-breasted Mountain-Tanagers, Golden-crowned, Silver-throated and Golden-naped Tanagers, Plush-capped Finch, Blue Seedeater, Scarlet-rumped Cacique.

Access

Getting on this road is tricky. Head south out of 'new' Quito on the Avenida Occidental through the tunnels and 'old' Quito. 5 km south of 'old' Quito turn west just after the CEPE gas station (on west side of road) into a small plaza. Turn south here. It is best to ask for directions to be on the safe side.

Anywhere along this road can be good. Torrent Duck occurs on the roadside stream before the village of Chriboga (42.5 km from Quito). There is some good subtropical forest below Chiriboga. There is some good forest below Chiriboga, especially 13 km below it, where Hoary Puffleg*, Toucan Barbet* and Plate-billed Mountain-Toucan* occur. The roadside and track north at km 90 (from Quito) is also a good area. Andean Emerald, Ruddy Foliage-gleaner, Scaled Fruiteater*, Black-chinned Mountain-Tanager and Yellow-collared Chlorophonia occur here. Torrent Duck and Andean Cock-of-the-Rock occur at the bridge where this road meets the new road to Santo Domingo.

PASOCHOA FOREST RESERVE **Map p. 209**

Over 100 species have been recorded in this small (400 ha) but important reserve, run by Fundacion Natura. It lies on the north flank of the extinct Pasochoa volcano between 2,700 m (8,858 ft) and 3,800 m (12,467 ft), 43 km south of Quito. The forest contains large stands of *Chusquea* bamboo, the favoured habitat of the superb Ocellated Tapaculo, a very hard bird to see elsewhere in South America. The adjacent fields are surrounded by shrubs which act as a magnet for over ten species of hummingbird (especially in May).

Specialities

Carunculated Caracara, Golden-breasted Puffleg, Giant Antpitta*, Ash-coloured and Ocellated Tapaculos, Plain-tailed Wren, Rufous-chested Tanager.

Others

Andean Condor, Red-backed Hawk, Black-and-chestnut Eagle*, Andean Guan, Andean Gull, Rufous-banded Owl, Green-fronted Lancebill, Mountain Velvetbreast, Collared Inca, Sword-billed Hummingbird, Glowing and Sapphire-vented Pufflegs, Black-tailed Trainbearer, White-bellied Woodstar, Crimson-mantled Woodpecker, White-browed Spinetail, Flammulated Treehunter, Undulated, Rufous and Tawny Antpittas, Unicoloured and Andean Tapaculos, Red-crested Cotinga, Rufous-headed Pygmy-Tyrant, White-banded Tyrannulet, Rufous-breasted Chat-Tyrant, Rufous-naped and Stripe-headed Bush-Tanagers, Scarlet-bellied Mountain-Tanager, Plush-capped Finch, Yellow-billed Cacique.

Access

Head south out of Quito on the Pan-American highway and turn east at Tambillo (32 km from Quito) towards Amaguana on the Sangolqui road. After 5 km turn south (just west of Amaguana) on to a rough 6-km track to the reserve. There is a small entrance fee.

Ocellated Tapaculo occurs along the trail behind the 'Bosque Protector Pasochoa' sign *en route* to the summit. Further up, on leaving the forest and beyond some Paramo, check the scrub either side of the trail for hummers.

Accommodation: camping (B), but no facilities except water.

COTOPAXI NP

Cone-shaped and snow-capped Cotopaxi, the world's highest active volcano (5,897 m/19,347 ft), and paramo, dominate this 34,000-ha park 65 km south of Quito. This is a good site for Chimborazo Hillstar, which was thought to be endemic to Ecuador until it was discovered in Colombia in 1991. Other goodies here include Carunculated Caracara, Noble Snipe and Rufous-bellied Seedsnipe. Take it easy and beware of altitude sickness in this high-altitude NP.

Specialities

Carunculated Caracara, Noble Snipe, Rufous-bellied Seedsnipe, Chimborazo Hillstar, Viridian Metaltail, Stout-billed Cinclodes, Many-striped Canastero.

Others

Andean Condor, Speckled Teal, Yellow-billed Pintail, Puna Hawk, Slate-coloured Coot, Andean Snipe, Andean Lapwing, Andean Gull, Black-winged Ground-Dove, Giant Hummingbird, Black-tailed Trainbearer, Andean Tit-Spinetail, Streak-backed Canastero, Tawny Antpitta, Tufted Tit-Tyrant, Subtropical Doradito, Black-billed Shrike-Tyrant, Spot-billed, White-browed (April-Sept) and Plain-capped Ground-Tyrants, Great Thrush, Paramo Pipit, Rufous-naped Brush-Finch, Plumbeous Sierra-Finch, Plain-coloured Seedeater.

Access

To bird Cotopaxi thoroughly requires a long day trip from Quito. However, it is possible to stay at the nearby Hosteria La Cienega (B), reached by turning west 9 km south of the Cotopaxi turn-off. Giant Hummingbird and Black-tailed Trainbearer occur in the grounds, and Subtropical Doradito occurs at the marsh 1 km from the turn off from the Pan-American highway.

Once you have found the Pan-American highway head south from Quito for 53 km, then turn east in response to the NP sign. From here it is 12 km to the entrance gate. Once in, take the right fork through the introduced conifers, past the small museum (on the right) to a campsite on the left which is worth checking, and a campsite on the right (also worth checking), then turn left to the Laguna de Limpiopungo (3,800 m/12,467 ft), where a trail around the right side of the lake is worth birding (Noble Snipe).

Back at the main NP track, those fit birders in search of Rufous-bellied Seedsnipe and Chimborazo Hillstar, and unconcerned about altitude sickness, should proceed up to the car park at 4,700 m (15,400 ft). The hillstar occurs alongside the track *en route*. As high as possible leave the track and walk the paramo in search of the seedsnipe. This well-camouflaged, confiding bird does not flush readily and is easy to miss.

Carunculated Caracara, Rufous-bellied Seedsnipe and Chimborazo Hillstar also occur on the flanks of the inactive Chimborazo volcano, the highest peak in Ecuador (6,310 m/20,702 ft), near **Riobamba.** In many ways this site is as good, if not better than Cotopaxi, for these species. Turn north 10 km south of Riobamba and follow signs to the volcano. From the turn-off it is 37 km to a refugio at 4,785 m (15,699 ft). Explore the trackside on the way up and the area around here. Subtropical Doradito occurs at **Lago Colta**, alongside the Riobamba–Cuenca road, 30 km south of Ríobamba.

Sangay NP (272,000 ha), some 200 km south of Quito, has not been fully explored by birders up to now but the rare Masked Mountain-Tanager* has been recorded here. It is one of the remotest and inaccessible parks in Ecuador. The high Eastern Cordillera (5,000 m/16,404 ft) forms the western boundary. Between there and the low eastern boundary (1 000 m/3 281 ft) is rugged, almost impenetrable, terrain. The best access is from Banos at the northern extremity of the NP. A number of trekking trails start here. There are also trails leading from the Hacienda Releche entrance, 2 km from Candelaria and 15 km from Penipe, a village approximately halfway between Riobamba and Banos. Only one road breaches the park boundaries, running southeast from Riobamba to Alao, from where a trail leads to Sangay volcano. Masked Mountain-Tanager* has been recorded here. Another road is under construction in the southern extremity of the park, southwest from Guamote towards Macas.

PAPALLACTA PASS–BAEZA ROAD

Map p. 222

The road to the Oriente runs through Papallacta pass (3,960 m/12,992 ft), 58 km east of Quito. This pass provides an excellent opportunity to see some high-shrub and *Polylepis* specialists such as Giant Conebill* and Black-backed Bush-Tanager, as well as such goodies as Sword-billed Hummingbird and the gorgeous Golden-crowned Tanager, not far from the capital, or *en route* to the Oriente. Furthermore, the lovely

Black-billed Mountain-Toucan* and Dusky Piha occur alongside the road to Baeza (below the pass).

Specialities

Curve-billed Tinamou, White-throated Hawk, Carunculated Caracara, Noble Snipe, Chimborazo Hillstar, Buff-winged Starfrontlet, Mountain Avocetbill, Black-billed Mountain-Toucan*, Stout-billed Cinclodes, Many-striped Canastero, White-chinned Thistletail, Dusky Piha, Giant Conebill*, Red-hooded Tanager, Black-backed Bush-Tanager.

Others

Andean Condor, Sickle-winged Guan, Giant Hummingbird, Shining Sunbeam, Mountain Velvetbreast, Great Sapphirewing, Sword-billed Hummingbird, Black-tailed Trainbearer, Blue-mantled Thornbill, Andean Tit-Spinetail, Tawny Antpitta, Red-crested Cotinga, Rufous-breasted Flycatcher, Tawny-rumped and White-banded Tyrannulets, Red-rumped Bush-Tyrant, Black-billed Shrike-Tyrant, Brown-bellied Swallow, Paramo Pipit, Rufous-naped Brush-Finch, Blue-backed Conebill, Black-eared Hemispingus, Hooded and Scarlet-bellied Mountain-Tanagers, Golden-crowned Tanager, Buff-breasted Mountain-Tanager.

PAPALLACTA PASS

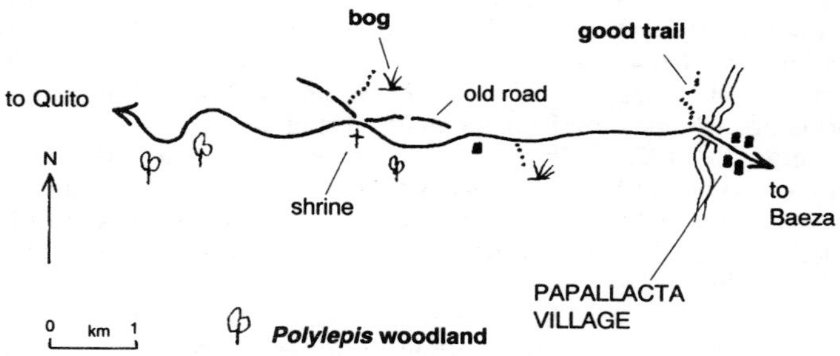

Access

Bird the roadside *Polylepis* patches a few km before the shrine at the pass (20 km east of Pifo) and around the marsh north of the shrine, the marsh on the south side of the road 3 km beyond the pass, the 3 km of scrub east of there to the village of Papallacta (Golden-crowned Tanager), the stone bridge area just before Papallacta (Mountain Avocetbill), and the gardens with *Datura* bushes around Papallacta (Sword-billed Hummingbird).

Dusky Piha and Red-hooded Tanager occur in the secondary Alder forest, between the road and the Río Maspa Chico, 15–20 km east of Papallacta.

Accommodation: Papallacta: Hotel Quitenita (C).

The quiet village of **Baeza**, 100 km east of Quito, is a good base from which to bird the road to Papallacta, the subtropical forest around Coca falls to the north, and the subtropical forest on the Huacamayo ridge, to the south.

Accommodation: Baeza: Hotel Samay (C); Hosteria El Nogal de Jumandi (C).

NORTHEAST ECUADOR

Northeast Ecuador encompasses east Andean slope temperate and subtropical forests, as well as west Amazonia, the richest place on earth for birds. Access is difficult and usually expensive, but well worth the effort and savings, for there are many spectacular birds here which are very difficult to see elsewhere. There are a number of birding sites, especially along the Río Napo which, apart from Cuyabeno Reserve, are all accessible by boat from Coca.

COCA (SAN RAFAEL) FALLS **Map p. 224**

The subtropical forest surrounding Ecuador's highest and most spectacular waterfall (150 m/500 ft), two hours from Baeza, supports some superb birds. It was here that the BBC filmed the Andean Cock-of-the-Rock lek for the epic series 'Flight of the Condor'. Situated on the east Andean slope at 1,280 m (4,200 ft) this site is renowned for its tanager flocks, here containing as many as 15 species, which often pass at eye-level. This is also a good site to look for such scarce birds as Coppery-chested Jacamar* and Golden-collared Honeycreeper.

Specialities

Semicollared Hawk*, Rufous-vented Whitetip, Coppery-chested Jacamar*, White-bellied Antpitta, Vermilion Tanager, Golden-collared Honeycreeper.

Others

Torrent Duck, Wattled* and Sickle-winged Guans, White-throated Quail-Dove, Red-billed Parrot, Band-bellied Owl, Green and Tawny-bellied Hermits, White-tipped Sicklebill, Blue-fronted Lancebill, White-tailed Hillstar, Booted Racket-tail, Wedge-billed Hummingbird, Gorgeted Woodstar, Golden-headed Quetzal, Blue-crowned Motmot, Red-headed Barbet, Yellow-vented and Crimson-crested Woodpeckers, Black-billed Treehunter, Sharp-tailed Streamcreeper, Rufous-rumped Antwren, White-backed Fire-eye, Rufous-vented Tapaculo, Amazonian Umbrellabird, Andean Cock-of-the-Rock, Golden-winged and White-bearded Manakins, Fulvous-breasted Flatbill, Ornate, Pale-edged and Lemon-browed Flycatchers, Barred Becard, Violaceous Jay, Spotted Nightingale-Thrush, Grey-mantled Wren, Olive Finch, Short-billed, Yellow-throated and Ashy-throated Bush-Tanagers, Rufous-crested Tanager, Bronze-green Euphonia, Yellow-throated, Golden-eared, Flame-faced, Golden-naped and Blue-browed Tanagers, Deep-blue Flower-piercer, Russet-backed Oropendola.

COCA FALLS

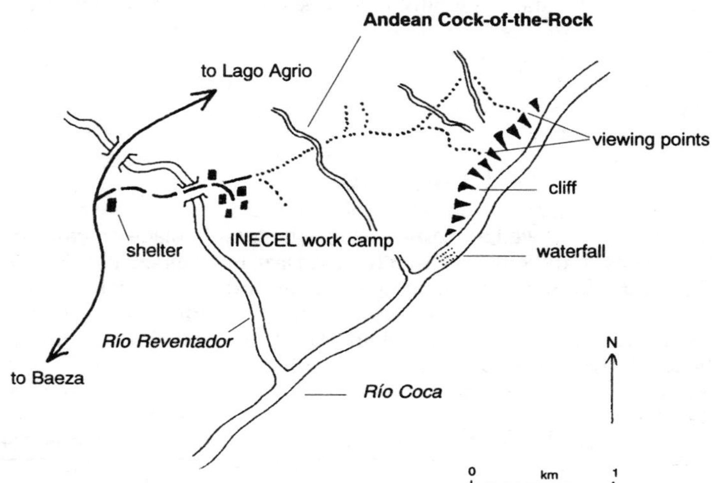

Access

The falls are 75 km northeast of Baeza on the road to Lago Agrio. Just before the bridge over the Río Reventador there is a concrete shelter on the right with an 'INECEL' (Electric Company) sign on it. From here there is a track to a workers' camp and a trail downhill to the falls. Andean Cock-of-the-Rock occurs approximately halfway along this trail, where it crosses a small river.

Accommodation: It is possible to stay at the INECEL work camp (C). Ask politely.

If continuing to **Lago Agrio** the grounds of the Gran Hostal El Lagro are worth a look. Scarlet-crowned Barbet and Lawrence's Thrush occur here. Those with spare time at Lago Agrio airport could check the marsh 200 m from the terminal building (turn left on facing airstrip), where Slate-coloured Hawk and Grey-breasted Crake occur.

BAEZA–TENA ROAD AND TENA

The remnant subtropical forest between Baeza and Tena is also good for spectacular east Andean slope tanager flocks, but the real goodies occurring along here are Andean Potoo, Moustached Antpitta*, until recently thought to be a very rare Colombian endemic, White-bellied Antpitta and, rarely, the stunning White-capped Tanager.

Specialities

Andean Potoo, Rufous-vented Whitetip, Spectacled Prickletail, Chestnut-winged Hookbill, Barred Antthrush, Moustached*, White-bellied and Slate-crowned Antpittas, Dusky Piha, Golden-winged Tody-Flycatcher, Plain-tailed Wren, White-capped, Vermilion, Yellow-

Four species of tody-flycatcher occur around the Hotel Auca in Tena, including the superb Golden-winged

throated and Metallic-green Tanagers, Golden-collared Honeycreeper.

Others

Black-and-chestnut Eagle*, Speckle-faced Parrot, White-throated Screech-Owl, Rufous-banded Owl, Lyre-tailed Nightjar, Tawny-bellied Hermit, Booted Racket-tail, Gorgeted Woodstar, Crested and Golden-headed Quetzals, Red-headed Barbet, Black-billed Mountain-Toucan*,

TENA

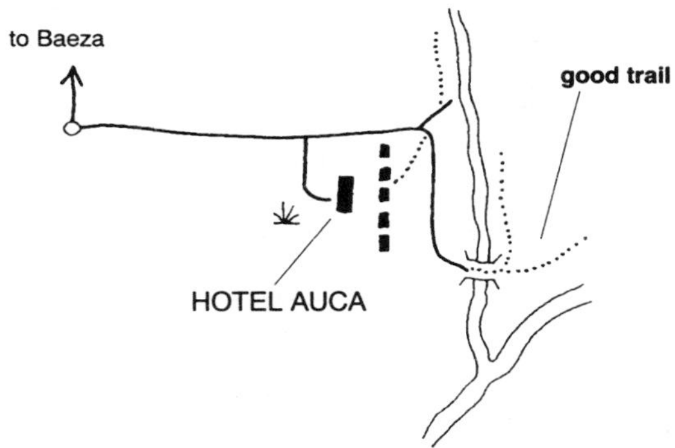

Black-mandibled Toucan, Lafresnaye's Piculet, Tyrannine Woodcreeper, Ornate Antwren, White-browed Antbird, Green-and-black Fruiteater, Rufous-headed Pygmy-Tyrant, Rufous-crowned Tody-Tyrant, Rusty-fronted Tody-Flycatcher, Lemon-browed Flycatcher, Andean Solitaire, Pale-eyed and Chestnut-bellied Thrushes, Rufous-crested Tanager, Bronze-green Euphonia, Golden-eared, Flame-faced, Golden-naped and Blue-browed Tanagers, Deep-blue and Bluish Flower-piercers, Subtropical and Mountain Caciques.

Access

Anywhere along this road can be productive, but the area approximately 65 km south of Baeza (40 km north of Tena), at a spur of the Andes known as the Cordillera de Guacamayos, is particularly good. Andean Potoo and Lyre-tailed Nightjar occur in the clearing around the house here at 1,900 m (6,234 ft). The extremely rare Moustached Antpitta*, has been recorded along the trail to the viewpoint at 1,700 m (5,577 ft). Dusky Piha and White-capped Tanager occur around the radio station at the pass (2,250 m/7,382 ft).

Some good birds may be seen around **Tena**, especially in the grounds of the Hotel Auca (Map p. 225), which would make an excellent base. Bird the trails on the far side of the river near this hotel and along the path down to the Río Tena from the cemetery. Look out especially for Napo Sabrewing, Spangled Coquette, Lemon-throated Barbet, Orange-fronted Plushcrown, Lined Antshrike, four species of tody-flycatcher including Rusty-fronted and Golden-winged, and Yellow-bellied Dacnis.

Nearby Misahualli, on the banks of the Río Napo, is one of the bases for tourist trips into the Napo region. Birders, however, are better off using Coca (see p. 223) as their gateway to this region.

LORETO ROAD

This road, between Tena and Coca, is very special. Constructed in 1988, it allows access to foothill forest on the east Andean slope (610 m/2,000 ft to 1,219 m/4, 000 ft), providing a very rare opportunity to look for a high number of rarities such as Orange-breasted Falcon*, Napo Sabrewing*, the superb Wire-crested Thorntail, Ecuadorian Piedtail*, Coppery-chested Jacamar*, Fiery-throated* and Scarlet-breasted* Fruiteaters, and Grey-tailed Piha.

Specialities

Orange-breasted Falcon*, Grey-breasted Crake, Maroon-tailed Parakeet, Spot-winged Parrotlet*, Rufescent Screech-Owl, Napo Sabrewing*, Wire-crested Thorntail, Ecuadorian Piedtail*, Chestnut-breasted Coronet, Coppery-chested Jacamar*, Lafresnaye's Piculet, Plain Softtail, Orange-fronted Plushcrown, Equatorial Greytail, Crested Foliage-gleaner, Lined Antshrike, Black Bushbird, Foothill, Plain-winged and Yellow-breasted Antwrens, Fiery-throated* and Scarlet-breasted* Fruiteaters, Grey-tailed Piha, Plum-throated Cotinga, Blue-rumped Manakin, Black-and-white Tody-Tyrant, Golden-winged Tody-Flycatcher, Ecuadorian Tyrannulet, Olive-chested Flycatcher, Yellow-cheeked Becard, Black-billed Peppershrike, Olivaceous Greenlet, Wing-banded Wren, Vermilion Tanager, Golden-collared Honeycreeper.

Others

Little Tinamou, Barred Hawk, Barred Forest-Falcon, Military Macaw, White-eyed Parakeet, Blue-winged Parrotlet, Little Cuckoo, Band-bellied Owl, Lyre-tailed Nightjar, White-bearded and Grey-chinned Hermits, Buff-tailed Sicklebill, Violet-headed, Many-spotted and Rufous-tailed Hummingbirds, Greenish Puffleg, Booted Racket-tail, Crested Quetzal, Red-headed Barbet, Cuvier's Toucan, Rufous-breasted Piculet, Little Woodpecker, Dark-breasted and Ash-browed Spinetails, Black-billed Treehaunter, Rufous-tailed Xenops, Stripe-chested and Rufous-winged Antwrens, Long-tailed, Blackish and Black Antbirds, White-backed Fire-eye, Thrush-like Antpitta, Amazonian Umbrellabird, Andean Cock-of-the-Rock, Golden-winged and Green Manakins, Bronze-olive Pygmy-Tyrant, Spectacled Bristle-Tyrant, Ornate and Cliff Flycatchers, Rufous-naped Greentlet, Grey-mantled Wren, White-thighed Swallow, Olivaceous Siskin, Yellow-browed Sparrow, Olive Finch, Black-faced Dacnis, Black-and-white Seedeater, Lesser Seed-Finch, Deep-blue Flower-piercer.

Access

Turn east 43 km north of Tena. Concentrate on the lower Loreto road, using Coca as a base, seeking out trails into the forest away from the roadside, and the upper part of the road, especially kms 13–18 from the Baeza–Tena road, using Tena as a base. Napo Sabrewing* and Fiery-throated Fruiteater* have been recorded along the trail south, 3 km from the Baeza–Tena road.

Accommodation: Coca: Hosteria La Mision (A/B). Tena: Hotel El Auca (C).

CUYABENO RESERVE

Map p. 229

This huge (655,000 ha) reserve, complete with lodges, allows access to some rarely seen species in lowland Amazonian rainforest (igapo, várzea and terra firme), lakes, marshes and *Mauritia* palm swamps, all accessed via a network of trails and boat. Over 400 species have been recorded in just a few years including many species originally thought to be rare, over 50 antbirds and the endemic Orange-crested Manakin. The avifauna is similar to La Selva (see p 000) but species found here and not there include the lovely Black-necked Red-Cotinga. The reserve is not usually accessible from January to March due to flooding.

The stunning Black-necked Red-Cotinga is one of the many special birds at Cuyabeno Reserve

Despite being set up to protect the forest, oil exploitation has led to the development of two oil towns: Cuyabeno and Tarapoa, on the Río Cuyabeno, and threatens the future of the reserve.

Endemics

Orange-crested Manakin.

Specialities

Zigzag Heron*, Nocturnal and Salvin's Curassows, Black-banded Crake, Sapphire-rumped Parrotlet, Rufous-vented Ground-Cuckoo, Oilbird, Fiery Topaz, White-eared and Purplish Jacamars, White-chested Puffbird, Brown Nunlet, Scarlet-crowned and Lemon-throated Barbets, Golden-collared Toucanet, Chestnut-throated Spinetail*, Orange-fronted Plushcrown, Chestnut-winged Hookbill, Rufous-rumped Foliage-gleaner, Undulated and Pearly Antshrikes, Short-billed, Río Suno, Spot-tailed, Dugand's and Ash-winged Antwrens, Black, Slate-coloured, Plumbeous, Sooty, Lunulated, Hairy-crested and Dot-backed Antbirds, Reddish-winged Bare-eye, Striated Antthrush, Ochre-striped Antpitta, Chestnut-belted Gnateater, Rusty-belted Tapaculo, Black-necked Red-Cotinga, Purple-throated Cotinga*, Double-banded Pygmy-Tyrant, Brownish Flycatcher, Citron-bellied Attila, Yellow-throated Flycatcher, Violaceous Jay, Masked Crimson Tanager, Yellow-bellied Dacnis, Casqued Oropendola.

Others

White-throated Tinamou, Agami Heron*, Slate-coloured Hawk, Harpy Eagle*, Blue-throated Piping-Guan, Sungrebe, Grey-winged Trumpeter, Violaceous Quail-Dove, Blue-and-yellow, Scarlet and Red-bellied Macaws, Festive Parrot, Hoatzin, Spectacled Owl, Band-tailed Nighthawk, Pavonine Quetzal, Black-throated and Blue-crowned Trogons, Rufous and Blue-crowned Motmots, Yellow-billed Jacamar, Rusty-breasted Nunlet, Lettered, Ivory-billed and Many-banded Aracaris, Yellow-throated Woodpecker, Barred, Ocellated and Elegant Woodcreepers, Curve-billed Scythebill, Striped Woodhaunter, Short-billed and Black-tailed Leaftossers, Rufous-tailed Xenops, Black-faced, Yellow-browed, Black-chinned, White-plumed and Spot-backed Antbirds, Rufous-capped Antthrush, White-browed Purpletuft, Spangled Cotinga, Wire-tailed Manakin, Yellow-browed Tody-Flycatcher, Slender-footed Tyrannulet, Southern Scrub-Flycatcher, Greyish Mourner, Swainson's, Short-crested and Sulphur-bellied Flycatchers, Dusky-capped and Tawny-crowned Greenlets, Collared Gnatwren, White-lored Euphonia, Paradise, Green-and-gold, Masked and Opal-rumped Tanagers.

Other Wildlife

Southern Tamandua, Brown-throated Sloth, Pygmy Marmoset, Black-mantled Tamarin, Squirrel Monkey, Giant Otter, Amazon River Dolphin, Manatee, Tapir, Jaguar, Peccary, Caiman, Anaconda, Giant Earthworm.

Access

Head for the Río Tarapoa bridge, north of Tarapoa (where it is necessary to pay an entrance fee), on the Tarapoa–Tipischa road. Canoes can be hired here to get to the lodges. New lodges are opening up here all the time. Those called Iripari (terra firme forest) and Imuya (várzea for-

CUYABENO RESERVE

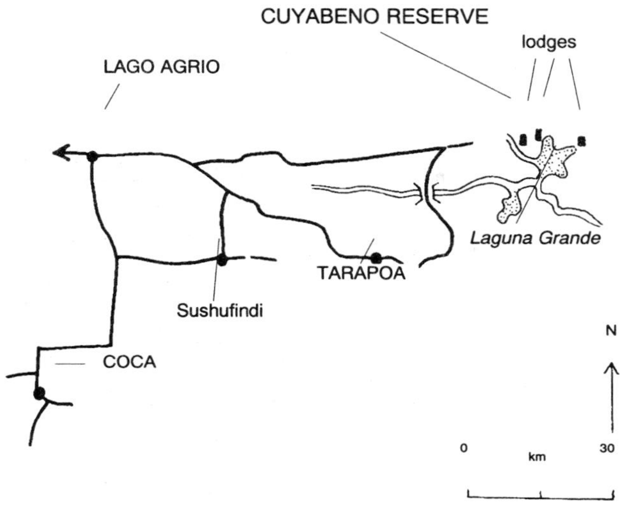

est), both along the Río Aguarico, are recommended. Book well in advance.

Several tour companies based in Quito operate expensive trips to the reserve which include flights to Lago Agrío or Tarapoa and boat trips to the reserve. Try Nuevo Mundo Expeditions, Amazonas 2468, (PO Box 402-A), Quito (tel: 552 617/816; fax: 552916), Neotropic Turis, Calle Robles 653 y Amazonas, Oficina 407, Quito (tel: 551477; fax: 554902), or Etnotur, Luis Cordero 1313 (tel: 230552).

Accommodation: Coca: Hosteria La Mision (A/B).

Primavera, 90 minutes downriver from Coca, supports similar birds to La Selva (see p. 230) including Purplish Jacamar, Golden-collared Toucanet and Orange-fronted Plushcrown.

Accommodation: Hacienda Primavera, run by Xetro, U Paez 229 y 18 de Septiembre, PO Box 653-A, Quito (tel: 541559).

Limoncocha (Map p. 230), 3–4 hours downriver from Coca, and a 2-hour walk from Pompeya, can also be reached by road from Lago Agrío. This is one of only a handful of sites (the only one in Ecuador) where Pale-eyed Blackbird occurs. Otherwise the species, of which over 300 have been recorded, are very similar to La Selva (see p. 230), and include Agami Heron*, Fiery Topaz, White-eared Jacamar, Yellow-billed Nunbird, Dark-breasted Spinetail, Plum-throated Cotinga, Orange-crested Manakin, Bicoloured Conebill, Yellow-bellied Dacnis, Band-tailed Oropendola and Ecuadorian Cacique. Bird the lake by hiring a boat with guide, northeast of the football pitch and the trails north of the disused runway, especially the one at the west end near a tributary of the Río Napo.

LIMONCOCHA

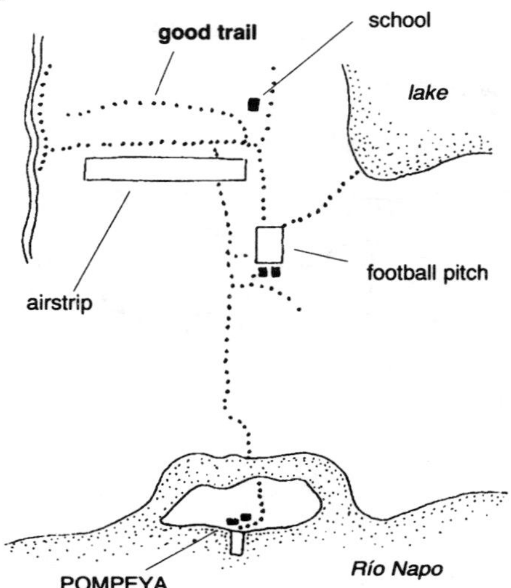

Accommodation: rooms (C) at the store and a small lodge.

LA SELVA

La Selva is, quite simply, one of the world's top birding sites. Only in Tambopata Reserve in Peru (see p. 335) have more birds (587) been recorded in such a small area. Nevertheless, the 530 species so far recorded in the Amazonian rainforest around the rustic La Selva lodge, three hours down river from Coca, include seven tinamous, 19 parrots, seven jacamars, 11 puffbirds, seven toucans, over 50 antbirds, eight cotingas and 34 tanagers. Specialities include Zigzag Heron*, Long-tailed Potoo, the endemic Orange-crested Manakin, and the endemic Cocha Antshrike*, which was rediscovered in 1990. However, many species are thin on the ground in the rainforest and very difficult to see, so do not expect to see more than 250 species here in a week.

Book well in advance and expect to pay a price to match the birding.

Endemics

Cocha Antshrike*, Orange-crested Manakin.

Specialities

Zigzag Heron*, Nocturnal and Salvin's Curassows, Chestnut-headed Crake, Grey-winged Trumpeter, Sapphire Quail-Dove, Maroon-tailed Parakeet, Scarlet-shouldered Parrotlet, Orange-cheeked Parrot, Long-tailed Potoo, White-eared, White-chinned and Purplish Jacamars,

White-chested Puffbird, Lemon-throated Barbet, Golden-collared Toucanet, Lesser Hornero, White-bellied Spinetail, Orange-fronted Plushcrown, Undulated, Castelnau's and Plain-winged Antshrikes, Black Bushbird, Río Suno Antwren, Black-and-white, Plumbeous, White-shouldered, Chestnut-crested and Dot-backed Antbirds, Reddish-winged Bare-eye, Striated Antthrush, Ochre-striped Antpitta, Chestnut-belted Gnateater, Rusty-belted Tapaculo, Purple-throated* and Plum-throated Cotingas, Cinnamon Tyrant-Manakin, Mottle-backed Elaenia, Lesser Wagtail-Tyrant, Lawrence's Thrush, Masked Crimson Tanager, White-lored Euphonia, Opal-crowned Tanager, Yellow-bellied Dacnis, Yellow-shouldered Grosbeak, Velvet-fronted Grackle.

Others

Great, Cinereous and Variegated Tinamous, Agami Heron*, Horned Screamer, Crested* and Harpy* Eagles, Black-and-white* and Ornate Hawk-Eagles, Black Caracara, Speckled Chachalaca, Spix's Guan, Blue-throated Piping-Guan, Marbled Wood-Quail, Azure Gallinule, Sungrebe, Collared Plover, Pied Lapwing, Large-billed Tern, Blue-and-yellow, Scarlet, Red-and-green, Chestnut-fronted and Red-bellied Macaws, Dusky-headed and Cobalt-winged Parakeets, Hoatzin, Tawny-bellied Screech-Owl, Mottled, Black-banded, Crested and Spectacled Owls, Great and Grey Potoos, Short-tailed and Sand-coloured Nighthawks, Ladder-tailed Nightjar, Pale-tailed Barbthroat, Blue-crowned Trogon, Rufous and Blue-crowned Motmots, Yellow-billed and Great Jacamars, Collared Puffbird, Lanceolated Monklet*, Lettered, Ivory-billed and Many-banded Aracaris, Yellow-ridged and Cuvier's Toucans, Scaly-breasted and Cream-coloured Woodpeckers, White-chinned, Long-billed, Cinnamon-throated, Striped, Ocellated and Elegant Woodcreepers, Speckled Spinetail, Point-tailed Palmcreeper, Olive-backed Foliage-gleaner, Spot-winged Antshrike, Pygmy, Plain-throated, Grey and Banded Antwrens, Black-faced, Silvered, White-plumed and Scale-backed Antbirds, White-browed Purpletuft, Screaming Piha, Spangled Cotinga, Bare-necked Fruitcrow, Amazonian Umbrellabird, Blue-crowned, Blue-backed and Striped Manakins, White-eyed Tody-Tyrant, Spotted and Yellow-browed Tody-Flycatchers, River Tyrannulet, Drab Water-Tyrant, Black-capped Donacobius, White-banded Swallow, Orange-headed Tanager, Fulvous Shrike-Tanager, Paradise and Green-and-gold Tanagers, Short-billed Honeycreeper, Yellow-rumped Cacique, Moriche Oriole.

Other Wildlife

Pygmy Marmoset, Squirrel Monkey, Dusky Titi, Saddle-backed Tamarin, Red Howler, Ocelot, Short-eared Dog,

Access

Book through La Selva Ltd, 6 de Diciembre 2816, PO Box 635, Quito (tel: 550995 or 554686; fax: 563814; telex: 2653). They also have an office in the USA, at Tumbaco Inc., PO Box 1036, Punta Gorda, Florida 33951 (tel: 800-247-2925). The high price includes flight to Coca, boat trip, sharp local guides, optional excursions to Panacocha and river islands, excellent food, and bungalows with private bathrooms. There is no hot water or electricity.

Orange-crested Manakin occurs along the boardwalk to the lodge, situated on Garzacocha lake which supports Agami* and Zigzag* Herons,

Cinnamon-throated Woodcreeper, Point-tailed Palmcreeper and Orange-crested Manakin. The 42-m (140-ft) canopy platform, 15 minutes walk from the lodge, is excellent for cotingas during the day, and owls at night, but was closed for safety reasons in late 1993. Azure Gallinule, Sungrebe, Cocha Antshrike* and Dot-backed Antbird occur on and around Mandicocha lake. The quiet Mandicocha trail is good for curassows. The Salado trail, on the other side of the Río Napo, has salt-licks which attract many parrots. Canoe and motor launch excursions along the Río Napo allow access to river-island specialities such as White-eared Jacamar, Lesser Hornero, White-bellied Spinetail, Castlenau's Antshrike, Black-and-white Antbird and Lesser Wagtail-Tyrant; one island is a roosting site for Amazonian Umbrellabird.

Yuturi Jungle Lodge, four to five hours down river from Coca, and a 30-minute walk, is owned by Yuturi Jungle Adventure, Amazonas 1022 y Pinto, Quito (tel and fax: 522133). Bartlett's Tinamou, Salvin's Curassow, White-eared Jacamar and Black-necked Red-Cotinga occur here.

PANACOCHA

Map below

Panacocha, six to seven hours down river from Coca, used to be the poor birder's La Selva, though there do not seem to be many poor birders left now, because few people have been here recently. This may be due to rumours that the once impressive Aguarico trail is now closed. In its day, this trail was described by some South American veterans as one of the best sites on the continent. Black-necked Red-Cotinga is one spectacular species found here but not at La Selva. The nearby lagoon of Panacocha on the Río Panayacu, at least, is still well worth exploring if you cannot afford La Selva (from where day-excursions are possible).

PANACOCHA

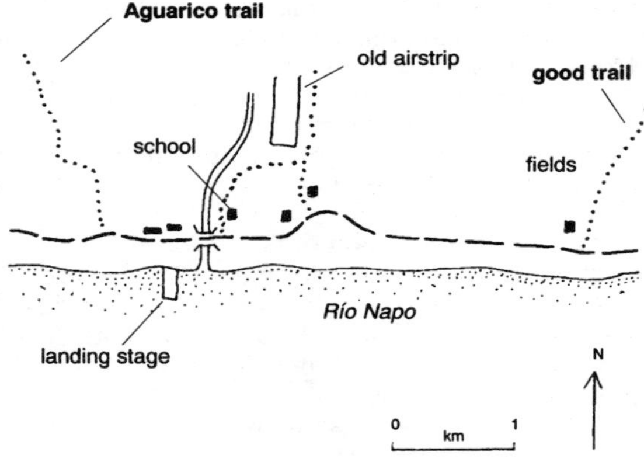

Specialities

Nocturnal and Salvin's Curassows, Grey-winged Trumpeter, Black-bellied Thorntail, White-eared Jacamar, White-chested Puffbird, Brown Nunlet, Lemon-throated Barbet, Golden-collared Toucanet, Dot-backed Antbird, Striated Antthrush, Chestnut-belted Gnateater, Black-necked Red-Cotinga, Plum-throated Cotinga, Masked Crimson and Opal-crowned Tanagers.

Others

White-throated and Undulated Tinamous, Blue-throated Piping Guan, Sungrebe, Festive Parrot, Dark-billed Cuckoo, Hoatzin, Spectacled Owl, Ladder-tailed Nightjar, Spangled Coquette*, Crimson Topaz, Blue-crowned Trogon, Green-and-rufous Kingfisher, Broad-billed and Rufous Motmots, Yellow-billed and Great Jacamars, Lanceolated Monklet*, Screaming Piha, Bare-necked Fruitcrow, Paradise and Green-and-gold Tanagers.

Access

The 17-km long Aguarico trail begins a short distance west of the landing stage, just beyond a large chicken coup. It is very wet, muddy and a favourite haunt of the Anaconda. Black-necked Red-Cotingas used to be seen after about one hour's walking along here. Also bird the area around the village and the drier trail 2 km east of the landing stage, where Lanceolated Monklet*, Dot-backed Antbird and Chestnut-belted Gnateater occur. Panacocha lagoon is north of Panacocha and is reached by boat via the Río Panayacu. Hire the boat and a guide to arrange accommodation. Grey-winged Trumpeter, Crimson Topaz and Plum-throated Cotinga occur here.

Accommodation: Cabanas Panacocha run by Xetro, U Paez 229 y 18 de Septiembre or PO Box 653-A, Quito (tel: 541 599).

Birding the Oriente

Birding the Oriente is a logistical nightmare for those with a small budget. Although there is some sort of passenger service between the major sites, it is very unreliable. Those with little time and money will need to hire their own canoe (expensive) for a return journey to their chosen site. Hiring the canoe one-way is not advisable since getting back will either involve a long wait or being charged an extortionate price.

Boat lovers may wish to arrange a trip on the Flotel Orellana, which travels down the Río Aguarico from Union Libre near Tarapoa, to Cofanes, and stops at a number of riverside trails and camps including Primavera, Limoncocha, Pacuya and Sacha Pacha. Book through Transturi, Orellana 1810 y 10 de Agosto (tel: 544963).

Yasuni NP, Ecuador's biggest NP (668,000 ha), is also in the Oriente. It covers a huge area of virtually inaccessible lowland rainforest south of the Río Napo (La Selva is situated on its northwestern boundary). Despite being designated as a Biosphere Reserve by UNESCO, oil exploitation seems imminent, and in 1990 there was just one lonely ranger. His station is at Puerto Rocafuerte. Some Coca guides may be persuaded to visit this NP, perhaps via the Río Tiputini, which could produce some very exciting birding.

SANTO DOMINGO

This large town, 130 km west of Quito, lies near two excellent birding sites, Tinalandia and Río Palenque Science Centre.

Accommodation: Hotel del Toachi (B); Hotel Zaracay (A+) (Scarlet-thighed Dacnis occurs in the excellent grounds).

TINALANDIA

Over 300 species have been recorded around this expensive rustic lodge and the adjacent golf course surrounded by forest (600 m/1,969 ft) just 15 km east of Santo Domingo and 14 km west of the lowland end of the Chiriboga Road (see p. 218). Specialities here include three forest-floor skulkers: Berlepsch's Tinamou, Rufous-fronted Wood-Quail and Indigo-capped Quail-Dove.

Non-residents must pay a high entrance fee.

Endemics

Pale-mandibled Aracari*.

Specialities

Berlepsch's Tinamou, Rufous-fronted Wood-Quail, Rufous-sided Crake, Pallid Dove, Indigo-capped Quail-Dove, Spectacled Parrotlet, Rose-faced Parrot, White-whiskered Hermit, Velvet-purple Coronet, Purple-bibbed Whitetip, White-eyed Trogon, Orange-fronted Barbet, Guayaquil Woodpecker, Esmeraldas Antbird, Broad-billed Manakin, Pacific Flatbill, Ecuadorian Thrush, Slate-throated Gnatcatcher, Golden-bellied Warbler, Yellow-green Bush-Tanager*, Ochre-breasted Tanager, Orange-crowned Euphonia, Yellow-collared Chlorophonia, Grey-and-gold and Rufous-throated Tanagers, Black-winged Saltator.

Others

Little Tinamou, Barred Hawk, Black-and-white Hawk-Eagle*, Crested Guan, Scarlet-fronted Parakeet, Bronze-winged Parrot, Black-and-white Owl, Band-tailed Barbthroat, White-tipped Sicklebill, Green Thorntail, Crowned Woodnymph, Rufous-tailed Hummingbird, Purple-crowned Fairy, Slaty-tailed, Black-tailed and Black-throated Trogons, Broad-billed and Rufous Motmots, White-whiskered Puffbird, Lanceolated Monklet*, Red-headed Barbet, Crimson-rumped Toucanet, Chestnut-mandibled Toucan, Black-cheeked Woodpecker, Barred and Spotted Woodcreepers, Red-billed Scythebill, Pale-legged Hornero, Slaty and Red-faced Spinetails, Spotted Barbtail, Lineated and Ruddy Foliage-gleaners, Black-tailed Leaftosser, Plain Xenops, Great and Russet Antshrikes, Plain Antvireo, Checker-throated, Slaty and Dot-winged Antwrens, White-backed Fire-eye, Chestnut-backed, Immaculate and Bicoloured Antbirds, Black-headed Antthrush, Green-and-black Fruiteater, Golden-winged Manakin, Ochre-bellied Flycatcher, Sooty-headed Tyrannulet, Marble-faced Bristle-Tyrant, Scale-crested Pygmy-Tyrant, Fulvous-breasted Flatbill, Masked Water-Tyrant, Rufous Mourner, Cinnamon and One-coloured Becards, Slaty-capped Shrike-Vireo, Spotted Nightingale-Thrush, Band-backed Wren, Tawny-faced Gnatwren, Buff-rumped Warbler, Orange-billed Sparrow, Yellow-throated and Ashy-throated Bush-Tanagers, Silver-throated Tanager, Black-faced Dacnis.

Access

Bird around the lodge, to the south of the Santo Domingo–Quito road, the golf course area and the main trail through the forest.

Accommodation: Lodge (A).

A little way east along the main road from the entrance there is a track running south which leads to some forest (a long 10-km trek), where Long-wattled Umbrellabird* has been recorded in the past.

RÍO PALENQUE SCIENCE CENTRE

Over 300 species have been recorded at this privately owned reserve (180 ha), 50 km south of Santo Domingo. It is a tiny oasis of Pacific lowland forest (100 ha) in an agricultural desert, and despite its small size, supports some localised species including Rufous-headed Chachalaca*, Dusky Pigeon and Barred Puffbird. There is an entrance fee for daily visitors.

Endemics

Pale-mandibled Aracari*.

Specialities

Plumbeous Hawk*, Rufous-headed Chachalaca*, Rufous-fronted Wood-Quail, Grey-breasted Crake, Brown Wood-Rail*, Dusky Pigeon, Ecuadorian Ground-Dove, Pacific Parrotlet, Bronzy Hermit, Violet-bellied Hummingbird, White-eyed Trogon, Barred Puffbird, Orange-fronted Barbet, Stripe-billed Aracari, Choco Toucan, Scarlet-backed and Guayaquil Woodpeckers, Slaty-winged Foliage-gleaner, Rufous-crowned Antpitta, Ochraceous Attila*, Scarlet-browed Tanager, Orange-crowned Euphonia, Blue-whiskered Tanager*, Crimson Finch-Tanager.

Others

Great Tinamou, Chestnut-fronted Macaw, Little Cuckoo, Band-tailed Barbthroat, Purple-crowned Fairy, Black-tailed Trogon, Broad-billed and Rufous Motmots, White-whiskered Puffbird, Olivaceous Piculet, Black-cheeked Woodpecker, Red-billed Scythebill, Pale-legged Hornero, Rusty-winged Barbtail, Striped Woodhaunter, Rufous-rumped and Ruddy Foliage-gleaners, Tawny-throated and Scaly-throated Leaftossers, Great and Western Slaty Antshrikes, Dot-winged Antwren, Dusky, Chestnut-backed and Spotted Antbirds, Black-headed Antthrush, Red-capped and White-bearded Manakins, Golden-crowned Spadebill, Lesser Greenlet, Whiskered and Bay Wrens, Southern Nightingale-Wren,, Dusky-faced and Tawny-crested Tanagers, Black-faced and Scarlet-thighed Dacnises, Blue Seedeater, Yellow-tailed Oriole.

(Old records include Black-tipped Cotinga* and Long-wattled Umbrellabird*).

The flashy Purple-crowned Fairy with snow white underparts and outer tail feathers, occurs at Río Palenque

Access

Head south out of Santo Domingo towards Quevedo and turn east 2 km south of the village of Patricia Pilar (45 km south of the Hotel Zaracay). Bird around the science centre and the 3 km of trails.

Accommodation: it is possible to stay at the science centre, which has six rooms for four people, showers, toilets and a kitchen (B). Book through Centro Cientifico Río Palenque, Casilla 95, Santo Domingo de los Colorados, Provincia de Pichincha (tel: 561 646).

Some 150 km southwest of Santo Domingo, *en route* to Machalilla NP, the roadside marshes and pools west of **Chone**, are worth a look. Pinnated Bittern, Masked Water-Tyrant and Dull-coloured Grassquit occur here.

MACHALILLA NP

Map opposite

This NP (60,000 ha) encompasses 40,000 ha of remnant dry and humid forest, as well as two offshore islands on Ecuador's west coast. Over 120 species have been recorded including the endemics Grey-backed Hawk* and Saffron Siskin*, as well as many specialities, not least Little Woodstar*, Elegant Crescent-chest and Grey-and-white Tyrannulet. The offshore islands, known as the 'poor man's (birder's) Galápagos', support some excellent seabirds including Waved Albatross*, which is otherwise endemic to Galápagos, Red-billed Tropicbird and Blue-footed Booby.

MACHALILLA NP

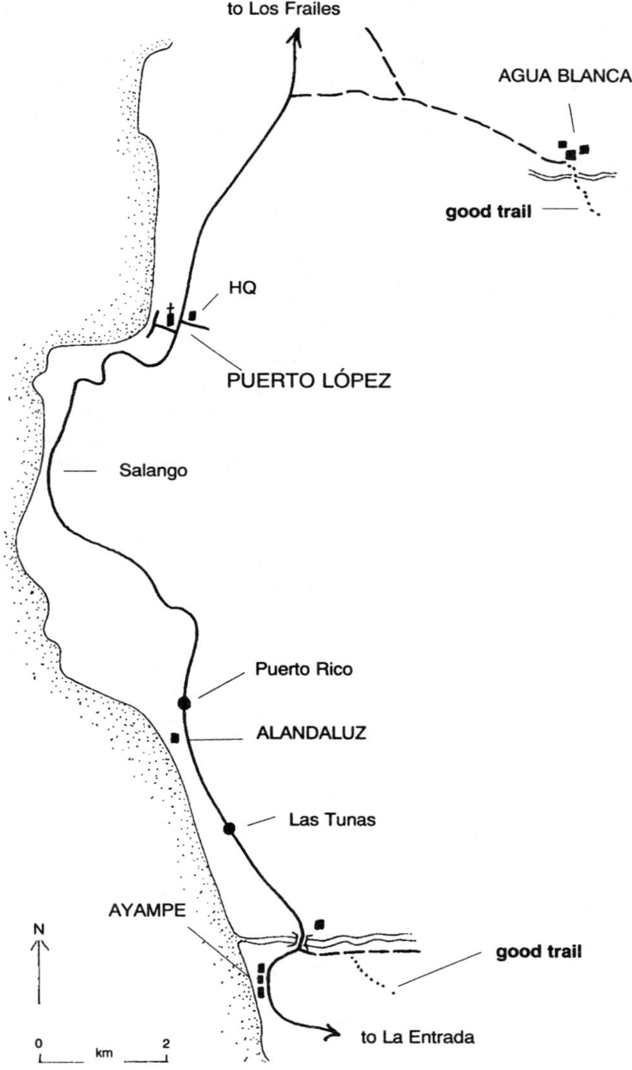

Endemics

Grey-backed Hawk*, Saffron Siskin*.

Specialities

Waved Albatross*, Pink-footed Shearwater*, White-vented Storm-Petrel, Red-billed Tropicbird, Great Frigatebird, Blue-footed, Masked and Red-footed Boobies, Rufous-headed Chachalaca*, Pallid Dove, Red-masked Parakeet*, Pacific Parrotlet, Scrub Nightjar, Little Woodstar*, Ecuadorian Piculet, Black-faced* and Necklaced Spinetails, Collared Antshrike, Watkins' Antpitta, Elegant Crescent-chest, Grey-and-white Tyrannulet, Pacific Elaenia, Grey-breasted* and Sooty-crowned Flycatchers, Ochraceous Attila*, Baird's Flycatcher, Plumbeous-backed Thrush, Fasciated and Superciliated Wrens, Grey-and-gold Warbler, Black-capped Sparrow, Collared Warbling-Finch, White-edged Oriole.

Other Wildlife

Humpback Whale, Southern Fur Seal, Mantled Howler, Ocelot, Tamandua.

Access

The NP HQ is in Puerto López. There are several areas worth birding: the maze of tracks through dry forest and scrub around Los Frailes (turn west 10 km north of Puerto López); the dry forest along the 5-km dirt road to **Agua Blanca** (turn east 6 km north of Puerto López); the river south of the museum there; the 15-km trail from there up to the humid forest at San Sebastian; the excellent humid forest at **Ayampe**, 19 km south of Puerto López, along the track running east from the village along the south side of the Río Ayampe (Grey-backed Hawk* and Saffron Siskin* occur here); and the roadside forest south of Ayampe to La Entrada.

To visit the islands (Isla de la Plata is best) arrange a boat trip at the museum of Fundacion Natura in Puerto López. The island is two hours away and the trip can be rough. There are two trails on Isla de la Plata; one, known as Puncta Machete, passes through a Masked Booby colony and leads on to Waved Albatrosses* and Red-billed Tropicbirds; the other goes to a Red-footed Booby colony. It is not possible to walk both on one visit, but it may be possible to camp overnight to save going twice. Little Woodstar* is also present.

Accommodation: Puerto López: Hotel Pacifico (B). Alandaluz: Ecological Tourist Centre (B/C), 10 km south of Puerto López (good birding).

After decades, the endemic Esmeraldas Woodstar* was rediscovered along the road south from Puerto López to the Santa Elena peninsula, in 1990. It occurs in the low, wet coastal hills along with Grey-backed Hawk*.

SANTA ELENA PENINSULA

West of Guayaquil, woodland gives way to arid scrub and cacti, at the northern tip of the Atacama desert, which supports a number of species restricted to southwest Ecuador and northwest Peru (an area known as Tumbesia). Particularly scarce specialities include Peruvian Pygmy-Owl, Sulphur-throated Finch and Parrot-billed Seedeater, whilst the Equasal salt pans near the beach resorts of La Libertad and Salinas support over 20 shorebird species.

Specialities

Peruvian Thick-knee, Ecuadorian Ground-Dove, Pacific Parrotlet, Grey-cheeked Parakeet*, Peruvian Pygmy-Owl, Scrub Nightjar, Amazilia Hummingbird, Short-tailed Woodstar, Ecuadorian Piculet, Necklaced Spinetail, Collared Antshrike, Elegant Crescent-chest, Grey-and-white Tyrannulet, Short-tailed Field-Tyrant, Baird's Flycatcher, Snowy-throated Kingbird, White-tailed Jay, Long-tailed Mockingbird, Superciliated Wren, Black-capped Sparrow, Crimson Finch-Tanager, Collared Warbling-Finch, Sulphur-throated Finch, Parrot-billed Seedeater, White-edged Oriole.

Others

Magnificent Frigatebird, Blue-footed Booby, White-cheeked Pintail, Chilean Flamingo, Pearl Kite, Wandering Tattler, Collared Plover, Grey and Grey-headed Gulls, Croaking Ground-Dove, Long-billed Starthroat, Tawny-crowned Pygmy-Tyrant, Band-tailed Sierra-Finch, Chestnut-throated Seedeater, Peruvian Meadowlark.

Access

Santa Elena is 130 km west of Guayaquil. Grey Gull occurs at San Pablo, a fishing village 15 km north of Santa Elena. South of this village there is a track inland to the village of **Cerro Alto**. Peruvian Pygmy-Owl and Short-tailed Woodstar occur in and around this village, whilst Peruvian Thick-knee and Short-tailed Field-Tyrant occur in the (usually) dry river bed behind.

The Equasal salt lagoons lie west of Punta Carnero, which is 17.5 km south of Santa Elena. From Punta Carnero head west on the beach road and take the north fork after 3.5 km to a maze of tracks through scrub, where Sulphur-throated Finch occurs, or continue west past the fork to the Equasal salt pans. Wandering Tattler occurs here and on the rocks at Salinas harbour, further west.

There are a few good birding places between here and Guayaquil to the east. Bird the two trails either side of the road to the west of the village of Zapotal, some 40 km east of Santa Elena (90 km west of Guayaquil), for desert scrub species. Pale-browed Tinamou, Peruvian Screech-Owl, Elegant Crecent-chest and White-tailed Jay occur near Progresso, some 60 km east of Santa Elena (70 km west of Guayaquil). Bird the track on the south side of the Santa Elena–Guayaquil road, 20 km west of Progresso, towards the low hills.

Accommodation: Punta Carnero Inn (A); Salinas: Miramar (A). La Libertad is cheaper (Hosteria Samarina (C)), and a better base for those without their own transport.

CERRO BLANCO (CHONGON HILLS)

A reserve has been established here, on the hills near Guayaquil, by Fundacion Natura, to protect the remaining dry, scrubby deciduous woodland which used to cover much of southwest Ecuador. Many 'Tumbesian' endemics occur here as well as other specialities, including Great Green Macaw. A permit is needed from the cement company who own this reserve.

Endemics

Grey-backed Hawk*, Saffron Siskin*.

Specialities

Pale-browed Tinamou*, Plumbeous Hawk*, Rufous-headed Chachalaca*, Brown Wood-Rail*, Ecuadorian Ground-Dove, Great Green Macaw, Red-masked Parakeet*, Pacific Parrotlet, Grey-cheeked Parakeet*, Scrub Nightjar, Amazilia Hummingbird, Short-tailed Woodstar, Ecuadorian Piculet, Scarlet-backed and Guayaquil Woodpeckers, Black-faced Spinetail*, Henna-hooded Foliage-gleaner*, Collared Antshrike, Elegant Crescent-chest, Pacific Elaenia, Sooty-crowned Flycatcher, Snowy-throated Kingbird, White-tailed Jay, Plumbeous-backed and Ecuadorian Thrushes, Long-tailed Mockingbird, Fasciated, Speckle-breasted, Stripe-throated and Superciliated Wrens, Grey-and-gold Warbler, Black-capped Sparrow, Crimson Finch-Tanager, White-edged Oriole, Scrub Blackbird.

Others

King Vulture, Pale-vented Pigeon, Croaking and Blue Ground-Doves, Bronze-winged and Red-lored Parrots, Short-tailed Swift, Streak-headed Woodcreeper, Pale-legged Hornero, Great and Barred Antshrikes, Yellow-bellied Elaenia, Tawny-crowned Pygmy-Tyrant, Yellow-olive and Bran-coloured Flycatchers, One-coloured Becard, Tropical Gnatcatcher, Masked Yellowthroat, Yellow-rumped Cacique, Yellow-tailed Oriole, Peruvian Meadowlark, Giant Cowbird.

Access

The reserve is on the north side of the Guayaquil–Santa Elena road, 4 km west of the major Puerto Maritimo/Daule junction just west of Guayaquil, altogether approximately 15 km west of Guayaquil. For more information write to Ing. Eduardo Aspiazu, Presidente, Fundacion Natura, Capitulo Guayaquil, Dolores Sucre 401 y Rosenda Aviles, Guayaquil, and for permission to visit write to: The Director, Cerro Blanco Reserve, La Cemento Nacional, Apartado Postal 09-01-04243.

In the event of problems it may be worth birding a nearby site. 5 km west of the reserve entrance, along the main Guayaquil–Santa Elena road, turn north to Insecadi and bird the fields and shrubs *en route*. After 7 km take the track to the right which allows access to some woodland.

Manta Real Reserve, west of La Merced on the Guyaquil–Cuenco road between La Troncal and Canar, supports El Oro Parakeet* and Long-Wattled Umbrellabird*.

CUENCA

Map p. 242

Colonial Cuenca (2,595 m/8,514 ft) is, arguably, one of the most beautiful cities in South America, and hence a good base from which to explore the remnant *Polylepis* woodland and paramo in Las Cajas NRA. It is also near the superb east Andean slope subtropical and temperate forests east of Gualaceo.

LAS CAJAS NATIONAL RECREATION AREA

Map p. 242

Much of this 28,808 ha NRA is lake-dotted, bleak paramo, although sheltered hollows harbour temperate forest and *Polylepis* woodland. This is the best site in Ecuador to look for the endemic Violet-throated Metaltail* and Tit-like Dacnis*, here at the northern edge of its range.

Endemics

Violet-throated Metaltail*.

Specialities

Carunculated Caracara, Rufous-bellied Seedsnipe, Chimborazo Hillstar, Rainbow Starfrontlet, Purple-throated Sunangel, Rainbow-bearded Thornbill, Stout-billed Cinclodes, Mouse-coloured Thistletail, Many-striped Canastero, Tawny Antpitta, Turquoise Jay, Giant Conebill*, Tit-like Dacnis*.

Others

Speckled Teal, Yellow-billed Pintail, Andean Condor, Black-chested Buzzard-Eagle, Puna Hawk, Andean Lapwing, Andean Pygmy-Owl, Sparkling Violet-ear, Shining Sunbeam, Mountain Velvetbreast, Great Sapphirewing, Sword-billed Hummingbird, Sapphire-vented Puffleg, Green-tailed Trainbearer, Tyrian Metaltail, Blue-mantled Thornbill, Masked Trogon, Grey-breasted Mountain-Toucan*, Andean Tit-Spinetail, Rufous Antpitta, Unicoloured Tapaculo, Red-crested Cotinga, Tawny-rumped and White-banded Tyrannulets, Tufted Tit-Tyrant, Rufous-breasted and Brown-backed Chat-Tyrants, Red-rumped and Smoky Bush-Tyrants, Black-billed Shrike-Tyrant, Spot-billed and Plain-capped Ground-Tyrants, Rufous-naped Brush-Finch, Blue-backed Conebill, Scarlet-bellied and Buff-breasted Mountain-Tanagers, Black Flower-piercer.

Other Wildlife

Spectacled Bear, Puma.

Access

To reach the Toreadora entrance, in the northern end of the NRA, head west out of Cuenca and continue on the Avenida Ordonez Lazo to the NRA entrance. From here it is a few km to a track south which leads to Laguna Llaviuco, where Grey-breasted Mountain-Toucan* (north side) and Sword-billed Hummingbird (southwest corner) occur. 20 km beyond the entrance is the Laguna Toreadora ranger station (3,870 m/12,697 ft) on the right. Violet-throated Metaltail* occurs in a *Polylepis* grove on the south side of the track a few km below the ranger station. Giant Conebill* and Tit-like Dacnis* occur in the *Gynoxys* thickets (grey-green shrubs) and *Polylepis* woodland to the left of the lake below

LAS CAJAS NRA

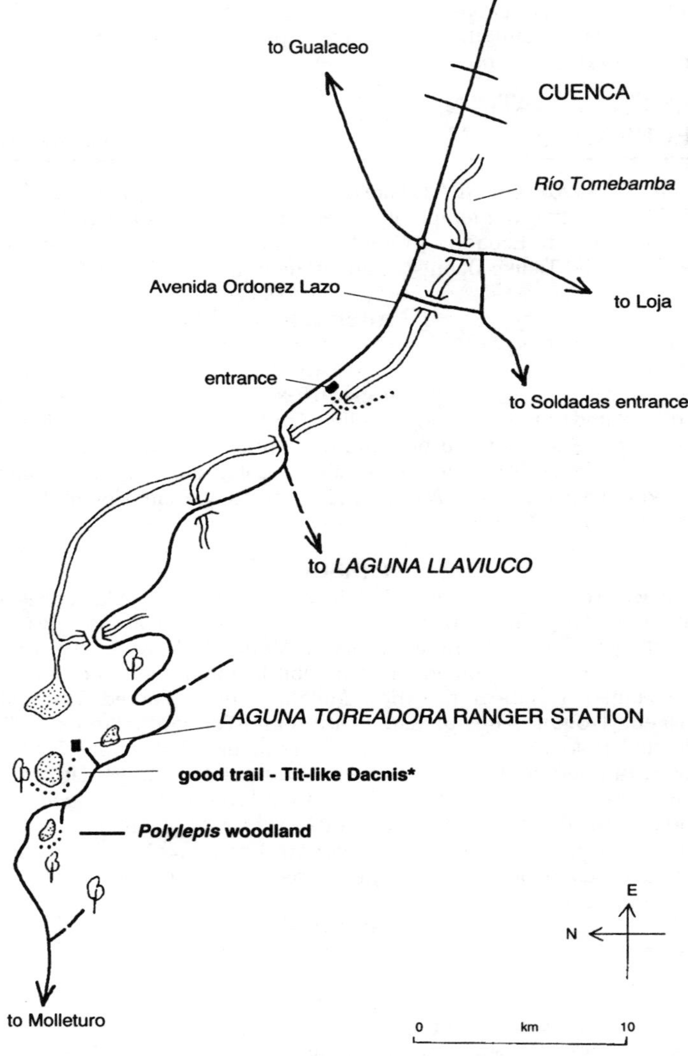

the ranger station. There is another large area of *Polylepis* worth exploring along a track to the left beyond Laguna Toreadora.

The south end of the park can be reached via the villages of Soldados (34 km from Cuenca) and Angas (56 km).

Accommodation: there is a refuge with four bunks and cooking facilities, bookable through the Ministry of Agriculture above the Banco Azuay, Bolivar 6-22 y Miguel (3rd floor), Cuenca. Camping is allowed.

A research station run by the **Río Mazan Project** also lies to the west of Cuenca. A letter of recommendation is required from the 'Amigos de Mazan' office at 38-40 Exchange Street, Norwich, NR2 1AX, England, U.K., or in Quito, at 7-84 Pres. Borrero Calle. Food and accommodation are basic, and birding tricky on poor trails, but Bearded Guan*, Golden-plumed Parakeet*, Violet-throated Metaltail* and Turquoise Jay occur here.

GUALACEO–MACAS ROAD

The superb elfin, temperate and subtropical forests on the rarely accessible east Andean slope, alongside the road between Gualaceo and Macas, is one of only a handful of sites where three rare and spectacular 'tanagers' have been seen. Birders in search of White-capped Tanager, Masked Mountain-Tanager* and Masked Saltator* must visit this road and/or the Loja area (see p.245).

It may be damp and chilly, especially in the early morning, but any one of these three birds will soon warm you up. If none of them appear, there is always the possibility of good compensation in the form of the scarce Black-billed Mountain-Toucan* or a tanager flock which may contain Grass-green Tanager, four 'mountain-tanagers', Yellow-throated and Golden-crowned Tanagers, and Black-backed Bush-Tanager. Even the hummers, for which this road is also famous, would struggle to surpass such a splash of colours.

Specialities

Carunculated Caracara, Golden-plumed Parakeet*, Buff-winged Starfrontlet, Tourmaline Sunangel, Viridian Metaltail, Rainbow-bearded Thornbill, Mountain Avocetbill, Black-billed Mountain-Toucan*, Mouse-coloured Thistletail, Equatorial Greytail*, Ash-coloured and Ocellated Tapaculos, Agile Tit-Tyrant, Orange-banded Flycatcher, Black-billed Peppershrike, Chestnut-breasted Wren, White-capped Tanager, Red-hooded Tanager, Masked Mountain-Tanager*, Yellow-throated Tanager, Black-backed Bush-Tanager, Masked Saltator*.

Others

Black-chested Buzzard-Eagle, Barred Parakeet, Andean Pygmy-Owl, White-collared Swift, Shining Sunbeam, Great Sapphirewing, Amethyst-throated Sunangel, Glowing Puffleg, Black-tailed Trainbearer, Grey-breasted Mountain-Toucan*, Crimson-mantled Woodpecker, White-browed Spinetail, Pearled Treerunner, Rufous Antpitta, Unicoloured Tapaculo, Red-crested Cotinga, Green-and-black and Barred Fruiteaters, Streak-necked Flycatcher, Bronze-olive Pygmy-Tyrant, Black-throated Tody-Tyrant, Rufous-crowned Tody-Tyrant, Highland Elaenia, Sulphur-bellied and White-banded Tyrannulets, Crowned and Slaty-backed Chat-Tyrants, Red-rumped, Streak-throated and Smoky Bush-Tyrants, Black-billed Shrike-Tyrant, Barred Becard, White-capped Dipper, Spectacled Redstart, Citrine and Black-crested Warblers, Pale-naped Brush-Finch, Blue-backed and Capped Conebills, Grass-green

The Gulaceo–Macas road is one of the best sites in the Andes to look for the beautiful but rare Black-billed Mountain-Toucan. It is one of only a few birds with sky-blue underparts*

Tanager, Black-headed Hemispingus, Hooded, Black-chested and Scarlet-bellied Mountain-Tanagers, Golden-crowned Tanager, Glossy, Black, Blue and Masked Flower-piercers, Yellow-billed Cacique.

Access

Head east out of Gualaceo (31 km east of Cuenca), and turn right 1.5 km west of the Chordeleg turn-off. From here it is 14 km to a good trail on the south side of the road, a further 9 km to the pass (3,350m/11,000 ft), and a further 17.5 km to a good river valley. Birding is also good for the next 8 km or so.

Masked Mountain-Tanager* and Black-backed Bush-Tanager occur in the high elfin forest zone, whilst the noisy nomad, White-capped Tanager, occurs in temperate forest lower down. Here the species are similar to those which occur on the Loja–Zamora road (see p. 245).

Since Gualaceo is 31 km east of Cuenca and the best birding some 20 km on from there, it is better to use pleasant Gualaceo as the base for this site.

Accommodation: Parador Turistico Gualaceo (B) (good birding) near town, Hosteria Uzhupud, northwest of town (good birding), and Gran Hostel (C) in town.

The arid valleys south of the village of **Ona**, (105 km south of Cuenca /122 km north of Loja), on the Cuenca–Loja road, are worth checking if travelling between these two places. Carunculated Caracara, Purple-throated Sunangel, Purple-collared Woodstar and Tawny Antpitta occur in these valleys, whilst Giant Hummingbird, White-browed Chat-Tyrant and the rare Drab Seedeater have been recorded along the Cuenca–Loja road.

The very rare, possibly extinct endemic, Pale-headed Brush-Finch*, is known only from around Ona and Giron, which is 43 km southwest of Cuenca on the road to Machala. This species may still inhabit scrubby areas in the arid Río Jubones valley.

Remnant temperate forest at **Acanama**, just north of Loja, supports Tawny-breasted Tinamou, Red-faced Parrot*, Buff-winged Starfrontlet, Crescent-faced Antpitta* and Orange-banded Flycatcher*. Just south of San Lucas, north of Loja, turn east. 7 km northeast of San Lucas take the track to the right and bird from 1 km onwards.

LOJA

Map p. 246

Loja is a big town in south Ecuador, situated at 2,100 m (6,890 ft), and a good base from which to bird two superb sites; the east Andean slope temperate and subtropical forests alongside the Loja–Zamora road, to the east, and the northern end of Podocarpus NP, to the south.

Accommodation: Vilcabamba International (B).

LOJA–ZAMORA ROAD

Map p. 246

The rough road, which is occasionally subject to landslides, east from Loja rises to a pass (2,850 m/9,350 ft) and then descends through relatively undisturbed temperate and subtropical forest where some 350 species have been recorded. Several bamboo specialists such as Chestnut-naped Antpitta occur near the pass, but the subtropical forest provides the very best in birding with big tanager flocks containing up to 15 species. This is also a good site for White-capped Tanager, and the best site in Ecuador to look for the rare endemic White-necked Parakeet*, which was only known from four specimens until it was rediscovered here in 1981.

Endemics

White-necked Parakeet*.

Specialities

Bearded Guan*, Buff-winged and Rainbow Starfrontlets, Tourmaline Sunangel, Rainbow-bearded Thornbill, Mouse-coloured Thistletail, Spectacled Prickletail, Chestnut-naped Antpitta, Plumbeous-crowned and Rufous-winged Tyrannulets, Turquoise Jay, Fasciated and Plain-tailed Wrens, White-capped, Red-hooded and Purplish-mantled Tanagers.

Others

Torrent Duck, Semicollared Hawk*, Solitary Eagle*, Sickle-winged Guan, Chestnut-collared Swift, Green Hermit, Green Violet-ear, Speckled Hummingbird, Bronzy Inca, Amethyst-throated Sunangel,

Glowing Puffleg, Booted Racket-tail, Rufous-capped Thornbill, Golden-headed Quetzal, Emerald Toucanet, Crimson-mantled Woodpecker, Olive-backed and Spot-crowned Woodcreepers, Ash-browed Spinetail, Rufous-rumped Antwren, Blackish Antbird, Rufous-vented Tapaculo, Andean Cock-of-the-Rock, Streak-necked and Rufous-breasted Flycatchers, Rufous-crowned Tody-Tyrant, Ashy-headed and Sulphur-bellied Tyrannulets, Fulvous-breasted Flatbill, Rufous-breasted Chat-Tyrant, Rufous Wren, Yellow-browed Sparrow, Short-billed Bush-Tanager, Black-headed Hemispingus, Rufous-crested, Yellow-throated, Golden-crowned, Orange-eared, Saffron-crowned, Flame-faced, Golden-naped, Metallic-green and Blue-necked Tanagers, Plush-capped Finch, Chestnut-bellied Seedeater, Lesser Seedfinch, Rusty and Glossy Flower-piercers, Russet-backed Oropendola, Mountain Cacique.

LOJA–ZAMORA ROAD

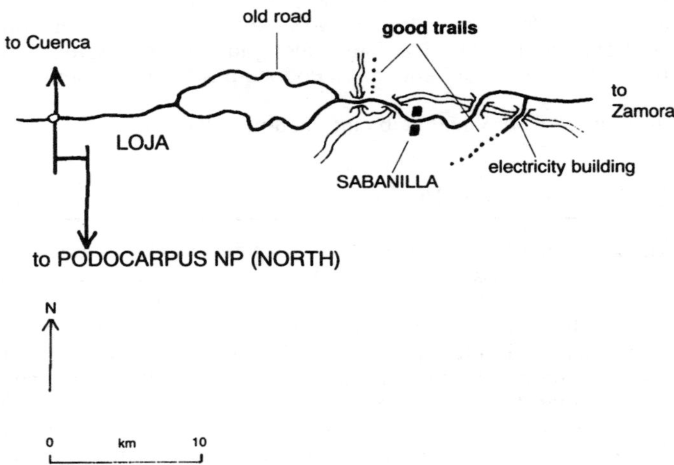

Access

From the pass 15 km east of Loja, the road descends via endless hairpin bends into good forest, best between 28 and 40 km from Loja. There is a good trail north of the road a few km beyond km 28, before the village of Sabanilla, and another south of the road beyond Sabanilla, which starts at an 'electricty building'; take the steps between two pipelines to get on to this trail (access not always allowed).

This road continues to Zamora (60 km east of Loja), the gateway to **Podocarpus NP (South)** (see p. 248 for the northern end) via the **Bombuscaro** entrance (Map p. 247), which is signposted from Zamora. Head south out of Zamora towards Romerillos and turn south on to a rough track, before the bridge over the Río Bombuscaro. This track, well worth birding, runs south to within 1 km of the park HQ. Bird the final km on foot and the trail beyond the HQ. White-necked Parakeet*, Ecuadorian Piedtail*, Coppery-chested Jacamar*, Lanceolated

PODOCARPUS NP (SOUTH): BOMBUSCARO ENTRANCE

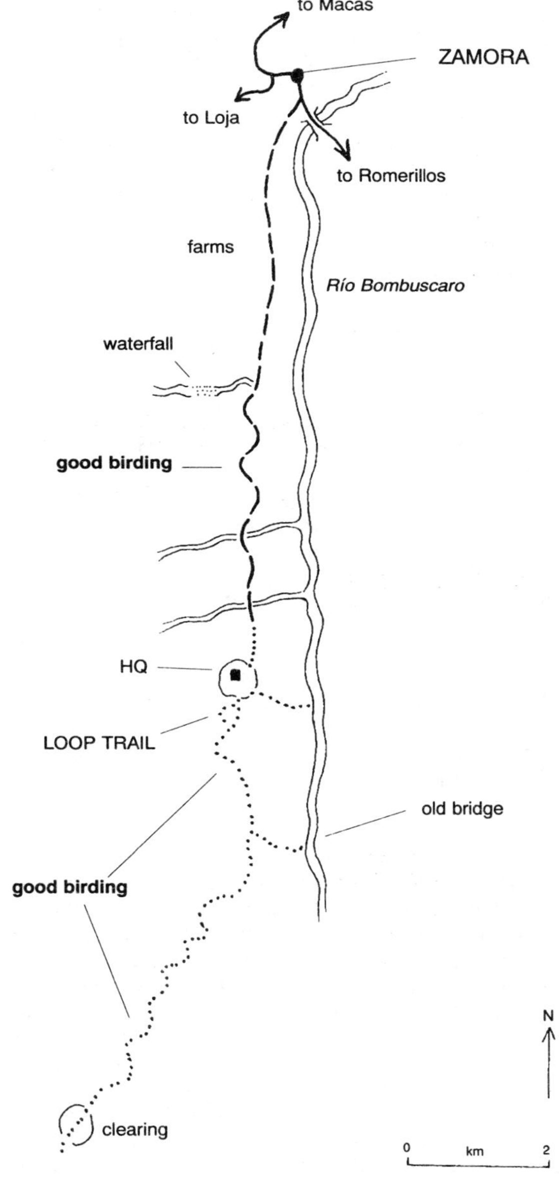

Monklet*, Crimson-bellied Woodpecker, Ash-browed Spinetail, Yellow-breasted Antwren, Shrike-like Cotinga*, Amazonian Umbrellabird, Andean Cock-of-the-Rock, Olivaceous Greenlet, and Paradise, Green-and-gold and Metallic-green Tanagers occur here. Torrent Duck and Sunbittern occur on the Río Bombuscaro, which can be reached by a number of side-trails.

The NP Information Office is on the right-hand side of the road just north of Zamora. Birding is also good at the Romerillos entrance 25 km south of Zamora.

Accommodation: Zamora: Hotel Maguna (C), Hotel Zamora (C).

Black-throated Brilliant and Napo Sabrewing have been recorded around the village of Guayzimi, two hours east of Zumbi, 45 km north-east of Zamora. Orange-throated Tanager* has been recorded near Shaime, 43 km southeast of Zamora

PODOCARPUS NP (NORTH) **Map p. 246 and opposite**

This huge (146,280 ha) NP is one of the richest in the world for birds. Some 540 species have been recorded so far, but many areas have yet to be fully explored, and the potential list could be as high as 800, which would make it second only to Manu NP in Peru (see p. 332).

Naturally, species diversity is directly related to habitat diversity, and Podocarpus NP was established to protect some of the richest habitats in the world: east Andean slope elfin, temperate and subtropical forests, as well as upper tropical forest with an Amazonian influence. Despite the importance and 'NP status' of these habitats they remain threatened by gold-mining, logging and hunting.

Although the species here are similar to those found along the Gualaceo–Macas road (see p. 243), and the specialities include Golden-plumed Parakeet*, Masked Mountain-Tanager* and Masked Saltator*, such rarities as Bearded Guan* and Imperial Snipe* also occur here. This can be a marvellous place to look for birds, amidst spectacular mountain scenery, but be prepared for grim weather since it is often misty, cloudy, damp, and downright wet.

Specialities

White-throated Hawk, Bearded Guan*, Imperial Snipe*, Golden-plumed Parakeet*, Oilbird, Buff-winged and Rainbow Starfrontlets, Chestnut-breasted Coronet, Tourmaline and Purple-throated Sunangels, Chestnut-naped Antpitta, Ash-coloured Tapaculo, Orange-banded Flycatcher*, Turquoise Jay, Plain-tailed Wren, Red-hooded Tanager, Masked Mountain-Tanager*, Yellow-throated and Yellow-scarfed Tanagers, Masked Saltator*.

Others

Brown Tinamou, Torrent Duck, Black-and-chestnut Eagle*, Rufous-breasted Wood-Quail, Blackish Rail, Sunbittern, Andean Pygmy-Owl, Buff-fronted Owl*, Chestnut-collared Swift, Shining Sunbeam, Mountain Velvetbreast, Collared Inca, Sword-billed Hummingbird, Glowing Puffleg, Crested and Golden-headed Quetzals, Masked Trogon, Blue-crowned Motmot, Crimson-rumped Toucanet, Grey-breasted Mountain-

PODOCARPUS NP (NORTH)

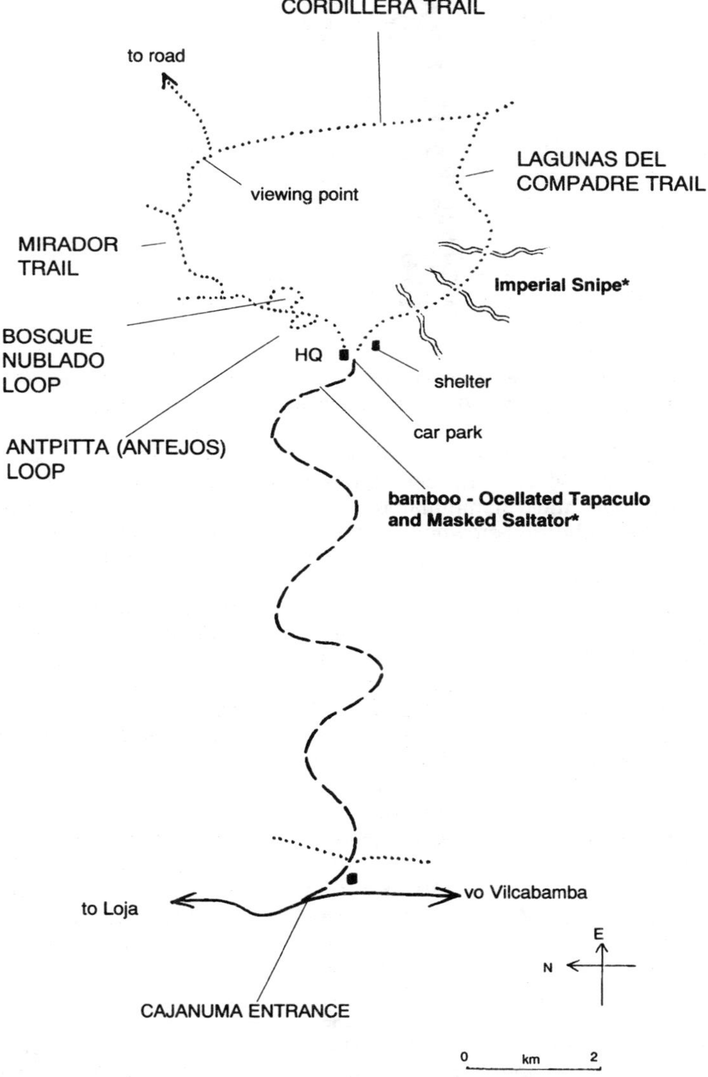

Look out for the striking orange, black and yellow bill of the Grey-breasted Mountain-Toucan in Podocarpus NP (North)*

Toucan*, Bar-bellied, Crimson-mantled and Powerful Woodpeckers, Tyrannine Woodcreeper, Pearled Treerunner, Streaked Tuftedcheek, Black-billed and Flammulated Treehunters, Barred Antthrush, Undulated, Chestnut-crowned, Rufous and Slate-crowned Antpittas, Unicoloured and Ocellated Tapaculos, Red-crested Cotinga, Barred Fruiteater, Amazonian Umbrellabird, Andean Cock-of-the-Rock, Rufous-headed Pygmy-Tyrant, Rufous-breasted Chat-Tyrant, Streak-throated Bush-Tyrant, Plain-capped Ground-Tyrant, Dusky-capped Flycatcher, Barred Becard, Slaty-backed Nightingale-Thrush, Rufous Wren, Pale-footed Swallow, Pale-naped, Rufous-naped and Stripe-headed Brush-Finches, Blue-backed Conebill, Grass-green Tanager, Grey-hooded Bush-Tanager, Black-capped and Black-headed Hemispinguses, Rufous-crested Tanager, Hooded, Black-chested and Scarlet-bellied Mountain-Tanagers, Golden-crowned Tanager, Chestnut-breasted Chlorophonia, Plush-capped Finch, Glossy Flower-piercer, Russet-backed Oropendola, Mountain Cacique.

Other Wildlife

Spectacled Bear, Mountain Tapir, Jaguar, Puma, Andean Fox, Giant Armadillo.

Access

The northern end of Podocarpus NP is accessible via the Cajanuma entrance. To reach this, head south out of Loja on the road to Vilcabamba and turn east in response to the park sign after 11 km. The 8-km track to the ranger station is worth prolonged birding. Most of the species listed above occur along here including Bearded Guan* and Masked Mountain-Tanager*, although this species is usually recorded much higher up. The trackside bamboo shortly before the HQ supports Ocellated Tapaculo and

Masked Saltator*. From the HQ bird:

(i) the Lagunas del Compadre trail which starts at the round shelter; Imperial Snipe*, first discovered in Ecuador at this site in 1990, occurs along the first 2 km of this trail, as well as Orange-banded Flycatcher*;

(ii) the very short Antpitta (Antejos) loop which starts just past the HQ (signposted);

(iii) the Bosque Nublado loop (signposted) further on from the Antpitta loop; Golden-plumed Parakeet* and Ocellated Tapaculo occurs along here;

(iv) the long Mirador trail, which the two loops above lead off; a cotinga, thought to be Bay-vented or a new species, was recorded here in 1989, at and beyond the viewpoint, on the Cordillera trail to the right and further along the Mirador trail. Imperial Snipe* has also been recorded along here.

Accommodation: Cabins.

Chestnut-crested Cotinga has been recorded at Quebrada Honda at the NP's extreme southwest boundary.

Maranon Thrush has been recorded around Chito, southeast of Zumba, 177 km south of Loja, near the disputed border with Peru. However, a large military presence in this area may present problems for the adventurous birder.

BUENAVENTURA (PIÑAS) **Map p. 252**

Buenaventura, on the west Andean slope at 914 m (3,000 ft), is the name given to an area where remnant Pacific slope upper tropical forest still exists. Over 250 species have been recorded at this small site including a curious mixture of 'Choco' and 'Tumbesian' endemics, and the highly localised endemic El Oro Parakeet*, which was described in 1985.

Endemics

Grey-backed Hawk*, El Oro Parakeet*, Pale-mandibled Aracari*.

Specialities

Rufous-headed Chachalaca*, Rufous-fronted Wood-Quail, Ochre-bellied Dove*, Red-masked Parakeet*, Blue-fronted Parrotlet, Rose-faced Parrot, White-whiskered and Tawny-bellied Hermits, Brown Inca, Gorgeted Sunangel, Purple-bibbed Whitetip, Violet-tailed Sylph, Little Woodstar*, Barred Puffbird, Choco Toucan, Guayaquil Woodpecker, Buffy Tuftedcheek, Uniform Treehunter, Chapman's and Uniform Antshrikes, Esmeraldas Antbird, Long-wattled Umbrellabird*, Club-winged Manakin, Rufous-winged Tyrannulet, Ochraceous Attila*, Slaty Becard*, Ecuadorian Thrush, Olive-crowned Yellowthroat, Grey-and-Gold and Three-banded Warblers, Tricoloured Brush-Finch, Ochre-breasted Tanager, Black-chinned Mountain-Tanager, Orange-crowned Euphonia, Rufous-throated Tanager, Black-winged Saltator.

BUENAVENTURA

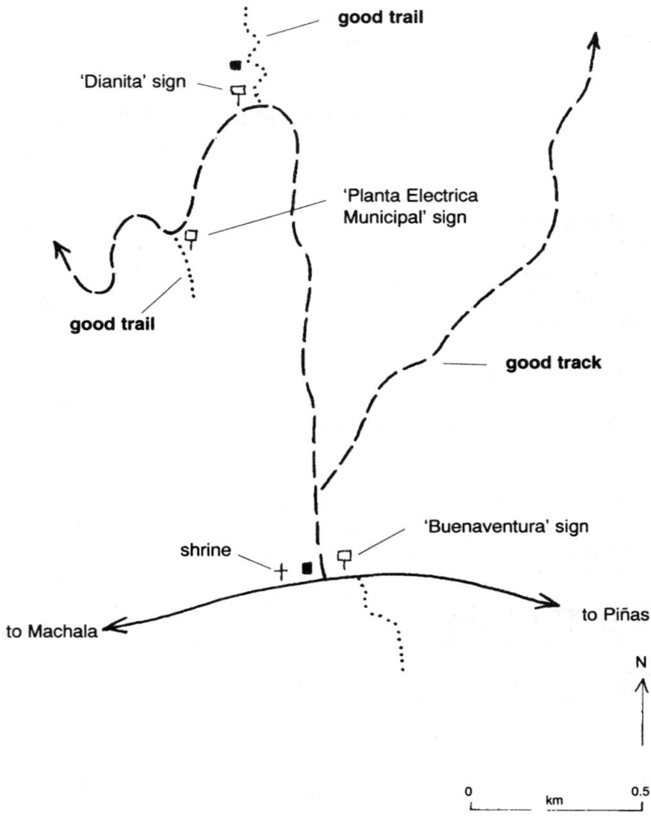

Others

Solitary Eagle*, Ornate Hawk-Eagle, Sickle-winged Guan, Plumbeous Pigeon, White-throated Quail-Dove, Bronze-winged Parrot, Black-and-white Owl, Little Nightjar, Lesser Swallow-tailed Swift, White-tipped Sicklebill, Crowned Woodnymph, Green-crowned Brilliant, Booted Racket-tail, Wedge-billed Hummingbird, Violaceous Trogon, Golden-headed Quetzal, Broad-billed and Rufous Motmots, Crimson-rumped Toucanet, Chestnut-mandibled Toucan, Olivaceous Piculet, Golden-olive Woodpecker, Plain-brown and Spotted Woodcreepers, Brown-billed Scythebill, Slaty Spinetail, Lineated and Scaly-throated Foliage-gleaners, Plain Xenops, Great and Russet Antshrikes, Rufous-rumped Antwren, Chestnut-backed and Immaculate Antbirds, Rufous-breasted Antthrush, Scaled and Ochre-breasted Antpittas, Scaled Fruiteater*, Golden-winged Manakin, Slaty-capped Flycatcher, Bronze-olive Pygmy-

Tyrant, Variegated Bristle-Tyrant, Scale-crested Pygmy-Tyrant, Ornate, Bran-coloured and Tawny-breasted Flycatchers, Black-and-white and One-coloured Becards, Lesser Greenlet, Whiskered, Bay and Song Wrens, Tropical Gnatcatcher, White-thighed Swallow, Yellow-bellied Siskin, Olive-crowned Yellowthroat, Buff-rumped Warbler, Orange-billed and Black-striped Sparrows, Silver-throated and Flame-faced Tanagers, Yellow-bellied Seedeater, Dull-coloured Grassquit, Subtropical Cacique.

Other Wildlife

Tayra.

Access

Head for the religious shrine at the pass, 9 km west of Piñas (15 km west of Zaruma via Portovelo, and 65 km east of Santa Rosa). Just east of the shrine, at the 'Buenaventura' sign, turn north on to a dirt road then bird the right and left forks after a few hundred metres. Either is good for El Oro Parakeet*. The muddy trail south of the main Santa Rosa–Piñas road, opposite the track, is also worth a look.

Accommodation: Piñas: Residencial Dumari (C). Zaruma: Finca Machay country resort (A) in Los Rosales de Machay, 12 km west of Zaruma; Hotel Municipal (C).

The arid valleys, woodland and cacti-clad slopes alongside the road from Zaruma to Catacocha, and southward to Celica, are worth stopping at. The specialities occurring here include Cloud-forest Screech-Owl, Henna-hooded Foliage-gleaner*, Elegant Crescent-chest, Plumbeous-backed Thrush, Black-and-white Tanager*, and Drab Seedeater.

Accommodation: Catacocha: Pension Guayaquil (C).

CELICA

This village in far south Ecuador near the Peruvian border lies next to virtually untouched west Andean slope subtropical and deciduous forests, where over 25 'Tumbesian' endemics occur, including Rufous-necked* and Henna-hooded* Foliage-gleaners and Grey-headed Antbird*, all of which are hard to find elsewhere. Some other specialities are at the northern edge of their mainly Peruvian ranges, and these include Black-crested Tit-Tyrant and the rare Black-and-white Tanager*.

Although this area requires a certain amount of pioneer spirit, it could reward the adventurous birder with some very special birds.

Endemics

Grey-backed Hawk*.

Specialities

Pale-browed Tinamou*, Rufous-headed Chachalaca*, Ochre-bellied Dove*, Red-masked Parakeet*, Pacific Parrotlet, Grey-cheeked Parakeet*, Cloud-forest Screech-Owl, Rainbow Starfrontlet, Purple-throated Sunangel, Ecuadorian Piculet, Black-faced* and Line-cheeked Spinetails, Rufous-necked* and Henna-hooded* Foliage-gleaners, Collared and Chapman's Antshrikes, Grey-headed Antbird*, Watkins' Antpitta, Elegant Crescent-chest, Rufous-winged Tyrannulet, Black-crested Tit-Tyrant, Jelski's Chat-Tyrant, Ochraceous Attila*, Sooty-crowned and Baird's Flycatchers, White-tailed Jay, Andean Slaty-Thrush, Plumbeous-backed and Ecuadorian Thrushes, Grey-and-gold and Three-banded Warblers, Tumbes and Black-capped Sparrows, Bay-crowned and White-winged Brush-Finches, Black-and-white* and Silver-backed Tanagers, Black-cowled Saltator.

Others

Black-and-chestnut Eagle*, Crested Guan, White-throated Quail-Dove, Striped Cuckoo, Green Hermit, Speckled Hummingbird, Golden-headed Quetzal, Black-tailed Trogon, Blue-crowned Motmot, Crimson-rumped Toucanet, Powerful Woodpecker, Tyrannine Woodcreeper, Pearled Treerunner, Streaked Tuftedcheek, Undulated, Scaled and Chestnut-crowned Antpittas, Rufous-vented Tapaculo, Yellow-bellied Elaenia, Tawny-crowned Pygmy-Tyrant, Piratic Flycatcher, Lesser Greenlet, Slaty-backed Nightingale-Thrush, Chiguanco Thrush, Golden-rumped Euphonia, Ash-breasted Sierra-Finch.

Access

Bird the 30-km road from Catacocha (93 km west of Loja) northwest to Lauro Guerrero, the 19 km road from here northwest to Orianga, and the road southwest to Guachanama.

Over 120 species have been recorded from the small Puyango Reserve (2,659 ha) near the village of Puyango, some 35 km northwest of Celica via Alamor, including Pale-browed Tinamou*, Rufous-headed Chachalaca*, Elegant Crescent-chest, Saffron Siskin* and Black-and-white Tanager*.

ADDITIONAL INFORMATION

Addresses

CECIA (Ornithological Society of Ecuador), PO Box 1717 906, Quito (tel and fax: 244734).

Fundacion Natura, Avenida America 5663, Quito (tel: 447341-4) (Ecuador's major non-governmental conservation organisation).

Books

Birds of Ecuador: Locational Checklist, Ortiz, F., Greenfield, P. and Matheus, J. 1990.

A Guide to the Birds of Colombia, Hilty, S, and Brown, W., 1986. Princeton UP.

ECUADOR ENDEMICS (12)	**BEST SITES**
Grey-backed Hawk*	West and southwest: Machalilla, Cerro Blanco, Buenaventura and Celica
El Oro Parakeet*	Southwest: Buenaventura and Manta Real

White-necked Parakeet*	South: Loja–Zamora road and Podocarpus NP (South)
Black-breasted Puffleg*	North: very rare on slopes of volcanoes Pichincha and Atacazo near Quito
Turquoise-throated Puffleg*	North: very rare. Known from Guayllabamba near Quito (and possibly Colombia)
Violet-throated Metaltail*	South: Cuenca; Las Cajas and Río Mazan
Esmeraldas Woodstar*	West: Machalilla–Santa Elena road
Pale-mandibled Aracari*	North and southwest: Nono–Mindo, Santo Domingo and Buenaventura
Cocha Antshrike*	East: La Selva
Orange-crested Manakin	East: Cuyabeno, Limoncocha and La Selva
Saffron Siskin*	West and southwest: Machalilla, Cerro Blanco and Puyango near Celica
Pale-headed Brushfinch*	South: very rare, possibly extinct. Cuenca–Loja road

North = Quito area.
East = Río Napo area.
West = Machalilla south to Santa Elena and inland to Guayaquil.
Southwest = inland to Zaruma and Macara.
South = Cuenca south to Loja and Podocarpus NP.

THE FALKLAND ISLANDS

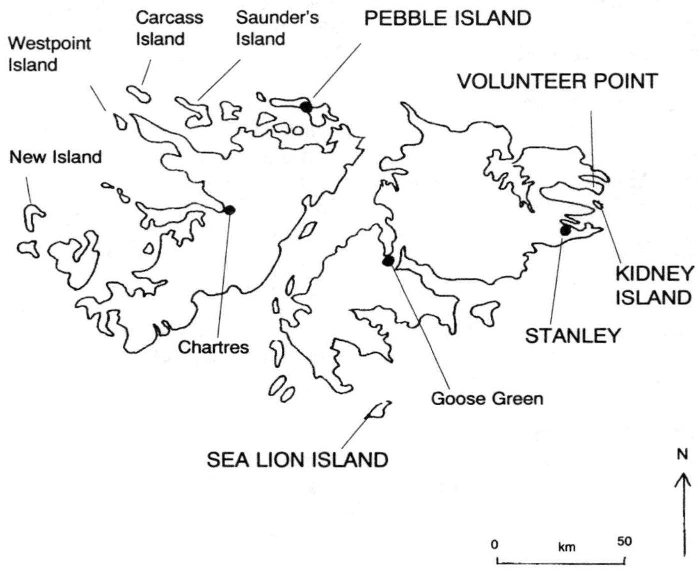

INTRODUCTION

Summary

This windy, rugged, treeless archipelago, 480 km east of Argentina, supports few birds by South American standards, and only 50 or so species may be seen in two weeks, even with plenty of flights and boat trips, which are necessary to reach the offshore islands. However, such star birds as King Penguin, and mainland South American rarities including Striated Caracara*, may justify the expense.

Size

This tiny archipelago (12,173 km^2), 258 km wide, is just a tenth the size of England and one fiftieth the size of Texas. However, it is still 1.5 times the size of the Galápagos Islands and 2.4 times larger than Trinidad and Tobago.

Getting Around

Most of the offshore islands are accessible via light aircraft run by the Falkland Islands Government Air Service (FIGAS), although flights are subject to seat availability and often affected by unpredictable weather.

A boat charter service caters for day trips out of Stanley, the capital. It is possible to hire cars, although the only real road is the 56-km one between Stanley and the airport. Land Rovers are more appropriate for tracks which criss-cross the boggy terrain, and are also available for hire.

Accommodation and Food

There are a few hotels and guest houses in Stanley and lodges on the larger offshore islands, which are all quite expensive by South American standards.

High quality 'meat and two veg.' is the order of the day.

Health and Safety

No problems with either of these on the friendly Falklands.

Climate and Timing

Situated at approximately the same latitude south as London, England or Vancouver, Canada are north, the Falklands are not so cold as one might expect with temperatures averaging 50° F in summer. Stiff westerly winds are more or less constant, although they tend to subside slightly during the southern summer, from November to March, so this is the best time to visit.

Habitats

Although over 700 islands make up the archipelago, there are two major islands split north to south by Falkland Sound.

Rolling moorland, dotted with lakes and sheep, rises to 712 m (2,312 ft), and dominates the landscape. It is encircled by the rugged coastline, broken here and there by wide, white beaches.

Conservation

The islanders are the conservationists here.

Bird Families

Only 28 of the 92 regularly occurring South American bird families are represented, the lowest total for the continent. There are no South American endemic families present and only two Neotropical endemic families are represented: furnariids and tapaculos.

Bird Species

185 species have been recorded on the Falkland Islands, which apart from the Galápagos total, is the lowest for South America. 58 species breed including 42 'waterbirds'.

Non-endemic specialities and spectacular species include King, Gentoo, Rockhopper and Macaroni Penguins, Wandering and Royal Albatrosses, Common Diving-Petrel, Ruddy-headed Goose*, Striated Caracara*, Snowy Sheathbill, Two-banded Plover, Rufous-chested Dotterel and Dolphin Gull. Passerines are few and far between but include Blackish Cinclodes, Dark-faced Ground-Tyrant, Austral Thrush, Correndera Pipit, Canary-winged Finch* and Long-tailed Meadowlark.

Endemics

Only one bird is endemic, Falkland Steamerduck, and this is quite easy to see.

Expectations

Although 185 species have been recorded, do not expect much above 50 species in two weeks.

STANLEY

The tiny capital, with a population of 1,700, sits at the northeast end of the archipelago. It is a good place to see the endemic Falkland Steamerduck, and the embarkation point for boat trips to a Rockhopper Penguin colony on Kidney Island and a King Penguin colony at Volunteer Point.

Endemics

Falkland Steamerduck.

Specialities

King, Gentoo and Magellanic Penguins, Upland Goose, Dolphin Gull, Canary-winged Finch*.

Others

Black-browed Albatross, Antarctic Giant Petrel, Sooty Shearwater, Rock Shag, Turkey Vulture, Red-backed Hawk, Peregrine, White-rumped Sandpiper, Two-banded Plover, Brown-hooded Gull, South American Tern, Austral Thrush, Correndera Pipit, Black-chinned Siskin, Long-tailed Meadowlark.

Access

Falkland Steamerduck occurs along the front. Check the wreck of *Charles Cooper* opposite the West Store for Rock Shag and Dolphin Gull. Black-chinned Siskin occurs near the Government House and Austral Thrush around the Official Residence.

Gentoo, Magellanic, and, occasionally, King Penguins are present at Penguin Walk, 5 km east of Stanley. Nearby Surf Bay is good for Two-banded Plover, with Black-browed Albatross offshore.

KIDNEY ISLAND

This small island reserve, 1.5 hours by boat north from Stanley, has been established to protect remnant tussac-grass habitat.

Specialities

Rockhopper Penguin, Blackish Cinclodes.

Others

White-chinned Petrel, Great and Sooty Shearwaters, Imperial Shag, Austral Thrush, Sedge and House Wrens, Black-chinned Siskin.

Other Wildlife

Southern Sealion.

Access

Permission from the government Lands Officer is needed to visit this island. The Tourist office arranges occasional trips. The Rockhopper Penguin colony, on the cliffs, is small but well worth seeing. Thousands of Sooty Shearwaters and some Great Shearwaters also breed on the island, along with White-chinned Petrel.

VOLUNTEER POINT

This headland at the northeast tip of the archipelago holds the world's most accessible King Penguin colony. However, landing is not always guaranteed.

Specialities

King, Gentoo and Magellanic Penguins, Wandering, Royal and Grey-headed Albatrosses, Common Diving-Petrel, Magellanic Oystercatcher, Dark-faced Ground-Tyrant.

Others

Black-browed Albatross, Southern Fulmar, Cape and White-chinned Petrels, Great and Sooty Shearwaters, Wilson's Storm-Petrel, Speckled Teal, Blackish Oystercatcher, Rufous-chested Dotterel.

Other Wildlife

Black-chinned (Peale's) and Commerson's Dolphins, Southern Fur Seal, Southern Elephant Seal.

Access

Permission from the landowner, Mr Osmond Smith at Johnson's Harbour, is needed to visit Volunteer Point. One or two of the 'big' albatrosses are occasionally seen on the sea crossing. About 120 pairs of King Penguin breed at the point, where it also worth scanning the rocks for seals.

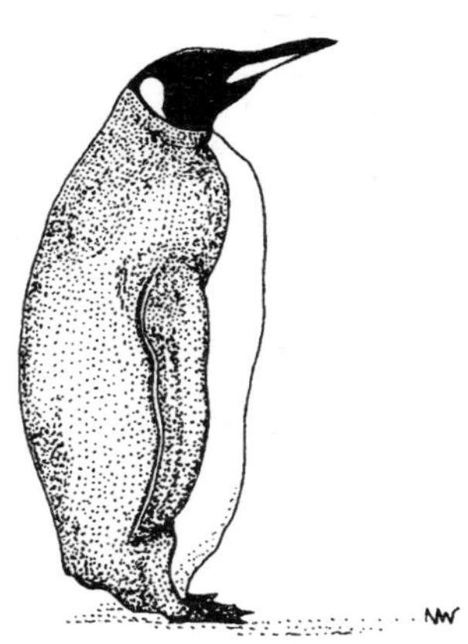

King Penguin is likely to be one of the highlights of a trip to the Falkland Islands

PEBBLE ISLAND

This large island situated in the northwest of the archipelago is the best place in the Falklands to see waterfowl including the rare Ruddy-headed Goose*, which is declining rapidly on mainland South America, as well as Striated Caracara*, one of the world's rarest raptors.

Specialities

Rockhopper and Macaroni Penguins, Ruddy-headed Goose*, Flying Steamerduck, Striated Caracara*, Snowy Sheathbill, Dolphin Gull, Canary-winged Finch*.

Others

White-tufted and Silvery Grebes, Imperial Shag, Black-necked Swan, Chiloe Wigeon, Yellow-billed Pintail, Silver Teal, Black-crowned Night-Heron, Crested Caracara, South American Snipe, Rufous-chested Dotterel, Brown-hooded Gull, South American Tern, Southern Skua, Sedge Wren, Black-chinned Siskin.

Other Wildlife

Southern Sealion.

Access

This island is accessible by FIGAS flights from Stanley (40 minutes). The Striated Caracara*, which is ridiculously tame, occurs along the cliffs and at the Rockhopper Penguin colony, which occasionally contains Macaroni Penguins as well. Snowy Sheathbills attend the Southern Sealion colonies.

Accommodation: Pebble Island Lodge (A); Chartres Lodge (A); Salvador Lodge (A); also a shanty (C), 19 km from the main centre of population.

WEST FALKLAND

The west coast of the Falklands archipelago is, arguably, the most rugged and beautiful part of the islands. Here, there are many offshore islands, some of which support the endemic Falkland Steamerduck, as well as a number of species absent in East Falkland and which are hard to see on mainland South America.

Endemics

Falkland Steamerduck.

Specialities

Gentoo, Rockhopper, Macaroni and Magellanic Penguins, Slender-billed Prion, Grey-backed Storm-Petrel, Common Diving-Petrel, Upland, Kelp and Ruddy-headed* Geese, Striated Caracara*, Magellanic Oystercatcher, Dolphin Gull, Blackish Cinclodes, Dark-faced Ground-Tyrant, Canary-winged Finch*.

Others

Black-browed Albatross, Antarctic Giant Petrel, Imperial and Rock

Shags, Crested Duck, Black-crowned Night-Heron, Turkey Vulture, Crested Caracara, Peregrine, Blackish Oystercatcher, South American Tern, Southern Skua, Austral Thrush, Sedge and House Wrens, Black-chinned Siskin, Long-tailed Meadowlark.

Other Wildlife

Peale's, Hourglass and Commerson's Dolphins, South American Fur Seal, Southern Sealion.

Access

Enquire in Stanley about possible cruises to West Falkland. There is a Black-browed Albatross colony on Saunder's Island. Carcass Island supports a Gentoo Penguin colony and is one of the last strongholds of the Ruddy-headed Goose*. Westpoint Island supports a large Rockhopper Penguin colony (4,000) and a Black-browed Albatross colony (1,000), which, in turn, attract Striated Caracara*. Slender-billed Prion breeds on New Island. Antarctic Cruises visit West Falkland (see that section, p. 403).

SEA LION ISLAND

On this large offshore island, which lies in the southeast of the archipelago, it is possible to see five species of penguin in a day.

Specialities

Gentoo, Rockhopper, Macaroni and Magellanic Penguins, Striated Caracara*, Snowy Sheathbill, Dolphin Gull.

Others

Antarctic Giant Petrel, Southern Skua, Short-eared Owl.

Other Wildlife

Southern Elephant Seal, Southern Sealion, Killer Whale.

Access

The island is a 35-minute flight from Stanley. Macaroni Penguins occasionally appear amongst the Rockhopper Penguin colonies, whilst King Penguins appear rarely amongst the Gentoo Penguin colony near the lodge.

Accommodation: Sea Lion Lodge (A).

SEA CROSSING TO URUGUAY

The ardent seawatcher may see some excellent seabirds on the three-day sea crossing to Montevideo, the capital of Uruguay.

Specialities

Wandering and Royal Albatrosses, Atlantic Petrel, Black-bellied Storm-Petrel.

Others

Black-browed and Shy Albatrosses, Antarctic Giant Petrel, Cape, Soft-plumaged and White-chinned Petrels, Cory's, Great and Sooty Shearwaters, Wilson's Storm-Petrel, Southern Skua, Long-tailed Jaeger.

ADDITIONAL INFORMATION

Addresses

Falkland Islands Tourist Board, John Street, Stanley (tel: 22216, or Falkland House, 14 Broadway, Westminster, London SW1H 0BH, England, U.K. (tel: 071 222 2542).

To hire boats in Stanley contact: South Atlantic Marine Services. (tel: 21145).

Books

Guide to the Birds of the Falkland Islands, Woods, R. 1988. Anthony Nelson.

FALKLAND ISLAND ENDEMICS (1)	BEST SITES
Falkland Steamerduck	Stanley and West Falkland

THE GALÁPAGOS ISLANDS

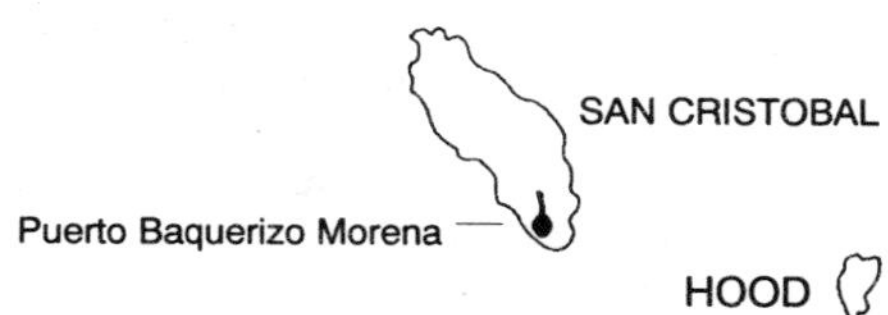

INTRODUCTION

Although not always considered to be geographically part of South America, the Galápagos Islands are administered by, and accessed via, Ecuador, so they have been included in this book.

Summary

To some these are the 'Enchanted Isles', to others they are a rather ugly sprawl of arid, barren islands, lying on the equator, 970 km west of Ecuador. Many of the birds, especially seabirds, are abundant and ridiculously tame, helping to provide an amazing birding experience. Although only 50 or so species are likely on a two-week trip, over half of these will be endemic or near-endemic. However, to 'clear up' the endemics on one trip could prove expensive and/or tricky since self-chartered tours rarely visit all the necessary sites, notably the very restricted mangrove habitat of the Mangrove Finch*.

Size

The islands span some 320 km and cover 7,845 km^2. The biggest island, Isabela, is 120 km long.

Getting Around

Travelling to these islands by boat can be expensive and time consuming. Services from Guayaquil, Ecuador are irregular so flying is advisable. Most flights go to Puerto Ayora on Santa Cruz, from where tours can be arranged. However, these are unlikely to satisfy the manic endemic-hunter since not all of the best birding sites are likely to be visited. Well prepared and patient budget-birders may be able to charter their own boat and arrange the itinerary with the captain beforehand. Otherwise, there are a number of specialist bird tour companies who guarantee most, if not all, of the endemics, at a price.

Accommodation and Food

There are plenty of hotels and eating places to choose from in Puerto Ayora. All accommodation (sometimes rather cramped) and food is provided on the cruises.

Health and Safety

No problems here except for sunburn and seasickness.

Climate and Timing

January to April are usually very hot and quite wet, although most rain falls in the highlands. The rest of the year is cooler so June to August is the most popular visiting time for general tourists and birders, since this is also when Waved Albatross and Great Frigatebird are breeding.

In 'El Nino' years it can be very wet. This warm current, which flows south from the Gulf of Panama, occasionally blocks the cold Humboldt current, which flows north from the Antarctic. Since the Humboldt current, provides the food for most of the seabirds breeding on Galápagos, an 'El Nino' year also causes a southward displacement of birds.

Habitats

The major terrestrial habitat on the Galápagos Islands is low, arid scrub, although older and larger islands with high ground support mangrove, *Opuntia* cacti scrub, lichen-covered scrub (200 m/656 ft), epiphyte-covered evergreen shrubs (400 m/1,312 ft) and a high zone of tree ferns and sedges.

The food-rich waters of the cold Humboldt current, which usually surround the islands, together with numerous rocky cliffs and islets, provide ideal habitat for masses of breeding seabirds.

The lovely Swallow-tailed Gull is virtually endemic to the Galápagos Islands

Conservation

The Galápagos Islands are a National Park and access is limited to 45 visitor sites. Introduced cats and rats cause many problems, an example being the demise of the Flightless Cormorant* at Tagus Cove on Isabela.

Five threatened species, of which four are endemic, occur on the islands. A further four endemics are near-threatened.

Bird Families

Only 38 of the 92 families which regularly occur in South America are represented, ten more than the total for the Falkland Islands. There are no Neotropical or South American endemic families present.

Bird Species

Only 136 species have been recorded, the lowest total for any country or archipelago in South America. Trinidad and Tobago, despite being smaller, has nearly 300 more species.

However, many of the species on the islands are hard, if not impossible, to see elsewhere. Non-endemic specialities and spectacular species include Waved Albatross*, Dark-rumped Petrel*, Wedge-rumped Storm-Petrel, Red-billed Tropicbird, Blue-footed, Masked and Red-footed Boobies, Paint-billed Crake, Swallow-tailed Gull, and Brown Noddy.

Endemics

The low species list does include 25 endemics, the fifth highest number of endemics for the 17 countries and archipelagos of South America. Only Brazil, Peru, Colombia and Venezuela have more. For such a small area this represents a very high degree of endemism.

One of the advantages of birding the Galápagos is that it needs only a single visit in order to see most, if not all, of the 25 endemics, providing it is a well-planned or, preferably, a birding company tour trip. These species are a penguin, Flightless Cormorant*, a heron, a hawk*, a rail, a gull, a dove, a flycatcher, four mockingbirds and 13 'Darwin's' finches, including the tool-using Woodpecker Finch.

Expectations

Cruises lasting two weeks, which visit most main islands, usually record around 50 species.

AT SEA

Some seabirds may be seen whilst travelling between the islands as they search for food before returning to their nests on the islands and islets.

Specialities

Waved Albatross*, Dark-rumped Petrel*, White-vented and Wedge-rumped Storm-Petrels, Blue-footed Booby.

Others

Audubon's Shearwater, Masked and Red-footed Boobies, Red-necked Phalarope, Brown Noddy.

Other Wildlife

Bottle-nosed Dolphin, Killer Whale, Californian Sealion, Galapagos Fur Seal, White-tipped Reef Shark, Green Marine Turtle.

Waved Albatross* breeds on the islands between May and August, and may be seen on land or at sea during that time. It is possible to miss this species at any other time of the year.

FERNANDINA

This young, barren island at the western edge of the archipelago, complete with mangroves, lagoons and sandy bays, is one of only two islands (the other is Isabela which is opposite) where the rare Flightless Cormorant* may be seen.

Endemics

Galapagos Penguin*, Flightless Cormorant*.

Others

Wandering Tattler, American Oystercatcher, Semipalmated Plover.

Other Wildlife

Marine Iguana, Green Turtle.

Flightless Cormorant* may be found around Punta Espinosa at the northeast corner of the island, opposite Isabela.

FLOREANA

This small island is the only one on which Medium Tree-Finch* occurs, and only one of two where Large Tree-Finch may be seen (the other is Santa Cruz). Furthermore, two of its offshore islets, at the southern edge of the archipelago, are the only sites where Charles Mockingbird* occurs.

Endemics

Galapagos Penguin*, Charles Mockingbird*, Large, Medium* and Small Tree-Finches.

Others

Greater Flamingo, Wandering Tattler, Brown Noddy, Short-eared Owl.

Access

Medium Tree-Finch* is endemic to Floreana. It occurs in the highlands, along with Large Tree-Finch, a two hour walk on a trail from Black Beach on the west coast. The lagoon at Punta Cormorant at the northern end is good for waterfowl and shorebirds. For those interested in underwater birding, snorkelling at Devil's Crown is excellent. Charles Mockingbird* occurs only on **Gardner** and **Champion Islets** to the east of the island.

HOOD

This tiny, 15-km long, flat island, at the archipelago's southeast corner is one of only two islands where Large Cactus-Finch occurs (the other is Tower), and the only island where Hood Mockingbird is found. Once on the island, it is not difficult to find the mockingbird since they seem to find shoelaces particularly interesting!

Endemics

Galapagos Dove, Hood Mockingbird, Common and Large Cactus-Finches.

Specialities

Waved Albatross*, Blue-footed Booby, Swallow-tailed Gull.

Others

Red-billed Tropicbird, Masked Booby.

Other Wildlife

Marine Iguana, Lava Lizard.

Access

One of the highlights of Hood is the large Waved Albatross* colony (over 400 birds), occupied from May to August. To reach it, walk carefully through the Blue-footed Booby colony from Punta Suarez.

ISABELA

The biggest island in the Galápagos is off the tourist circuit but supports no less than ten of the endemics, including Flightless Cormorant*,

Large Ground-Finch (difficult elsewhere except Tower), and the very rare Mangrove Finch*.

Endemics

Galapagos Penguin*, Flightless Cormorant*, Galapagos Hawk*, Galapagos Flycatcher, Galapagos Mockingbird, Large, Medium and Small Ground-Finches, Small Tree-Finch, Woodpecker and Mangrove* Finches.

Others

Brown Noddy.

Access

The coastal mangroves near Punta Tortuga, on the west side of the island opposite Fernandina, are the best place in Galápagos to look for Mangrove Finch*. Large ships may be unable to penetrate the preferred habitat of dense mangroves so a smaller vessel would be more useful here. Flightless Cormorant* also occurs around Punta Tortuga.

(Some authorities believe Mangrove Finch* is extinct, having hybridised completely with Woodpecker Finch).

JAMES

This island in the centre of the archipelago has a large intertidal area which supports Galapagos Heron, and two offshore islets, one of which supports a big Blue-footed Booby colony.

Endemics

Galapagos Penguin*, Galapagos Heron, Sharp-beaked Ground-Finch, Common Cactus-Finch.

Specialities

Blue-footed Booby.

Others

Red-billed Tropicbird, Magnificent Frigatebird, Masked Booby, White-cheeked Pintail, Greater Flamingo, Yellow-crowned Night-Heron, Wandering Tattler, Brown Noddy, Southern Martin.

Other Wildlife

Galapagos Fur Seal, Lava Lizard, Sally Lightfoot Crab.

Access

The Galapagos Heron feeds on the abundant crabs at Puerto Egas on the west side of the island. The 2-km trail from here up to the crater of Sugarloaf volcano (395 m/1,296 ft) is good for 'Darwin's' finches. Greater Flamingo occasionally occurs at Espumilla beach, 5 km north of Puerto Egas. Finches also occur along the trail which leads inland from here.

It is possible to swim with Galapagos Penguins* around **Bartholemew** Islet, off the southeast corner of James. **Daphne Major Islet** to the southeast of here, between James and Santa Cruz, supports Red-billed Tropicbirds, a big Blue-footed Booby colony and Masked Booby.

SAN CRISTOBAL

This island is relatively well vegetated compared to many others. Apart from cacti and poisonous *Manzanillo* trees, there are even some introduced coffee and orange trees here. The island lies at the eastern edge of the archipelago and is the only island on which San Cristobal Mockingbird occurs.

Endemics

Galapagos Dove, San Cristobal Mockingbird, Woodpecker Finch.

Others

Paint-billed Crake, Vermilion Flycatcher, Yellow Warbler.

Access

San Cristobal Mockingbird occurs in the plantations *en route* to the 'highlands', accessible by bus from Puerto Baquerizo Moreno in the southwest corner of the island. Frigatebird Hill, 2 km east of Puerto Baquerizo Moreno, is also worth a look.

It is possible to visit **Santa Fe** island between San Cristobal and Santa Cruz. Galapagos Hawk* and Common Cactus-Finch, which is tricky elsewhere, occur here.

SANTA CRUZ

This is the archipelago's second largest island, and the base from which to cruise around the other surrounding islands. The area around Puerto Ayora, where flights from Ecuador arrive and depart, at the southern end of the island, supports a number of arid-scrub species and presents the best opportunity in Galápagos to get to grips with Galapagos Rail*, Vegetarian Finch, Large Tree-Finch (also present but scarce on Floreana) and Woodpecker Finch, as well as Giant Tortoise.

Endemics

Galapagos Heron, Galapagos Rail*, Lava Gull*, Galapagos Flycatcher, Galapagos Mockingbird, Common Cactus-Finch, Vegetarian Finch, Large Tree-Finch, Woodpecker Finch, Warbler Finch.

Specialities

Blue-footed Booby, Paint-billed Crake.

Others

Magnificent Frigatebird, Brown Pelican, White-cheeked Pintail, Great Blue Heron, Great and Cattle Egrets, Common Moorhen, Wandering Tattler, Brown Noddy, Dark-billed Cuckoo, Barn Owl, Belted Kingfisher, Vermilion Flycatcher, Yellow Warbler.

Other Wildlife

Giant Tortoise, Californian Sealion.

Access

Puerto Ayora harbour is a good place to look for Lava Gull*. The Prickly

Pears and Candleabra Cacti around the Darwin Research Station, 20 minutes walk northeast of town, supports Common Cactus-Finch and Vegetarian Finch. The Giant Tortoise Reserve at Santa Rosa, 12 km west of Bellavista (7 km north of Puerto Ayora), supports Paint-billed Crake, Galapagos Rail*, Vegetarian Finch, Large Tree-Finch and Woodpecker Finch. The rails also occur along the path which leads north from Bellavista to Cerro Crocker, and on Mt Media Luna.

North Seymour Islet to the north of Santa Cruz supports Magnificent Frigatebird, Blue-footed Booby and Californian Sealion. **South Plaza Islet** just off the east coast supports breeding Red-billed Tropicbird and Swallow-tailed Gull, as well as Land Iguana. **Barrington Island** to the southeast of Santa Cruz also supports Land Iguana as well as Galapagos Hawk*, Manta Ray and White-tipped Shark.

TOWER

This tiny, low, dry and remote island at the northeast edge of the archipelago is one of the best places to see Large Ground-Finch (which also occurs on Isabela), Sharp-beaked Ground-Finch, and Large Cactus-Finch, which occurs only here and on Hood Island, as well as Great Frigatebird.

Endemics

Large, Small and Sharp-beaked Ground-Finches, Large Cactus-Finch, Warbler Finch.

Specialities

Wedge-rumped Storm-Petrel, Great Frigatebird, Swallow-tailed Gull.

Others

Red-billed Tropicbird, Magnificent Frigatebird, Masked and Red-footed Boobies, Wandering Tattler, Short-eared Owl.

Access

There are Great Frigatebird (May to August), Red-footed Booby and Swallow-tailed Gull colonies in Darwin Bay at the southern end of the island. Wedge-rumped Storm-Petrels can be seen in daylight over their colony at the lava fields on the far side of the island, at the end of a path from Prince Philip's steps. A short walk inland should be enough to find Large Cactus-Finch.

(The Sharp-beaked Ground-Finches here are considered, by some 'authorities', to be a long-billed form of Small Ground-Finch).

ADDITIONAL INFORMATION

Books

A Field Guide to the Birds of The Galapagos, Harris, M. 1982. HarperCollins.

GALÁPAGOS ENDEMICS (25)	BEST SITES
Galapagos Penguin*	Fernandina, Floreana, Isabela and James
Flightless Cormorant*	Fernandina and Isabela
Galapagos Heron	Uncommon on most islands except James
Galapagos Hawk*	Uncommon on most islands
Galapagos Rail*	Santa Cruz
Lava Gull*	Uncommon on most islands except Santa Cruz
Galapagos Dove	Most islands
Galapagos Flycatcher	Most islands
Galapagos Mockingbird	Most islands
Charles Mockingbird*	Floreana: Champion and Gardner Islands
Hood Mockingbird	Hood
San Cristobal Mockingbird	San Cristobal
Large Ground-Finch	Isabela and Tower
Medium Ground-Finch	Most islands
Small Ground-Finch	Common on most islands
Sharp-beaked Ground-Finch	Tower
Common Cactus-Finch	Santa Cruz and Santa Fe
Large Cactus-Finch	Hood and Tower
Vegetarian Finch	Santa Cruz
Large Tree-Finch	Floreana and Santa Cruz
Medium Tree-Finch*	Floreana
Small Tree-Finch	Most islands
Woodpecker Finch	Santa Cruz
Mangrove Finch*	Fernandina and Isabela
Warbler Finch	Most islands

GUYANA

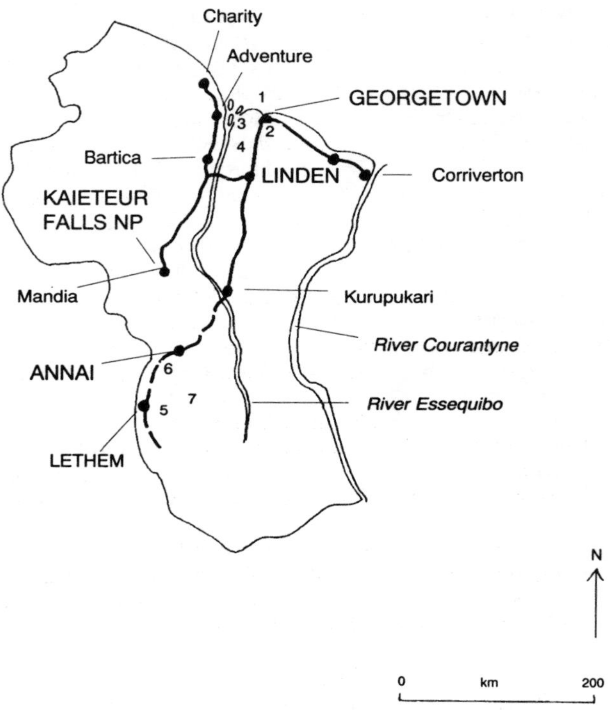

1	Georgetown	4	Timberhead
2	Emerald Towers	5	Kanuku Mountains
3	Shanklands	6	Rupunini Savanna
		7	Iwokrama Forest

INTRODUCTION

Summary

Guyana has recently realised the potential of ecotourism and new lodges in prime birding habitat are appearing fast. Although many of these need to be booked in advance, and some of the other best birding areas are still subject to red-tape, the patient non-budget birder could reap the rewards of a well prepared trip to Guyana since the country is full of cotingas.

Size

At 214,969 km^2 Guyana is a relatively small South American country. It is only twice the size of England, a third the size of Texas and slightly smaller than Ecuador.

Getting Around

Apart from the coast road the few other roads are poor and short of public transport, although 'Maxi' taxis serve coastal towns and Linden, and Bedford trucks are available for extensive inland expeditions. There is a small range of scheduled flights (including one to Lethem) but most planes must be chartered, at considerable expense. Tour companies in Georgetown, most of which are based in the major hotels, arrange trips into the interior (see Additional Information). Permission from the Ministry of Home Affairs, Brickdam, is required to travel in the interior, where it is also necessary to register with the local police or village captain when staying overnight.

Accommodation and Food

Outside of Georgetown, where virtually all accommodation is expensive, basic government guest houses are the commonest type of accommodation. Creole, Chinese and seafood dominate the menus.

Health and Safety

Immunisation against typhoid, yellow fever and malarials recommended. Do not drink the water, beware of electric eels, piranhas and stingrays in rivers, and walk around Georgetown during night at your peril.

Climate and Timing

This is a hot, humid tropical country with two main wet seasons, from late April to early August, and from November to January, so the best times to visit are January to February and August to September.

Habitats

The primarily reclaimed coastal plain varies in width from 10 to 70 km, giving way to sandy hills and a highland plateau carpeted in remote rainforest (which covers a remarkable 70–75% of the country), rising in the west to the Pakaraima Mountains. The forest is broken in the south by the huge Rupununi savanna.

Conservation

Guyana has only one National Park (Kaieteur Falls NP), although the Kanuku Mountains, near Lethem, are also being proposed. Together with a willingness to at least experiment with sustainable forestry, and encourage ecotourism, the government of Guyana seem to be heading in the right direction.

One threatened species occurs in Guyana.

Bird Families

77 of the 92 families which reguarly occur in South American are represented, a few less than the bigger countries such as Venezuela but more than Bolivia. 20 of the 25 Neotropical endemic families and five

of the nine South American endemic families are represented. There are no rheas, seriemas, Magellanic Plover, seedsnipes or tapaculos.

Well-represented families include antbirds and cotingas.

Bird Species

825 species have been recorded in Guyana, although this figure probably only represents 70–80% of the real total.

Non-endemic specialities and spectaculars species include Zigzag Heron*, Grey-bellied Goshawk*, Rufous Crab-Hawk, Yellow-knobbed Curassow, Sunbittern, Grey-winged Trumpeter, Spot-tailed Nightjar, Racket-tailed Coquette, Blood-coloured Woodpecker, Rio Branco Spinetail*, Rio Branco Antbird, Guianan Red-Cotinga, Dusky Purpletuft, Purple-breasted and Pompadour Cotingas, Crimson Fruitcrow, Capuchinbird, Bearded Bellbird, Guianan Cock-of-the-Rock, Tepui Manakin, Saffron-crested Tyrant-Manakin, Painted Tody-Flycatcher, Bearded Tachuri*, Crested Doradito, Cayenne Jay, Guianan Gnatcatcher, Finsch's Euphonia, and Great-billed Seed-Finch*.

Endemics

There are no endemics.

Expectations

A two-week trip is likely to produce in the region of 350–400 species.

GEORGETOWN

The remnant forest in and around Guyana's capital supports some excellent birds including the bright Blood-coloured Woodpecker, which is restricted to the Guianan lowlands, and the rare Bearded Tachuri*.

Specialities

Plain-bellied Emerald, Blood-coloured Woodpecker, White-bellied and Ferruginous-backed Antbirds, Bearded Tachuri*, Yellow Oriole.

Others

Pearl and Snail Kites, Red-shouldered Macaw, Brown-throated Parakeet, Black-headed, Yellow-crowned and Orange-winged Parrots, Striped Cuckoo, Reddish Hermit, Brown and Paradise Jacamars, Swallow-wing, Point-tailed Palmcreeper, Black-crested and Barred Antshrikes, Painted Tody-Flycatcher, Burnished-buff Tanager, Blue-black Grassquit.

Access

Blood-coloured Woodpecker, Point-tailed Palmcreeper and White-bellied Antbird occur in Georgetown's Botanic Gardens. Head for the zoo end of the gardens near downtown Georgetown, go straight across the mini-roundabout then turn left onto a track for the best area.

The old glass factory and racing track area around Timerhi airport, 45 km south of Georgetown, supports Brown-throated Parakeet, Reddish Hermit, Brown and Paradise Jacamars, and Ferruginous-backed Antbird.

The best birding near Georgetown is along the **Linden highway**.

Turn right some 5 km north of the airport *en route* to Georgetown. This road, which is paved for the most part, passes through savanna and gallery forest, and on, as a gravel track, beyond Linden (70 km south of Timerhi airport), to remnant rainforest. Black-headed Parrot, Striped Cuckoo, Swallow-wing and the rare Bearded Tachuri* occur along here. The latter is best looked for along side-tracks into the savanna, the best of which is the one which leads to Emerald Towers (see next site), to the left near the start of the highway. Do not go down the tracks marked 'GDF' since these lead to military bases.

At the south end of Georgetown, there is a pontoon bridge across the Demerarra river. If it is in use cross it, turn right at the first junction, and bird the paddies on the left-hand side of this road. Long-winged Harrier, Yellowish Pipit and Grassland Yellow-Finch occur here. Further on, turn left 100 m before the police station, by a bus-stop, on to a gravel track leading to Haig dam. The paddies alongside this road are good for crakes.

Scarlet Ibis and Rufous Crab-Hawk are two typical species which may be seen from the coast road. The area east of the Essequibo river is the most accessible.

Accommodation: Georgetown: Forte Creste, Hotel Tower; Woodbine Hotel; Ariantze Hotel; Rima Guest House. Linden Highway: Emerald Towers (A) (see next site).

Eskimo Curlew* has been recorded, in the dark distant past, from Guyana's coast.

EMERALD TOWERS

This new, expensive lodge near Timehri airport, is set in some excellent forest where the specialities include the lovely Pompadour Cotinga.

Specialities

Little Chachalaca, Black-necked Aracari, Red-billed Toucan, Ferruginous-backed Antbird, Pompadour Cotinga, Cayenne Jay, Plumbeous Euphonia.

Others

Little Tinamou, Agami Heron*, Green Ibis, Grey-lined Hawk, Marbled Wood-Quail, Rufous-sided Crake, Grey-fronted Dove, Red-bellied and Red-shouldered Macaws, Black-bellied Cuckoo, Grey Potoo, Long-tailed Hermit, White-necked Jacobin, Fork-tailed Woodnymph, Blue-crowned Motmot, Paradise Jacamar, Swallow-wing, Channel-billed Toucan, Chestnut and Ringed Woodpeckers, Lineated Woodcreeper, Mouse-coloured Antshrike, Black-headed Antbird, Blue-backed and White-bearded Manakins, Painted Tody-Flycatcher, White-lored Tyrannulet, Helmeted Pygmy-Tyrant, Greyish Mourner, Yellow-throated Flycatcher, Lemon-chested Greenlet, Black-capped Donacobius, Coraya Wren, Neotropical River Warbler, Pectoral Sparrow, Fulvous-crested Tanager, White-vented Euphonia.

Access

Bird the few trails, the tower and the creek.

Accommodation: Lodge (A).

SHANKLANDS

This new lodge, west of Georgetown, on a hill overlooking the River Essequibo, lies next to some excellent forest, where the goodies include Black Curassow and Racket-tailed Coquette.

Specialities

Black Curassow, Grey-winged Trumpeter, Racket-tailed Coquette, Black-necked Aracari, Red-billed Toucan, Chestnut-rumped Woodcreeper, Double-banded Pygmy-Tyrant, Cayenne Jay, Buff-cheeked Greenlet.

Others

Little Tinamou, Red-throated Caracara, Southern Lapwing, Large-billed Tern, Mealy and Red-fan Parrots, Grey Potoo, Lesser Swallow-tailed Swift, Reddish Hermit, White-chinned Sapphire, Black-tailed Trogon, Paradise Jacamar, Swallow-wing, Channel-billed Toucan, Red-necked Woodpecker, Mouse-coloured, Dusky-throated and Cinereous Antshrikes, Brown-bellied, Grey and Ash-winged Antwrens, Grey, Black-headed and Spot-winged Antbirds, Rufous-capped Antthrush, Screaming Piha, White-crowned Manakin, Painted Tody-Flycatcher, Helmeted Pygmy-Tyrant, Sulphur-rumped Flycatcher, Greyish Mourner, Ashy-headed and Tawny-crowned Greenlets, Chestnut-bellied Seedeater.

Access

To reach Shanklands head west out of Georgetown to the River Essequibo where, at Roed-en-Rust, it is possible to hire a motor-boat up river to the lodge. There is a maze of tracks and trails worth birding, although the best one is probably to Sand Hills and Timberhead (see next site). Most of the specialities listed above occur along here.

TIMBERHEAD

This lodge is surrounded by marshes, palm stands and forest where scarce birds such as Spot-tailed Nightjar and White-fronted Manakin occur.

Specialities

Little Chachalaca, Spot-tailed Nightjar, Green-tailed Goldenthroat, Black-necked Aracari, Red-billed Toucan, Chestnut-rumped Woodcreeper, Ferruginous-backed Antbird, White-fronted Manakin, Double-banded Pygmy-Tyrant, Cayenne Jay, Buff-cheeked Greenlet, Finsch's Euphonia.

Others

Great, Little, Undulated and Variegated Tinamous, Long-winged

The fabulous bright orange Guianan Cock-of-the-Rock graces the forest around the magnificent 228-m high waterfall in Kaieteur Falls NP

Harrier, Grey-lined Hawk, Marbled Wood-Quail, Ash-throated Crake, Sungrebe, Sunbittern, Southern Lapwing, Black-bellied Cuckoo, Grey Potoo, Reddish Hermit, Grey-breasted Sabrewing, Black-throated Trogon, Blue-crowned Motmot, Paradise Jacamar, Ringed and Red-necked Woodpeckers, Plain-crowned Spinetail, Point-tailed Palmcreeper, Amazonian, Dusky-throated and Cinereous Antshrikes, Rufous-bellied Antwren, Grey, Black-headed, Spot-winged and Black-throated Antbirds, White-crowned Manakin, Painted Tody-Flycatcher, Yellow-margined Flycatcher, White-vented Euphonia.

Access

From Sand Hills on the River Demerara head down river to Pokeraro Creek. Head up creek here to the lodge. Point-tailed Palmcreeper occurs in the palms around the lodge, and Sungrebe and Sunbittern up-creek.

Georgetown hotel tour companies arrange expensive one-day or 4–5 day trips to the spectacular **Kaieteur Falls NP** (577 km^2) in west Guyana. The waterfall, 228 m (748 ft) high and 100 m (328 ft) wide, is on a par, if not even more spectacular than, Angel Falls in Venezuela. Guianan Cock-of-the-Rock occurs here, as well as McConnell's Flycatcher and Red-shouldered Tanager.

Accommodation: Government guest house.

LETHEM

Lethem, in southwest Guyana, on the Brazilian border, is accessible by air from Georgetown, and is the gateway to a number of superb birding sites.

Accommodation: Government rest house.

KANUKU MOUNTAINS

Only 15–20 km south of Lethem, the steep slopes of these mountains are cloaked in primary forest where a number of spectacular cotingas occur, including Crimson Fruitcrow, its only known locality in Guyana.

Specialities

Tufted Coquette, Purple-breasted Cotinga, Crimson Fruitcrow, Guianan Cock-of-the-Rock, Finsch's Euphonia.

Others

Harpy Eagle*, Black-and-white* and Ornate Hawk-Eagles, Blue-chinned Sapphire, Blue-tailed Emerald, Amethyst Woodstar, Amazonian Umbrellabird, Cliff Flycatcher, Pink-throated Becard.

Other Wildlife

Jaguar, Ocelot, Tapir, Black Spider-monkey, Red Howler Monkey, Bearded Saki.

Access

It is possible to hire vehicles or horses in Lethem, although good habitat is within walking distance. Access is easiest from Macushi village or Nappi.

RUPUNINI SAVANNA

Much of this huge area of dry grassland, dotted with termite mounds, a few pools and gallery forest, has been converted to grazing land, but it still supports some rare and scarce birds, notably Rio Branco Spinetail* and Rio Branco Antbird, two birds thought, until very recently, to be localised Brazilian endemics. Crested Doradito and Great-billed Seed-Finch* also occur here.

Specialities

Sharp-tailed Ibis, Rio Branco Spinetail*, Rio Branco and White-bellied Antbirds, Saffron-crested Tyrant-Manakin, Smoky-fronted Tody-Flycatcher, Crested Doradito, Black-hooded and Bare-eyed Thrushes, Bicoloured Wren, Great-billed Seed-Finch*.

Others

Undulated Tinamou, Muscovy Duck, Capped Heron, Buff-necked and Green Ibises, Maguari Stork, Jabiru, Lesser Yellow-headed Vulture, Savanna and Black-collared Hawks, Double-striped Thick-knee, Pied and Southern Lapwings, Large-billed and Yellow-billed Terns, Red-shouldered Macaw, Least, Band-tailed and Nacunda Nighthawks, Blue-tailed Emerald, Long-billed Starthroat, Rufous-tailed and Green-tailed Jacamars, Spotted Puffbird, Straight-billed and Striped Woodcreepers, Pale-legged Hornero, White-browed Antbird, Plain-crested Elaenia, Pale-tipped Tyrannulet, Pale-eyed Pygmy-Tyrant, Rusty-margined Flycatcher, Ashy-headed Greenlet, Buff-breasted Wren, Long-billed

Gnatwren, Yellowish Pipit, Red-capped Cardinal, Burnished-buff Tanager, Troupial.

Other Wildlife

Jaguar, Puma, Giant Anteater, Crab-eating and Grey Foxes,

Access

To look for Rio Branco Spinetail and Rio Branco Antbird*, it is necessary to arrange a boat trip along the River Ireng, between Lethem and Karanambo.

Sunbittern, Spotted Puffbird, Saffron-crested Tyrant-Manakin and Smoky-fronted Tody-Flycatcher occur at the Karanambo Ranch, 80 km northeast of Lethem, which caters for guests.

Annai, some 150 km northeast of Lethem, is another good base from which to bird the savanna, as well as semi-deciduous forest on the surrounding hills and the lowland forests of Iwokrama (see next site). Permission from the Ministry of Regional Development is required to visit this area.

Accommodation: rest house in Annai.

IWOKRAMA FOREST

Designated as a reserve for research into sustainable forestry, this site supports such rarities as Grey-bellied Goshawk* and Yellow-knobbed Curassow, as well as six species of jacamar, over 20 species of antbird and Capuchinbird. Ecotourism is one aim of the project and access, which is currently restricted, may be made easier in the future.

Specialities

Grey-bellied Goshawk*, Yellow-knobbed Curassow, Sun Parakeet, Red-fan Parrot, Todd's Antwren, Capuchinbird, Guianan Cock-of-the-Rock.

Others

Crested Eagle*, Sunbittern, Grey-winged Trumpeter, Blue-and-yellow, Scarlet and Red-and-green Macaws, Blue-crowned Motmot, Short-tailed Antthrush, Spangled and Pompadour Cotingas, Bearded Bellbird.

Other Wildife

Black Spider-monkey, White-faced Saki, Brown and Weeper Capuchins, Tapir, Two-toed Sloth, Giant Anteater, Giant Otter, Tayra, Grey Fox.

Access

A research station is planned for Surama. The Guyana–Brazil highway runs through the site. Check details of visiting arrangements with tourist agencies in Georgetown.

ADDITIONAL INFORMATION

Addresses

Ministry of Home Affairs, 6 Brickdam, Georgetown, Guyana.
Ministry of Regional Development, Water Street, Georgetwon, Guyana.

Tropical Adventures, Forte Crest Pegasus Hotel, PO Box 101147, Georgetown (tel: 52856; fax: 60532).
Torong Guyana, 56 Coralita Avenue, Bel Air Park East, Georgetown (tel: 65298).
Demerara Safari Tours, Hotel Ariantze, 176 Middle Street, Georgetown (tel: 65363).

Books

A Guide to the Birds of Venezuela, Meyer de Schauensee, R and Phelps, W. 1978. Princeton UP.(All but a few species are illustrated).
The Birds of Guyana, Snyder, M. 1966. Salem: Peabody Museum. (Old, and hard to track down, but interesting).

GUYANA ENDEMICS

None.

GUYANE

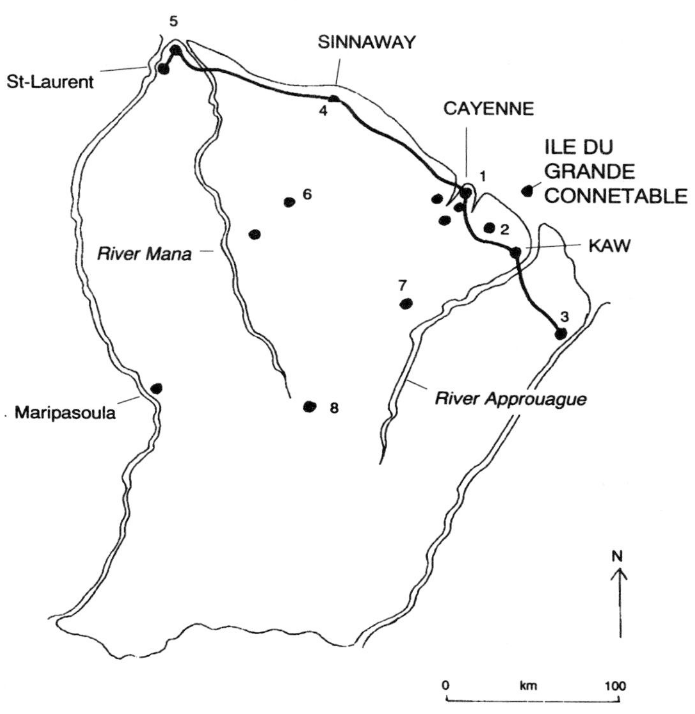

1	Cayenne	5	Mana
2	Kaw	6	Saint Elie
3	Saint Georges	7	Nouragues
4	Sinnamary	8	Saul

INTRODUCTION

Summary

Guyane is an expensive French colony where ecotourism is poorly developed and shooting birds is somewhat more popular than watching them. However, birders who are prepared to make extensive preparations before a visit (prior permission is required to visit the best sites), may be rewarded with a fine selection of scarce and spectacular species, especially cotingas, without a great deal of time spent in transit.

Size

At 91,000 km^2 Guyane is the smallest mainland country in South America, one eighth the size of Texas and only one third the size of Ecuador.

Getting Around

There is a severe lack of surface public transport in Guyane, and although cars may be hired, the few good roads are restricted to the coast. Fortunately, the internal air network accesses most of the best birding sites, which are inland. Rivers are also fully utilised.

Accommodation and Food

Most accommodation and food (creole and sea), mainly of a high quality, is very expensive by South American standards.

Health and Safety

Immunisation against malaria, typhoid and yellow fever is recommended.

Climate and Timing

Guyane is a hot, humid tropical country. Over four metres of rain usually fall in northeast Guyane every year, especially in the Kaw area, with May being the wettest month. The best time to visit is the dry season, from August to October.

Habitats

Guyane is a surprisingly low-lying country, with the land mostly below 300 m (984 ft) and rarely above 800 m (2,625 ft). There are a few offshore islands, such as Grande Connetable, where some seabirds breed, and a narrow coastal belt of mangroves and savanna, but the rest of the country (over 95%) is almost completely covered by rainforest. Most of the numerous rivers run south to north and gallery forest lines these rivers as they pass through the coastal savanna.

Conservation

Guyane has just one protected area and, as one might expect, hunting and shooting birds is a national pastime. Even Scarlet Ibis has been seen on one local menu. There are a lot of spectacular birds in Guyane and they look much better in the wild than on a plate. Hopefully, the government will realise this soon, get ecotourism off the ground, and make Guyane the popular destination it deserves to be.

One threatened species, the only endemic, occurs in Guyane.

Bird Families

74 of the 92 families which regularly occur in South American are represented, three less than Guyana to the west. 19 of the 25 Neotropical endemic families and four of the nine South American endemic families (screamers, Hoatzin, trumpeters and gnateaters) are represented, less than most of the bigger countries.

Well-represented families include seven tinamous, eight puffbirds, seven toucans, 45 antbirds and 12 cotingas.

Bird Species

710 species have been recorded in Guyane. Although this total is over 100 fewer than nearby Guyana, it reflects the lack of visiting birders

rather than a paucity of birds, since a further 119 species occur near the country's borders and are quite likely to be found in Guyane, making a possible total of 829.

Non-endemic specialities and spectacular species include Scarlet Ibis, Rufous-thighed Kite, Rufous Crab-Hawk, Crested* and Harpy* Eagles, Black Curassow, Spotted Rail, Grey-winged Trumpeter, Sooty Barbthroat, Toco Toucan, Blood-coloured Woodpecker, White-plumed Antbird, Spotted Antpitta, Guianan Red-Cotinga, Crimson Fruitcrow, Capuchinbird, White Bellbird, Guianan Cock-of-the-Rock, Sharpbill, Guianan Gnatcatcher, and Blue-backed Tanager.

Eskimo Curlew* was seen regularly on migration (September) in the early 1900s, usually in the company of American Golden-Plovers in the coastal savanna.

Endemics

The sole endemic, Cayenne Nightjar*, is known only from the type specimen, a male collected at Degrad Tamanoir on the Río Iracoubo (northwest Guyane) in 1917.

Expectations

A two-week trip to Guyane, which takes in the best sites, is likely to produce over 300 species. As many as 400 species may be seen on a longer trip.

CAYENNE

Situated on an island on the north coast, Cayenne, the small capital of Guyane, is a good base from which to explore a few interesting birding sites.

The Paradise Jacamar, like a bee-eater, preys on flying insects with sweeps and swoops from exposed branches and is a familiar sight in Guyane

From the sea-wall at the front of the town it is possible to see Magnificent Frigatebird and Plumbeous Kite whilst the mangroves support the localised Spotted Rail and Green-throated Mango. The old port is also a good site, and the rare Orange-breasted Falcon* has been recorded along the airport road.

The secondary forest and savanna around **Montsinery** and **Tonnegrande** are worth a look, the reservoir area near **Le Rorota** on Mt Mahury and the archaelogical site, Vidal, hold some forest species such as trogons, manakins (including Crimson-hooded Manakin) and tanagers, and **Mount Matoury** a few more than Rorota and Vidal including White-bellied Piculet and the striking Blood-coloured Woodpecker, which is restricted to the Guianan lowlands.

The small island of **Ile du Grande Connétable,** 40 km from Cayenne, supports a good breeding seabird colony (May to August), which includes Magnificent Frigatebird, Sooty Tern (250 pairs), and Brown Noddy (100–150 pairs). The island, Guyane's only nature reserve, is a restricted area for tourists although some tour operators (see Additional Information) may be persuaded to visit by persistent birders.

KAW

The village of Kaw, built on an 'island' amidst wet savanna, is 83 km southeast of Cayenne. Whilst the wetlands hold some interesting species, the nearby mountain (Mount Kaw), with superb forest, holds the two major specialities, Capuchinbird and Guianan Cock-of-the-Rock. Most birds are uncommon however, owing to hunting.

Specialities

Rusty Tinamou, Rufous-thighed Kite, Black-bellied Cuckoo, Black-necked Aracari, Capuchinbird, Guianan Cock-of-the-Rock.

Others

Horned Screamer, Muscovy Duck, Pinnated Bittern, Scarlet Ibis, Black-collared Hawk, Crested* and Harpy* Eagles, Grey-necked Wood-Rail, Azure Gallinule, Sungrebe, Wattled Jacana, Blue-and-yellow and Red-and-green Macaws, Little Cuckoo, Hoatzin, Green-tailed Jacamar, Spotted Puffbird, Toco Toucan, Purple-breasted and Pompadour Cotingas, Bare-necked Fruitcrow, White-headed Marsh-Tyrant, Black-capped Donacobius, Golden-sided Euphonia.

Access

Mount Kaw lies between Roura and Kaw.

Accommodation: Kaw: basic available (tel: 31-88-15). Roura: Hotel Relias de Patawa, 36 km southeast of Cayenne beyond Roura; the owners are entomologists and run guided tours.

SAINT GEORGES

This settlement, accessible by road and air from Cayenne, on the border with Brazil in east Guyane is a good base from which to see some savanna and wetland species including Sooty Barbthroat.

Specialities
Ash-throated Crake, Sooty Barbthroat, White-bellied Spinetail, Smoky-fronted Tody-Flycatcher.

Others
Grey-necked Wood-Rail, Southern Lapwing, Large-billed Tern, White-tailed Nightjar, Toco Toucan, Plain-crowned Spinetail, Long-tailed Tyrant, Chestnut-bellied Seedeater.

Access
Birding involves exploring the general area.

Accommodation: Chez Modestine (B).

SINNAMARY

This small town, some 100 km west of Cayenne on the north coast, is surrounded by savanna and lies near a large estuary which is an internationally important shorebird site. Although these habitats offer good birding the best birding near here is on the track to Saint Elie (see p. 286).

Specialities
Rufous Crab-Hawk, Rufous-crowned Elaenia.

Others
Scarlet Ibis, King Vulture, Snail Kite, Savanna Hawk, Upland Sandpiper, Mangrove Cuckoo, Ruby-topaz Hummingbird, Green-tailed Goldenthroat, Lesser Elaenia, Yellowish Pipit, Plumbeous Seedeater.

Access
Birding involves exploring the local area.

Accommodation: Hotels Sinnarive and Eldo Grill.

MANA

West of Sinnamary, 290 km from Cayenne, in the northwest of the country, is the pleasant town of Mana. The primary and secondary forest and savanna here supports some excellent birds, notably a couple of guans, toucans, woodpeckers, White-plumed Antbird, and cotingas.

Specialties
Marail Guan, Black Curassow, Spotted Rail, Green-throated Mango, Racket-tailed Coquette, Black-necked Aracari, Red-billed Toucan, Golden-collared and Waved Woodpeckers, Crimson Fruitcrow, Capuchinbird.

Others
Little Tinamou, Lesser Yellow-headed Vulture, Long-winged Harrier, Great Black-Hawk, Savanna Hawk, Plumbeous Pigeon, Golden-winged Parakeet, Black-headed, Blue-headed, Dusky and Orange-winged Parrots, Ladder-tailed Nightjar, Long-tailed, Straight-billed and Reddish Hermits, Blue-chinned Sapphire, White-tailed Trogon, Yellow-billed and

Paradise Jacamars, White-necked Puffbird, Channel-billed Toucan, Lineated and Red-necked Woodpeckers, Plain-brown, White-chinned, Long-tailed and Wedge-billed Woodcreepers, Short-billed Leaftosser, White-flanked and White-fringed Antwrens, Warbling, Spot-winged, White-plumed and Rufous-throated Antbirds, Thrush-like Antpitta, Screaming Piha, Spangled Cotinga, Golden-headed and White-crowned Manakins, Cinnamon Attila, Greyish Mourner, Pink-throated Becard, Musician Wren, Violaceous Euphonia, Turquoise Tanager, Red-legged Honeycreeper, Blue-black Grassquit, Lined Seedeater, Large-billed Seed-Finch*, Red-rumped Cacique, Epaulet Oriole.

Access

Just east of St Laurent turn north onto road D9 to Mana. 100 ha of primary and secondary forest, surrounded by savanna, remains between Crique Rouge and Nouveau Camp, 14 km along this road. It is possible to stay at Crique Rouge (reached via track between km posts 7 and 8). Bird around here, around Jan Rodriquez's Farm, 2.5 km north of here on the east side of the D9 (Yellow-billed Jacamar), around the farm opposite Godebert, also on the east side of the D9 south of Crique Rouge, and the roadside between km posts 8 and 10 on the D9 itself.

Scarlet Ibis, Limpkin and Collared Plover as well as Leatherback Turtles (June–July) occur at Les Hattes, west of Mana on the Suriname border.It is possible to take boat trips up the Maroni river from St Laurent into excellent forest.

SAINT ELIE

Along with Nouragues, this is one of Guyane's best sites. The road to St Elie south of Sinnamary is lined with superb forest, where Sooty Barbthroat, Spotted Antpitta, and a number of rare cotingas, including the incredible Crimson Fruitcrow, occur.

Specialities

Slaty-backed Forest-Falcon, Lilac-tailed and Sapphire-rumped Parrotlets, Black-bellied Cuckoo, Tawny-bellied Screech-Owl, Sooty Barbthroat, Green Aracari, Spotted Antpitta, Dusky Purpletuft, Crimson Fruitcrow, Capuchinbird, Double-banded Pygmy-Tyrant, Glossy-backed Becard, Blue-backed Tanager, Red-and-black Grosbeak.

Others

Little Tinamou, Scarlet Ibis, Bicoloured and Black-faced Hawks, Harpy Eagle*, Black-and-white* and Ornate Hawk-Eagles, Collared Forest-Falcon, Spectacled Owl, Grey Potoo, Brown and Great Jacamars, Spotted Puffbird, Yellow-tufted and Red-necked Woodpeckers, Rufous-tailed Xenops, White-plumed, Wing-banded and Spot-backed Antbirds, Pompadour Cotinga, Purple-throated Fruitcrow, Cinnamon-crested Spadebill, Coraya and Musician Wrens, Rose-breasted Chat, Red-billed Pied Tanager, Paradise and Opal-rumped Tanagers.

Access

Contact the Tourist Office and/or travel agents who may be able to arrange a visit to this site.

Further inland from St Elie is **La Trinite** where there is a naturalist's lodge set amongst excellent preserved forest and savanna.

NOURAGUES

This remote inland site, some 100 km south of Cayenne, is, along with Saint Elie, one of the best birding sites in Guyane. Unfortunately, access is restricted. Goodies recorded from this site, and not from Saint Elie, include Guianan Red-Cotinga and Boat-billed Tody-Tyrant*.

Specialities

Rufous-thighed Kite, Slaty-backed Forest-Falcon, Grey-winged Trumpeter, Lilac-tailed and Sapphire-rumped Parrotlets, Tawny-bellied Screech-Owl, Amazonian Pygmy-Owl, Green Aracari, MacConnell's Spinetail, Band-tailed Antshrike, Guianan Red-Cotinga, Dusky Purpletuft, Crimson Fruitcrow, Capuchinbird, White-fronted Manakin, Boat-billed Tody-Tyrant*, Glossy-backed Becard, Guianan Gnatcatcher, Fulvous Shrike-Tanager, Blue-backed Tanager, Golden-sided Euphonia.

Others

Cinereous and Little Tinamous, Tiny, Bicoloured and Black-faced Hawks, Crested Eagle*, Black-and-white* and Ornate Hawk-Eagles, Collared Forest-Falcon, Little Chachalaca, Scaled Pigeon, Blue-and-yellow Macaw, Crested and Spectacled Owls, Grey Potoo, Reddish Hermit, Great Jacamar, Spotted Puffbird, Scaly-breasted Woodpecker, Striped Woodcreeper, Rufous-tailed Xenops, Spot-winged Antshrike, White-plumed and Scale-backed Antbirds, Spangled Cotinga, Bare-necked Fruitcrow, Sharpbill, Snethlage's Tody-Tyrant, Cinnamon-crested, White-throated and Golden-crowned Spadebills, Black-capped Becard, Red-billed Pied Tanager, Grey-headed and Red-shouldered Tanagers, Red-legged Honeycreeper, Epaulet Oriole.

Access

Contact the Tourist Office and/or travel agents who may be able to arrange a visit to this site.

SAUL

This remote gold-mining settlement, virtually in the middle of Guyane and accessible by air from Cayenne, is the gateway to some excellent forest. Although a number of species are uncommon owing to hunting, this is another site in Guyane full of cotingas.

Specialities

Rufous-thighed Kite, Black Curassow, Lilac-tailed Parrotlet, Amazonian Pygmy-Owl, Sooty Barbthroat, Band-tailed Antshrike, Dusky Purpletuft, Crimson Fruitcrow, Capuchinbird, White Bellbird, Blue-backed Tanager.

Others

Cinereous Tinamou, Rufescent Tiger-Heron, Black-faced Hawk, Crested Eagle*, Collared Forest-Falcon, Grey-necked Wood-Rail, Scaled Pigeon, Little Cuckoo, Crested and Spectacled Owls, White-tailed Nightjar, Black-throated Trogon, Great Jacamar, Spotted Puffbird, Yellow-tufted and Little Woodpeckers, Chestnut-crowned Foliage-gleaner, Rufous-tailed Xenops, Great and Spot-winged Antshrikes, Blackish Antbird,

Cinnamon-crested Spadebill, Long-tailed Tyrant, Black-capped Becard, Red-billed Pied Tanager, Grey-headed Tanager, Chestnut-bellied Seedeater, Epaulet Oriole.

Access

The well-maintained 90-km network of trails allows access to the forest.

Accommodation: Gite d'Etape (A).

West of Saul on the Suriname border is Maripasoula, accessible by air from Cayenne, a departure point for river trips into the forest. Birding here may be very interesting.

ADDITIONAL INFORMATION

Addresses

The Tourist Office, Jardin Botanique (Botanical Gardens), BP 801, or 12 rue Lalouette, Cayenne.
JAL Voyages (tel: 38-23-70) and Guyane Excursions, Centre Commercial Simarouba (tel: 32-05-41).

Books

Oiseaux de Guyane, Tostain, O. 1992. SEO.
A Guide to the Birds of Venezuela, Meyer de Schauensee, R. and Phelps, W., 1978. Princeton UP.

GUYANE ENDEMICS (1)	**BEST SITES**
Cayenne Nightjar*	Known only from a specimen collected in 1917, although a bird recorded at Saul in 1982 showed a number of this species' characteristics

NETHERLAND ANTILLES

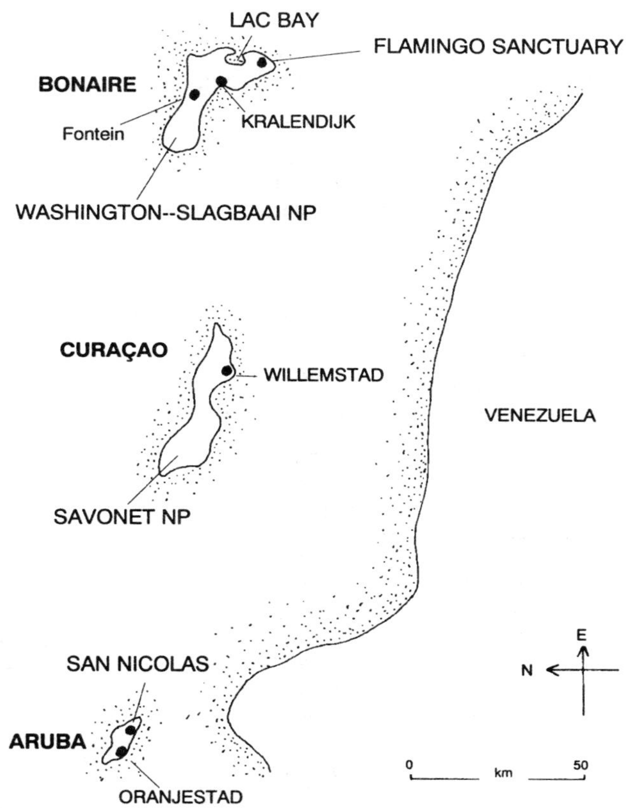

INTRODUCTION

Summary

The three main islands which comprise the (southern) Netherland Antilles lie just off the north Venezuela coast, and are usually considered to be geographically a part of South America rather than the Caribbean.

These mainly arid islands are a popular tourist destination and support few birds, although three species occur only here within South

America, and one, the Yellow-shouldered Parrot*, is more likely to be seen here than on mainland South America.

Size

At only 920 km^2 the Netherland Antilles are the smallest archipelago in South America. The largest of the three major islands, Curaçao, is only 60 km long and a few km wide.

Getting Around

Taxis are abundant though often expensive; there are buses on Curaçao, and cars may be hired.

Accommodation and Food

There are plenty of hotels to choose from, although these are expensive during the October–April high season. Seafood is a speciality, but the food is generally cosmopolitan.

Health and Safety

Beware of the sun and sandflies. All three islands are almost completely free of tourist-related crime.

Climate and Timing

Any time is a good time to visit since the temperature is more or less constantly hot. However, there is a light rainy season from October to January.

Habitats

The islands are low and arid with moon-like landscapes on Bonaire, and plenty of cacti and scrub on all three islands.

Conservation

There are NPs on Bonaire and Curaçao, and a few marine parks. Much of the thorny woodland which once covered the islands has been lost to tourist complexes and agriculture.

One threatened species occurs in the Netherland Antilles.

Bird Families

40 of the 92 families which regularly occur in South American are represented. None of these are endemic to the Neotropical region or South America itself.

Bird Species

236 species have been recorded on the Netherland Antilles, of which nearly 50% are passage and winter visitors. Although a low total, it is over 50 more than that for the Falkland Islands, and 100 more than that for the Galápagos.

Non-endemic specialities and spectacular species include Scaly-naped Pigeon, Caribbean Elaenia and Pearly-eyed Thrasher, all of which occur only here within South America, Brown and Black Noddies, Yellow-shouldered Parrot*, which is rare and hard to find on mainland South America, Ruby-topaz Hummingbird, Black-faced Grassquit and Yellow Oriole.

Endemics

There are no endemic species.

Expectations

A two-week visit during the winter months, November to March, when northern migrants such as shorebirds and warblers are present, may produce around 100 species.

ARUBA

This low, dry, barren, densely populated island, rising to just 188 m (617 ft), is some 30 km long. It is the most westerly of the three main islands, 68 km west of Curaçao and 24 km north of Venezuela. The island was once covered with thorny woodland, but the tourist complexes, cultivation and the consequent spread of semi-desert has left little. Remnant scrub may be found in the southeastern hills. Good birds include Ruby-topaz Hummingbird and Caribbean Elaenia, and there is an important tern colony on offshore islands near San Nicolas.

Specialities

Scaly-naped and Bare-eyed Pigeons, Caribbean Elaenia, Black-faced Grassquit.

Others

Magnificent Frigatebird, Brown Booby, Brown Pelican, Reddish Egret, Tricoloured Heron, Osprey, Caribbean Coot, Roseate, Bridled and Sooty Terns, Brown and Black Noddies, Black Skimmer, Brown-throated Parakeet, Burrowing Owl, White-tailed Nightjar, Ruby-topaz Hummingbird, Grey Kingbird, Yellow Oriole, Troupial.

Ruby-topaz Hummingbird occurs in the relatively well-vegetated dry river beds. Terns breed on the small islands off the south coast at San Nicolas. The fish ponds behind the tourist complex at Bubali attract herons, shorebirds and terns.

CURAÇAO

This, the middle of the three main islands, is the largest. It is 68 km east of Aruba, 48 km west of Bonaire and 64 km north of Venezuela. Virtually all of the thorny woodland which once covered much of the island has been lost, leaving remnant patches of cacti and scrub amidst the numerous fruit plantations. What woodland remains is mostly protected on the slopes of the hills (375 m/1,230 ft) within Savonet NP, in the northwest of the island. Good birds include Scaly-naped Pigeon and Ruby-topaz Hummingbird.

Specialities

Scaly-naped and Bare-eyed Pigeons, Black-faced Grassquit.

Others

Least Grebe, Magnificent Frigatebird, Tricoloured Heron, Yellow-crowned Night-Heron, Caribbean Coot, Wilson's Plover, Ruby-topaz

Hummingbird, Northern Scrub-Flycatcher, Black-whiskered Vireo, Tropical Mockingbird, Grasshopper Sparrow, Yellow Oriole, Troupial.

Scaly-naped Pigeon occurs in Savonet NP. Magnificent Frigatebirds occur at Saint Anna Bay near Willemstad (the capital). Coastal mangroves, mudflats and salt pans, mostly along the south coast, support breeding herons and attract wintering shorebirds.

BONAIRE

This low, arid island, rising to 239 m (784 ft), is the most easterly of the archipelago, 48 km east of Curaçao and 80 km north of Venezuela. The large Washington-Slagbaai NP (3,800 ha) protects the wooded hills, lagoons and coast of the northwest, and supports most of the best birds on the Netherland Antilles, including the rare Yellow-shouldered Parrot*. Bonaire is also the only island in the archipelago where Pearly-eyed Thrasher may be found. Although this species occurs throughout the West Indies, this is the only known site for it in South America. Ruby-topaz Hummingbird also occurs on Bonaire, helping to make it the best of the three islands for birding.

Specialities

Scaly-naped Pigeon, Yellow-shouldered Parrot*, Pearly-eyed Thrasher.

Others

Magnificent Frigatebird, Brown Booby, Greater Flamingo, Reddish Egret, Tricoloured and Great Blue Herons, White-tailed Hawk, Brown-throated Parakeet, Ruby-topaz Hummingbird, Northern Scrub-Flycatcher, Brown-crested Flycatcher, Grey Kingbird, Troupial.

Access

The best place to look for Yellow-shouldered Parrot* is within Washington-Slagbaai NP, although it has also been recorded from the rugged plains inland from the north and east coasts. The skulking Pearly-eyed Thrasher occurs in dense scrub, usually near water, amongst the fruit plantations around Fontein, in the north. Magnificent Frigatebirds occur at Kralendijk (the capital) and along with herons and shorebirds at Lac Bay, in the southeast. Greater Flamingo breeds at the Flamingo Sanctuary (55 ha) in the south of the island.

ADDITIONAL INFORMATION

Books

Birds of the Netherland Antilles, Voous, K. 1983. Curaçao. Privately published.

NETHERLAND ANTILLES ENDEMICS

None.

PARAGUAY

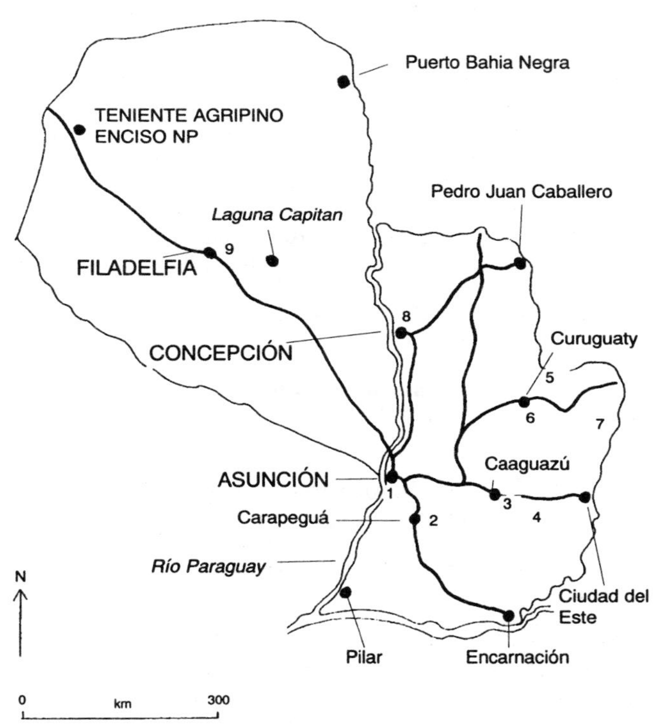

1	Asunción	6	La Golondrina PNR
2	Ybicuí NP	7	Estancia Itabo PNR
3	Estancia la Golondrina PNR	8	Concepción
4	Estancia San Antonio PNR	9	Filadelfia
5	Reserva Natural del Bosque Mbaracayú		

INTRODUCTION

Summary

Although prior permission is needed to visit Paraguay's best sites, this is a small, friendly country where local conservationists are doing their best to protect some fine tracts of Atlantic forest and open them up to ecotourism. The birder prepared to spend some extra time organising a trip to this country may see some of South America's most endangered birds, in these forests, as well as many chaco specialities.

Size

Paraguay (406,752 km^2) is three times the size of England, but considerably smaller than Texas.

Getting Around

The excellent bus services utilise the extensive paved road system, although this is susceptible to unpredictable flooding. Cars are available for hire. There is a good internal air network and a ferry service along the Río Paraguay between Asunción and Concepción. The few trains are slow.

Accommodation and Food

A wide variety of accommodation is available near the main routes, but scarce elsewhere. Bread, beef and soup are the major types of food.

Health and Safety

Immunisation against hepatitis, typhoid, tetanus, malaria, tuberculosis and even rabies is recommended. Dysentery is also present, but the major health problem seems to be Hookworm.

Paraguay is a very friendly country and crime is rare.

Climate and Timing

It is very hot and humid from January to March, cool and dry from April to September and wet from October to April. The austral spring, September to November, is the best time to visit.

Habitats

Paraguay is a very flat country where the land rises to a maximum of only 400 m (1,312 ft). The country is split into two by the southbound Río Paraguay. Northwest of the river lies the vast expanse of the chaco: a flat, hot, dry scrubby wilderness covering some 60% of the land surface. The southeast portion of the chaco, known as the 'low' chaco, is mainly marshy although palm forest is also present. The 'mid' chaco around Filadelfia is composed mainly of scrub and low woodland. In the far northwest, the 'high' chaco, towards Bolivia, more cacti appears as the land becomes remote semi-desert.

East of the Río Paraguay, the Oriental region consists largely of marshes in the southwest, campo and cerrado in the northeast and, most significantly, remnant humid, subtropical forest known as Atlantic forest, in the east and southeast. Much of this forest, of which little remains in Brazil, has been lost to agriculture but the remaining pockets support many endangered birds, endemic to this forest-type, which reaches its western limit here.

Conservation

Since 95% of Paraguay is in private ownership the Fundacion Moises Bertoni (FMB), a non-governmental conservation organisation, has set up a Private Nature Reserves (PNRs) scheme. Up to 1992 they had already established 42 PNRs, on private ranches with sympathetic owners. There are also over ten well-managed National Parks.

22 threatened bird species occur in Paraguay.

Bird Families

69 of the 92 families which regularly occur in South America are represented, over ten behind most of the bigger countries. 18 of the 25 Neotropical endemic families and four of the nine South American endemic families are represented. There are no Sunbittern, trumpeters, Magellanic Plover, seedsnipes, Hoatzin, Oilbird or barbets.

Well-represented families include rails and tyrants.

Bird Species

635 species have been recorded in Paraguay, nearly 200 more than Chile but over 600 fewer than Bolivia.

Non-endemic specialities and spectacular species include Solitary Tinamou*, Greater Rhea*, Ringed Teal, Crowned Eagle*, Black-fronted Piping-Guan*, Giant Wood-Rail, Black-legged Seriema, Blue-winged Macaw*, Red-spectacled* and Vinaceous* Parrots, Long-tailed Potoo, Plovercrest, Toco Toucan, Helmeted Woodpecker*, Great Rufous Woodcreeper, Canebrake Groundcreeper*, Lark-like Brushrunner, Rufous Gnateater, White-tipped Plantcutter, Bare-throated Bellbird*, Dinelli's Doradito*, Sao Paulo Tyrannulet*, Russet-winged Spadebill*, Strange-tailed Tyrant*, Ochre-breasted Pipit*, Green-chinned Euphonia*, Marsh* and Grey-and-chestnut* Seedeaters, and Saffron-cowled Blackbird*.

Endemics

There are no birds endemic to Paraguay, although many of those mentioned above and more are restricted to east Paraguay, southeast Brazil and northeast Argentina.

Expectations

Whilst it is possible to see around 150 species in a single day in the wet 'low' chaco, a ten-day trip is unlikely to produce much more than 250 species.

ASUNCIÓN

Map p. 296

The small, pleasant capital of Paraguay lies near some excellent birding sites, where the goodies include Great Rufous Woodcreeper, Ochre-breasted Pipit* and Grey-and-chestnut Seedeater*.

Specialities

Spotted Nothura, Giant Wood-Rail, Nanday Parakeet, Pale-crested Woodpecker, Great Rufous Woodcreeper, Black-capped Antwren, Plush-crested Jay, Creamy-bellied Gnatcatcher*, Ochre-breasted Pipit*, White-rimmed Warbler, Red-crested and Yellow-billed Cardinals, Red-crested Finch, Tawny-bellied, Dark-throated* and Grey-and-chestnut* Seedeaters, Golden-winged Cacique, Scarlet-headed Blackbird.

Others

Red-winged Tinamou, Whistling Heron, Maguari Stork, Savanna and Black-collared Hawks, Yellow-headed Caracara, Southern Lapwing, Large-billed and Yellow-billed Terns, Picazuro and Pale-vented Pigeons, Picui Ground-Dove, Peach-fronted and Monk Parakeets, Blue-winged Parrotlet, Canary-winged Parakeet, Guira Cuckoo, Black-throated Mango, Glittering-bellied Emerald, Gilded Sapphire, White-barred Piculet, Campo Flicker, White, Little and, Green-barred Woodpeckers, Planalto and Narrow-billed Woodcreepers, Rufous Hornero, Chotoy, Sooty-fronted and Pale-breasted Spinetails, Greater Thornbird, Firewood-gatherer, Great Antshrike, Pearly-vented Tody-Tyrant, Suiriri Flycatcher, Large Elaenia, Tawny-crowned Pygmy-Tyrant, Grey and White Monjitas, Yellow-browed Tyrant, Rufous Casiornis, Creamy-bellied Thrush, Chalk-browed Mockingbird, Masked Gnatcatcher, Flavescent Warbler, Saffron-billed Sparrow, Chestnut-vented Conebill, Hooded Tanager, Purple-throated Euphonia, Black-capped Warbling-Finch, Great Pampa-Finch, Greyish Saltator, Chestnut-capped and Chopi Blackbirds.

Access

ASUNCIÓN AREA

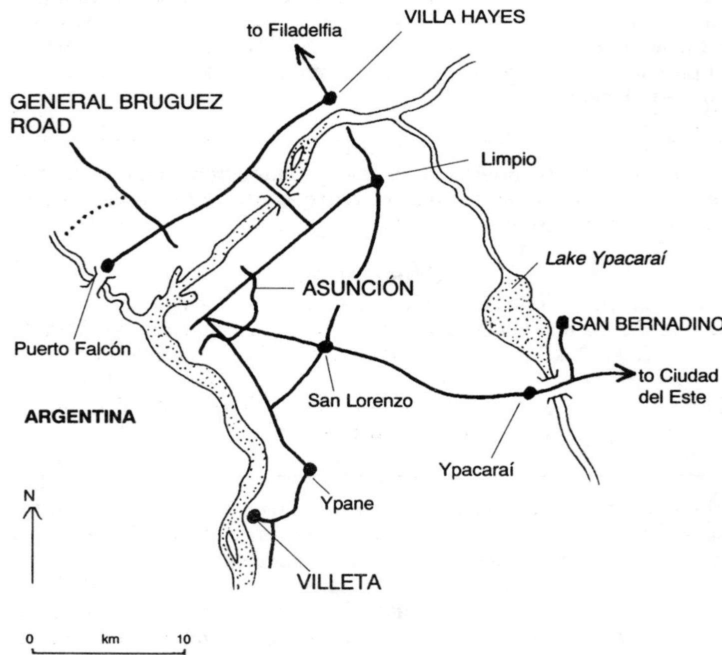

Bird the waterfront, the botanical gardens on Premier Presidente, 6 km northeast of the city centre and the Parque Nu Guazú nearby. Wetland species can be found on the northern side of Asunción Bay, reached by boat or via the road along the northeast side.

The small town of **Villeta** on the banks of the Río Paraguay, 27 km south of Asunción, is a good base from which to explore the surrounding area which supports some scarce birds. The track south from here is good for seedeaters.

The surroundings of **San Bernardino**, a popular resort town on the shores of the sterile Lake Ypacaraí, 56 km east of Asunción, supports some interesting birds. Head east out of Asunción on the Mariscal Estigarribia highway towards San Lorenzo. After 40 km a branch road leads off to San Bernardino.

There is some good chaco alongside the **General Bruguez road**, where Great Rufous Woodcreeper and Black-capped Warbling-Finch occur. Head northwest (opposite the Chaco-í road), 5 km northeast of Puerto Falcón, and bird along this road and to the south.

More chaco specialities occur in the scrub, palm savanna, woods, ponds and marshes around Villa Hayes. Take the ferry from Asunción across the Río Paraquay to Chaco-í. Bird the Pilcomayo road (route 12), the road to **Villa Hayes** and the road to Bolivia northwest from there, especially past Benjamin Aceval (9 km from the junction of Chaco 1 and Villa Hayes roads). These roads are often impassable in the wet season so it is best to visit between May and October.

Adventurous birders may wish to take the three-day boat trip (or a flight) from Asunción to Puerto Bahia Negra in the northeast chaco, where species akin to those of the Pantanal may be seen (see that site, in Brazil, p. 137).

YBICUÍ NP

Map p. 298

This small NP, 130 km southeast of Asunción, protects remnant humid forest with waterfalls and a reservoir, where the birds are similar to those of Iguazú/Iguaçu falls (see that site, in Argentina and Brazil, p. 62 and p. 112).

Specialities

Rufous-thighed Kite, Pheasant Cuckoo, Great Dusky Swift, White-spotted Woodpecker, Scaled Woodcreeper, Rufous Gnateater, Blue Manakin, Southern Antpipit, Southern Bristle-Tyrant*, Purplish and Plush-crested Jays, White-rimmed Warbler, Chestnut-headed, Ruby-crowned and Sayaca Tanagers, Red-crested Finch.

Others

Brazilian Teal, Chimango Caracara, Limpkin, Southern Lapwing, Grey-fronted Dove, Canary-winged Parakeet, Surucua Trogon, White-barred Piculet, Green-barred Woodpecker, Buff-browed, Ochre-breasted and White-eyed Foliage-gleaners, Sharp-tailed Streamcreeper, Variable Antshrike, Red-ruffed Fruitcrow, Wing-barred Manakin, Ochre-faced Tody-Flycatcher, Yellow Tyrannulet, Rufous-crowned Greenlet, Rufous-bellied and Creamy-bellied Thrushes, Black-goggled Tanager, Red-crowned Ant-Tanager, Violaceous Euphonia, Blue-naped Chlorophonia, Burnished-buff Tanager, Red-rumped Cacique, Epaulet

Oriole.

Access

Head southeast from Asunción towards Carmen. Turn left at Carapeguá and continue to the end of the road at Santa Rosa. Turn left here on to

YBICUÍ NP

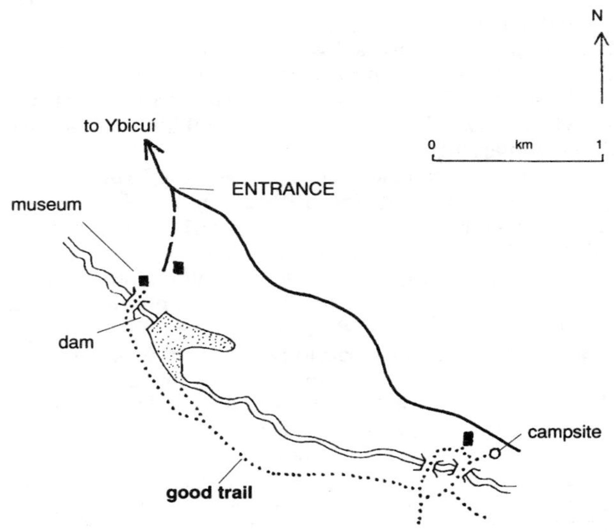

a track which leads to the NP. Bird the many trails.

Accommodation: camping. The nearest town with hotels is Ybicuí, 30 km away.

Collared Plover and Large-billed Tern occur on the Río Tebicuary at Villa Florida south of Carapeguá. Tufted Tit-Spinetail, Strange-tailed Tyrant*, Marsh Seedeater* and Saffron-cowled Blackbird* occur south and southwest of Villa Florida. They may be found along the road to Pilar, or the 50-km road to Ayolas, which leads south from the Asunción–Encarnación road, 260 km south of Asunción. From Ayolas it is possible to visit the Isla Yacyretá in the Río Parana. The Hotel El Tirol (B), 20 km northeast of Encarnación, lies in excellent forest with many trails worth exploring for Plovercrest, Spot-backed Antshrike, Dusky-tailed Antbird and Chestnut-headed Tanager.

CAAGUAZÚ

This town, approximately 180 km east of Asunción, on the main route east to Ciudad del Este lies near two fine birding sites, both of which require prior permission to visit.

ESTANCIA LA GOLONDRINA PNR

.Over 60% (15,477 ha) of this cotton and cattle ranch (24,077 ha) has been designated as a Private Nature Reserve, affording some protection to the island of Atlantic forest that still remains on the ranch. Prior permission from the FMB is required to visit this reserve, where over 200 species have been recorded, including such rarities as Long-tailed Potoo and Canebrake Groundcreeper*.

Specialities

Solitary Tinamou*, Pileated Parrot*, Long-tailed Potoo, Robust Woodpecker, Canebrake Groundcreeper*, Southern Bristle-Tyrant*, Sao Paulo* and Bay-ringed Tyrannulets*, Creamy-bellied Gnatcatcher*.

Others

King Vulture, Grey-headed Kite, Rusty-margined Guan, Plovercrest, Surucua Trogon, Rufous-capped Motmot, Rusty-breasted Nunlet, Rufous Gnateater.

Other Wildlife

18 species of large mammal, including Jaguar.

Access

Contact FMB for details of this site.

ESTANCIA SAN ANTONIO PNR

This small ranch (3,500 ha) supports 1,000 ha of Atlantic forest, where over 200 species have been recorded including the very rare Black-fronted Piping-Guan* and Helmeted Woodpecker*. Prior permission from FMB is required to visit this ranch.

Specialities

Black-fronted Piping-Guan*, Pileated* and Blue-fronted Parrots, Saffron* and Spot-billed Toucanets, Red-breasted Toucan, Helmeted* and Robust Woodpeckers, Dinelli's Doradito*, Southern Bristle-Tyrant*, Bay-ringed Tyrannulet*, Creamy-bellied Gnatcatcher*, Blackish-blue Seedeater*.

Others

Grey-headed Kite, Rusty-margined Guan, Chestnut-eared Aracari, Toco Toucan, Rufous Gnateater, Red-ruffed Fruitcrow.

Access

Contact FMB for details of this site.

The small Reserva Florestal, near Ciudad del Este, 327 km east of Asunción, on the Brazilian border, is worth a visit if in the area. Red-breasted Toucan, Robust Woodpecker, White-shouldered Fire-eye, Eared Pygmy-Tyrant, Green-backed Becard, and Chestnut-bellied Euphonia occur here. The reserve is 12 km west of Ciudad del Este. There is a pond surrounded by some forest with several trails.

*Birders prepared to make the effort involved in visiting the remote Atlantic forest reserves in east Paraguay may see the rare Helmeted Woodpecker**

CURUGUATY

This town, approximately 300 km northeast of Asunción on the road to Salta del Guaíra, is the nearest to three fine Atlantic forest reserves, all of which support the rare Black-fronted Piping-Guan* and Helmeted Woodpecker*.

RESERVA NATURAL DEL BOSQUE MBARACAYÚ

This big, remote, reserve (57,715 ha) contains the largest remaining area of Atlantic forest in Paraguay, as well as cerrado. Illegal logging, hunting (legal in the case of the native Ache indians) and marijuana cultivation are all problems, and yet over 250 species have been recorded, including Solitary Tinamou*, Crowned Eagle*, Vinaceous Parrot*, and the rare Russet-winged Spadebill*. Prior permission to visit is required from FMB.

Specialities

Solitary Tinamou*, Crowned Eagle*, Black-fronted Piping-Guan*, Pileated*, Blue-fronted and Vinaceous* Parrots, Sooty Swift, Saffron* and Spot-billed Toucanets, Red-breasted Toucan, Helmeted Woodpecker*, Purplish Jay, Bare-throated Bellbird*, Southern Bristle-Tyrant*, Bay-ringed Tyrannulet, Russet-winged Spadebill*, Creamy-bellied Gnatcatcher*, Green-chinned Euphonia*.

Others

King Vulture, Rusty-margined Guan, Red-and-green Macaw, Plovercrest, Surucua Trogon, Rufous-capped Motmot, Rusty-breasted Nunlet, Chestnut-eared Aracari, Toco Toucan, Rufous Gnateater, Red-ruffed Fruitcrow.

Other Wildlife

Jaguar, Tapir, Black Howler, Southern River Otter.

Access

The site is 40 km northeast of Curuguaty. The road in can be impassable in the wet season. Bird the track running east from the Jejuí-mi guardpost towards the Lagunita guardpost, parallel to the Río Jejuí. Solitary

Tinamou* and Green-chinned Euphonia* occur along here. The forest at the edge of the cerrado, which surrounds the Lagunita guardpost, holds Crowned Eagle* and Black-fronted Piping-Guan*, as well as Jaguar.

LA GOLONDRINA PNR

This large (54,000 ha) cattle ranch, of which part has been designated as a Private Nature Reserve, supports nearly 20,000 ha of Atlantic forest which, unfortunately, is still subject to logging and hunting. However, over 220 species have been recorded from here including Giant Snipe and Great Dusky Swift.

Specialities

Black-fronted Piping-Guan*, Giant Snipe, Great Dusky Swift, Saffron* and Spot-billed Toucanets, Red-breasted Toucan, Helmeted Woodpecker*, Bare-throated Bellbird*, Creamy-bellied Gnatcatcher*.

Others

Greater Rhea*, King Vulture, Sungrebe, Red-and-green Macaw, Surucua Trogon, Rufous-capped Motmot, Rusty-breasted Nunlet, Chestnut-eared Aracari, Toco Toucan, Red-ruffed Fruitcrow.

Other Wildlife

19 species of large mammal including Jaguar, Tapir and Southern River Otter.

Access

Contact FMB for details of this site.

ESTANCIA ITABO PNR

This reserve (8,000 ha) is an excellent model of sustainable forest use. The *palmito* 'heart of palm', which only grows in primary forest, is harvested here. Unfortunately, even sustainable forestry is beset by problems, which in this case involves 'palm-bandits', known locally as grilleros.

FMB are trying to encourage ecotourism at this site as part of their conservation package, but it is tricky to reach from the main road and prior permission is needed from FMB. However, those who make it may see the best selection of rare Atlantic forest endemics in Paraguay including the rare and restricted Red-spectacled* and Vinaceous* Parrots.

Specialities

Solitary Tinamou*, Black-fronted Piping-Guan*, Blue-winged Macaw*, Red-spectacled*, Blue-fronted and Vinaceous* Parrots, Least Pygmy-Owl, Great Dusky Swift, Saffron* and Spot-billed Toucanets, Red-breasted Toucan, Helmeted Woodpecker*, Bare-throated Bellbird*, Southern Bristle-Tyrant*, Sao Paulo* and Bay-ringed* Tyrannulets, Creamy-bellied Gnatcatcher*, Saffron-cowled Blackbird*.

Others

King Vulture, Black-and-white Hawk-Eagle*, Rusty-margined Guan, Plovercrest, Surucua Trogon, Rufous-capped Motmot, Rusty-breasted Nunlet, Chestnut-eared Aracari, Rufous Gnateater, Red-ruffed Fruitcrow.

Other Wildlife

Jaguar, Tapir.

Access

Contact FMB for details of this site.

Some of the birds present in Reserva Natural del Bosque Mbaracayú and Estancia Itabo PNR also occur in four small reserves run by Itaipu Binacional, a private company. These are **Itabo**, **Mbaracayú**, **Limoy** and **Tati Yupí Biological Reserves**, all of which require prior permission to visit from IB.

The little known **Cerro Corá NP** protects campo and cerrado, amidst the low mountains of north-central Paraguay near the Brazilian border. This park is 20 km west of Pedro Juan Caballero, accessible by air (or a very long drive) from Asunción and may be worth birding. There are trails near the HQ.

Accommodation: camping or hotels in Pedro Juan Caballero.

CONCEPCIÓN

The pleasant port of Concepción, 312 km north of Asunción, lies on the east bank of the Río Paraguay. There is some good chaco on the virtually uninhabited west bank where Crowned Eagle* occurs.

Specialities

Crowned Eagle*, Giant Wood-Rail, Nanday Parakeet, Checkered Woodpecker, Great Rufous Woodcreeper, Purplish and Plush-crested Jays, Red-crested Cardinal, Tawny-bellied Seedeater, Golden-winged Cacique, Screaming Cowbird.

Others

Southern Screamer, Whistling Heron, Rufescent Tiger-Heron, Maguari Stork, Plumbeous Ibis, Muscovy Duck, Snail Kite, Long-winged Harrier, Crane Hawk, Great Black-Hawk, Limpkin, Wattled Jacana, Black Skimmer, Scaly-headed Parrot, Guira Cuckoo, Nacunda Nighthawk, Blue-crowned Trogon, Toco Toucan, White-barred Piculet, White and Green-barred Woodpeckers, Red-billed Scythebill, Rufous Hornero, Sooty-fronted Spinetail, Common Thornbird, Tawny-crowned Pygmy-Tyrant, White Monjita, Rufous Casiornis, Red-crowned Ant-Tanager, Black-capped Warbling-Finch, Double-collared Seedeater, Solitary Cacique, Bay-winged Cowbird.

Access

The best place to bird on the west bank, accessible by bridge, is the small military post of Puerto Militar. There is a track into the chaco here. Boats can be hired at the port.

Accommodation: Hotel Victoria.

Concepción is directly accessible from Asunción by boat. The drive is very, very long.

FILADELFIA

The 472-km road through the chaco northwest from Asunción to this slightly whacky and neat German Mennonite agricultural town (304 km from Bolivia), offers an excellent chance to see many typical chaco species, with over 150 species possible in a day. Goodies include Paint-billed Crake and Black-legged Seriema.

Specialities

Spotted Nothura, Ringed Teal, Chaco Chachalaca, Rufous-sided Crake, Giant Wood-Rail, Paint-billed Crake, Spotted Rail, Black-legged Seriema, Checkered and Cream-backed Woodpeckers, Scimitar-billed Woodcreeper, Chaco Earthcreeper, Crested Hornero, Lark-like Brushrunner, Warbling Doradito, Purplish Jay, Yellow-billed Cardinal, Many-coloured Chaco-Finch, Tawny-bellied Seedeater.

Others

Greater Rhea*, White-tufted Grebe, Southern Screamer, Rosy-billed Pochard, Stripe-backed Bittern, Maguari Stork, Jabiru, Spot-flanked Gallinule, Red-legged Seriema, Nacunda Nighthawk, Gilded Sapphire, White Woodpecker, Chotoy Spinetail, White-tipped Plantcutter, Suiriri Flycatcher, Rufous Casiornis, White-naped Xenopsaris.

Access

The best places to bird along this long road are the bridges which cross the wetlands. Around Filadelfia, bird the side-tracks around town and nearby Loma Plata, especially to the south of Filadelfia, the rubbish dump, Parque Trebol, 5 km east of town, and Laguna Capitan, east of Loma Plata.

Accommodation: Filadelfia: Hotel Florida (B); Mennonite Hotel, Loma Plata (B).

The remote **Defensores del Chaco NP** and **Teniente Agripino Enciso NP**, which are hard to reach in wet weather, are little known and would no doubt produce something special for the adventurous birder with some time to spare.

ADDITIONAL INFORMATION

Addresses

FMB: Fundacion Moises Bertoni, Av. Gaspar Rodriquez de Francia 770, CC 714, Asunción (tel: 440 328 or 444 253; fax: 440 239). This is a non-governmental organisation keen on encouraging ecotourism.
Museo Nacional de Historia Natural del Paraguay, Ministerio de Agricultura y Granaderia, Sucursal 19, San Lorenzo. This is a department of the governmental Direccion de Parques Nacionales y Vida Silvestre.

IB: Itaipu Binacional, Calle De la Residenta 1075, Asunción (tel: 207161).

Books

A Birder's Field Checklist of the Birds of Paraguay, Hayes, F., Scharf, P. and Loftin, H., 1991.
Birds of Argentina and Uruguay: A Field Guide, Narosky, T and Yzurieta, D. 1989.(Covers 95% of birds occuring in Paraguay.)
Project Canopy 1992: BirdLife International Study Report No. 57, Brooks, T. *et al.*

PARAGUAY ENDEMICS

None.

PERU

1	Lima	7	Yungay
2	Lima–Oroya Road	8	Paracas NP and Pisco
3	Carpish Pass	9	Abancay
4	Tingo Maria	10	Machu Picchu
5	Iquitos	11	Abra Malaga
6	Trujillo–Chiclayo Circuit	12	Cuzco–Manu Road

13	Manu NP: Amazonia Lodge	16	Arequipa
14	Manu NP: Manu Lodge	17	Arequipa–Puno Road
15	Tambopata Reserve		

INTRODUCTION

Summary

Although terrorism prevents access to some areas of Peru, two of the best habitats in the world for birds, the east Andean slope temperate-subtropical forest belt, and west Amazonian lowland rainforest, are safely accessible and full of superb antbirds, cotingas and tanagers. These habitats exist together in Manu NP which, not surprisingly, supports more birds than any other NP in the world. It is also safe to look for high-Andes specialities, including Diademed Sandpiper-Plover*. Quite simply, Peru has (almost) got the lot. So, despite the dangers, the lack of a field guide and the limited, often remote, ranges of some of the most spectacular endemics, Peru is still a destination any birder should seriously consider.

Size

Peru is ten times the size of England and twice as large as Texas, a total of 1,285,216 km^2. However, since two-thirds of the country is virtually out-of-bounds for reasons of safety and topography, travelling distances between birding sites are relatively small.

Getting Around

Even though most of the roads are poor, thanks mainly to the terrain and climate, there is an excellent, extensive bus network. In the Amazon region travel is mainly restricted to boats and planes. There is an extensive internal air network but it is somewhat unreliable. The trains are slow but interesting, especially the one that takes visitors up to Machu Picchu from Cuzco (a good way to see Torrent Duck). Keep your passport handy ready for the numerous spot-checks, and beware of problems with cars at high altitude.

Accommodation and Food

Basic hotels can be found in most towns but beware of the best ones being fully booked before fiestas and market-days. Try to book in advance. Menus are dominated by various mixtures of rice, noodles, chicken and fish. The beer is good.

Health and Safety

Immunisation against hepatitis, typhoid, tuberculosis, malaria and rabies is recommended, and beware of altitude sickness in the high Andes.

Peru is a troubled country. The Sendero Luminoso (Shining Path) is a ruthless terrorist organisation which 'takes no prisoners', especially tourists. Do not enter an area controlled by them, and there are plenty of these. These 'no-go zones' usually lie beyond Tingo Maria, but the situation is always changing. Check with the embassy in your country. Over-zealous birders, including a friend of mine, have entered these areas and not lived to tell their tales of glorious birds. Birds are special, but so is life.

Having said that, many birders have enjoyed and survived travelling and birding around the south, and rarely felt threatened.

Climate and Timing

The coast is hot, sunny and humid between January and March. For the rest of the year it is usually foggy and cool, especially from June to October.

While the west Andes are hot and sunny all year, the central Andes are dry from May to September, and the east Andes and Amazon region generally wet all year, especially from January to April. June to September is the best time to visit this region.

High in the Andes it is very cold at night.

Habitats

Peru is blessed with a great range of habitats, from the cold Humboldt current flowing along its coast, with its attendant seabirds, across the coastal desert dotted with riverine oases, over the magnificent peaks (rising to a maximum at Huascaran of 6,768 m (22,205 ft)) of the Andes above the altiplano and lakes of central Peru, to the dripping wet east Andean elfin, temperate and subtropical forests, down to the lowland Amazonian rainforest in the north and east.

Conservation

Although many of Peru's habitats have been seriously degraded over the years and much forest has been lost especially at the eastern base of the Andes in the centre and north of the country, much of the remaining lowland forest in the east and north has so far escaped destruction, owing to its remoteness. Long may these forests remain inaccessible, at least to anyone not interested in sustainable ecotourism.

64 threatened species, of which no less than 31 are endemic, occur in Peru. A further 19 endemics are near-threatened.

Bird Families

86 of the 92 families which regularly occur in South America are represented, the greatest diversity of all the countries on the continent. These include 23 of the 25 Neotropical endemic families, equal highest with Bolivia and Brazil, and seven of the nine South American endemic families. Only seriemas and Magellanic Plover are absent.

Well-represented families include over 25 tinamous, some 50 parrots, over 100 hummingbirds, nearly 100 antbirds, 30 cotingas and 225 'tanagers'.

Bird Species

With a list of around 1,700 species (1,678 at the end of 1980 since when at least 12 have been added) Peru is virtually the richest country in the world for birds; only Colombia has more species. However, there are only 150 more species than Ecuador, which is one fifth the size of Peru.

Non-endemic specialities and spectacular species include Lesser Rhea*, Short-winged Grebe, Andean* and Puna* Flamingos, Zigzag Heron*, Andean Condor, Nocturnal Curassow, Sunbittern, Pale-winged Trumpeter, Rufous-bellied Seedsnipe, Peruvian Thick-knee, Andean Avocet, Diademed Sandpiper-Plover*, Inca Tern, Hoatzin, Peruvian Sheartail, Chestnut Jacamar, Ash-throated Gnateater, Black-necked Red-Cotinga, Chestnut-crested Cotinga, Band-tailed and Scarlet-breasted* Fruiteaters, White-browed Purpletuft, Purple-throated*, Plum-

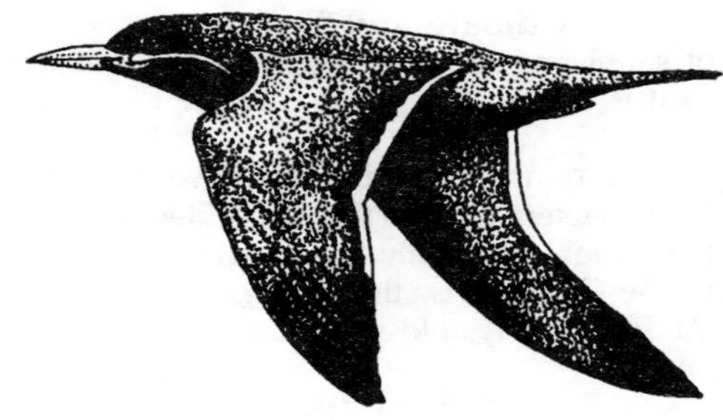

The unique Inca Tern occurs only along the coasts of south Peru and north Chile

throated and Spangled Cotingas, Amazonian Umbrellabird, Andean Cock-of-the-Rock, White-eared Solitaire, Pearly-breasted* and Tamarugo* Conebills, Grass-green, White-capped, Golden-collared and Yellow-scarfed Tanagers, Tit-like Dacnis*, Moustached Flowerpiercer, and Slender-billed Finch*.

Endemics

Over 100 (104) birds are unique to Peru. Only Brazil has more, and Colombia has nearly 50 fewer.

Many of Peru's endemics which include 17 hummingbirds, 26 furnariids, four cotingas and 21 'tanagers', are rare, endangered and little known, so very few are likely to be seen on a short trip to the south. However, possibilities include Bearded Mountaineer, Scarlet-hooded Barbet*, Red-and-white Antpitta, Black-faced Cotinga* and Inca Wren.

The treats awaiting the first birders to visit the centre and north, 'at the end of the Shining Path', include Long-whiskered Owlet, Marvellous Spatuletail*, Maranon Crescent-chest*, Masked Fruiteater and Golden-backed Mountain-Tanager*.

Expectations

A three-week trip to south Peru, running from the coast to the Amazon lowlands, is likely to produce around 600 species, perhaps even over 650.

LIMA

It is possible to see some excellent birds in and around Peru's often foggy capital. Amazilia Hummingbird and Chestnut-throated Seedeater can be seen in the parks and gardens of the city, whilst seabirds such as Inca Tern are possible at the shoreline, and Tawny-throated Dotterel has been recorded from local fields.

Specialities

Peruvian Thick-knee, Tawny-throated Dotterel, Grey Gull, Inca Tern, Croaking Ground-Dove, Amazilia and Oasis Hummingbirds, Short-tailed Field-Tyrant, Long-tailed Mockingbird, Chestnut-collared Swallow,

Slender-billed Finch*, Collared Warbling-Finch, Parrot-billed, Drab and Chestnut-throated Seedeaters.

Others

White-tufted Grebe, Peruvian Booby, Guanay Cormorant, Peruvian Pelican, White-cheeked Pintail, Plumbeous Rail, Least Seedsnipe, Wren-like Rushbird, Southern Beardless-Tyrannulet, Many-coloured Rush-Tyrant, Dark-faced Ground-Tyrant (April–Sep), Yellowish Pipit, Shiny Cowbird.

Access

Lying adjacent to the Humboldt current, Lima is a good place to see seabirds, especially from Miraflores, Chorillos and Villa beaches as well as Callao waterfront, where boats can be arranged to tour San Lorenzo Island. Parks in the centre of town as well as at San Isidiro, Miraflores and Independencia hold most of the landbirds.

Least Seedsnipe, Peruvian Thick-knee and Tawny-throated Dotterel all occur at **Puente San Pedro**, 34 km south of Lima. Head south out of Lima on the Pan-American highway and turn off towards the Pachacamac ruins at km 34. Turn right before the ruins, cross the highway and scan the fields before the bridge for Tawny-throated Dotterel, and after the bridge for Least Seedsnipe. Seabirds can also be seen around the island out to sea here.

Dark-faced Ground-Tyrant (April–Sep), Slender-billed Finch* and Parrot-billed Seedeater occur at **San Antonio**, 70 km southeast of Lima, along the Pan-American highway. Check the riverine scrub 1 km west of San Antonio (3 km from the highway).

Shorebirds sometimes occur in large numbers at **Ventanilla**, 35 km northwest of Lima. From Ventanilla head for Playa Ventanilla. Good pools lie alongside the road to the coast, 3 km away.

Accommodation: Lima: Hostal Senor Roial in Miraflores.

CENTRAL HIGHWAY

Map p. 310

This is the road (4,550 m/14,928 ft) which leads inland from Lima, ascending to 4,550 m/14,928 ft before dropping into the Amazonian lowlands near Pucallpa. The Shining Path operate sections of this road, especially beyond Tingo Maria. Check to see if it is safe before travelling.

LIMA–LA OROYA ROAD

Map p. 310

A few sites along this 200-km stretch of road support a number of endemics including Bronze-tailed Comet* and White-cheeked Cotinga*, and Milloc Bog, near Marcapomacocha, is one of the few known sites for the almost mythical Diademed Sandpiper-Plover*.

Endemics

Black-breasted Hillstar, Bronze-tailed Comet*, Black Metaltail, Dark-winged Miner, Striated Earthcreeper, White-bellied Cinclodes*, Rusty-

CENTRAL HIGHWAY

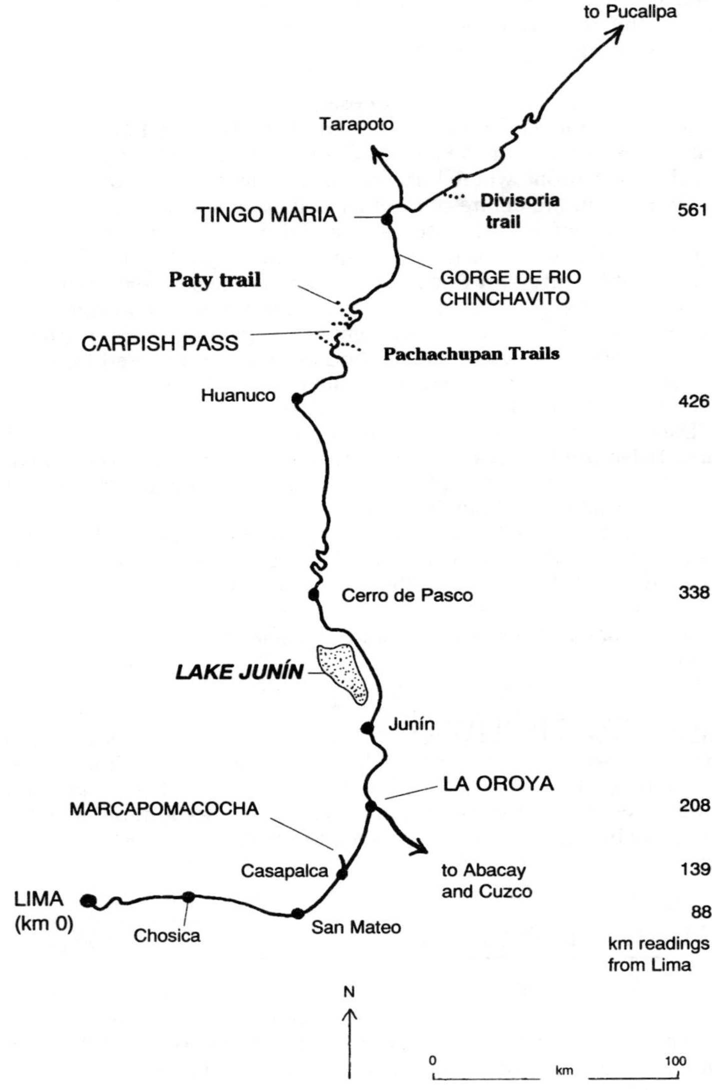

crowned Tit-Spinetail, Canyon and Junin Canasteros, White-cheeked Cotinga*, Rusty-bellied Brush-Finch, Great Inca-Finch, Rufous-breasted Warbling-Finch*.

Specialities

Giant Coot, Diademed Sandpiper-Plover*, Olivaceous Thornbill, Oasis Hummingbird, Peruvian Sheartail, Plain-breasted Earthcreeper, Streak-throated Canastero, Pied-crested Tit-Tyrant, D'Orbigny's Chat-Tyrant, Giant Conebill*, Collared Warbling-Finch.

Others

Ornate and Puna Tinamous, Silvery Grebe, Puna and Black-faced Ibises, Andean Condor, Puna Snipe, Grey-breasted Seedsnipe, Andean Lapwing, Bare-faced Ground-Dove, Mountain Parakeet, Andean Swift, Giant Hummingbird, Common and Slender-billed Miners, White-winged Cinclodes, Andean Tit-Spinetail, Stripe-headed Antpitta, Yellow-billed and Tufted Tit-Tyrants, White-browed Chat-Tyrant, Rufous-webbed Tyrant, Black-billed Shrike-Tyrant, Plain-capped and White-fronted Ground-Tyrants, Andean Negrito, Brown-bellied Swallow, Correndera Pipit, Black Siskin, Mourning and Ash-breasted Sierra-Finches, White-winged Diuca-Finch, Bright-rumped Yellow-Finch, Black-throated Flower-piercer, Golden-billed Saltator.

Access

Check the shrubby roadsides at km 70 for Mountain Parakeet, Oasis Hummingbird and Collared Warbling-Finch. Stop at **San Mateo** (km 88) (accommodation: Hotel Andino), and bird the trails above and below the hotel. Black Metaltail, Rusty-crowned Tit-Spinetail, Rufous-webbed Tyrant and Rusty-bellied Brush-Finch occur here.

From San Mateo head up to Casapalca (km 139). Turn north off the central highway shortly after here to Chinchan and take the rough road towards Marcapomacocha. Ignore the right turning to Marcapomacocha after 28km (but look for the rare endemic White-bellied Cinclodes* at this junction) and continue to the next rise. Below this rise to the right is **Milloc Bog** (Map p. 312). Diademed Sandpiper-Plover* occurs alongside the streams running into this bog, on the far side (try to stem your excitement and walk over slowly for this is high-altitude birding). Puna Tinamou, Andean Condor, Black-breasted Hillstar, Dark-winged Miner, Striated Earthcreeper and Junin Canastero also occur here. Giant Coot occurs on two lakes over the hill from the bog in the direction of Milloc and on the lake at **Marcapomacocha** village (Map p. 312).

Accommodation: Chosica: Hostal Hans.

The central highway runs north from La Oroya to Junín and, further north, passes close to the eastern shore of **Lake Junín** (4,080 m/13 385 ft), through wet fields full of birds. The rare Puna Grebe* is endemic to this lake. It is now threatened by declining water quality in the lake and seems to be heading for extinction. The best place to look for it is near Ondores. Turn off the central highway into the town of Junín then take the road west, then northwest around the southern shores of the lake to Ondores. In Ondores take the north fork closest to the lake shore. Check

MILLOC BOG AND MARCAPOMACOCHA

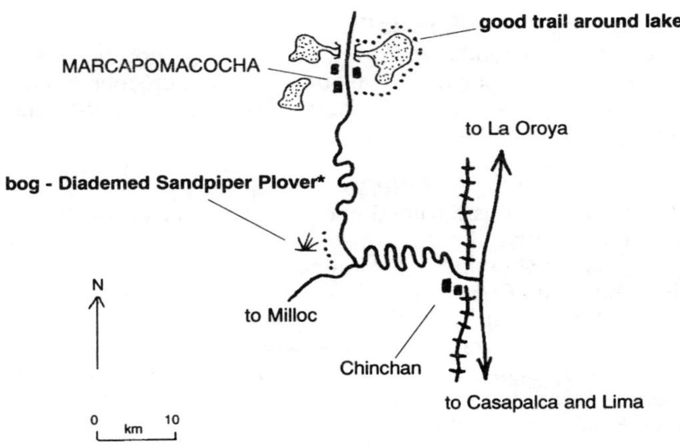

the lake between 3 and 4 km along here where the water is usually deep and relatively reed-free. A 'scope or, to be certain of seeing this bird, a boat, may be required. Black-breasted Hillstar also occurs here.

The central highway then climbs steeply beyond Cerro de Pasco towards Huanuco, which makes stopping difficult, but try, because Brown-flanked Tanager occurs in the dense thickets near the road. The fields around **Ungymaran** are a good place to look.

Striated Earthcreeper, Stripe-headed Antpitta, Pied-crested Tit-Tyrant and Giant Conebill* occur in the remnant *Polylepis* woodland (if it still remains) on the east side of the road just before the village of **La Quinua** (km post 320). Walk up the irrigation channels into the *Polylepis*.

The rare endemic Rufous-backed Inca-Finch has also been recorded near the road between Junín and Huanuco.

Accommodation: Huanaco: Hotel Turistas.

CARPISH PASS (HUANUCO–TINGO MARIA ROAD)

Map opposite

Just one hour north of the arid Huanuco area, the central highway enters east Andean slope temperate forest with stands of *Chusquea* bamboo. Many star birds occur here, not least the endemic Masked Fruiteater, but the real dazzler, Golden-backed Mountain-Tanager*, occurs only way above the road, and a mini-expedition is necessary to look for it. Those that manage it could be rewarded with one of South America's most fabulous birds.

Endemics

Bay Antpitta, Masked Fruiteater, Inca Flycatcher, Rufous-browed Hemispingus*, Brown-flanked Tanager, Golden-backed Mountain-Tanager*, Pardusco.

Specialities

Fasciated Tiger-Heron*, White-throated Quail-Dove, Golden-plumed Parakeet*, Speckle-faced Parrot, Chestnut-breasted Coronet, Emerald-bellied Puffleg, Chestnut-crested Cotinga, Band-tailed Fruiteater, Black-throated Tody-Tyrant, Ochraceous-breasted Flycatcher, White-collared Jay, White-eared Solitaire, Giant Conebill*, Drab Hemispingus, Yellow-scarfed Tanager, Chestnut-bellied Mountain-Tanager, Golden-collared Honeycreeper, Moustached Flower-piercer.

CARPISH PASS

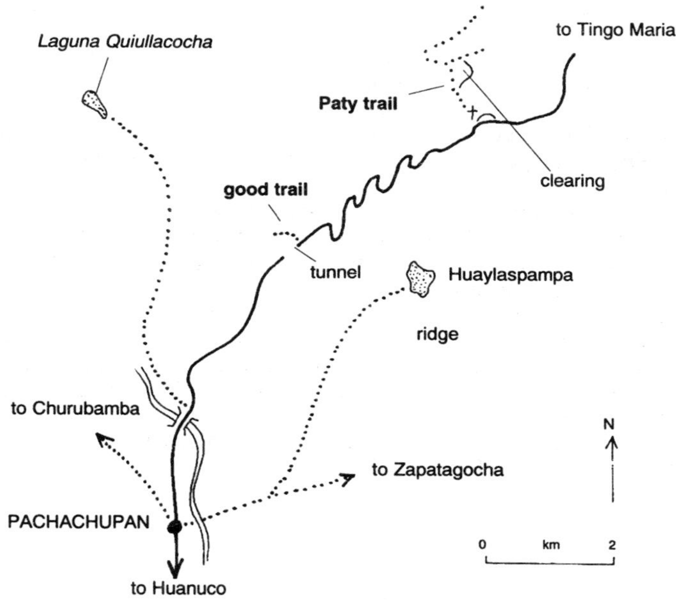

Others

Brown Tinamou, Sickle-winged Guan, Scaly-naped Parrot, Buff-tailed Sicklebill, Speckled Hummingbird, Mountain Velvetbreast, Violet-throated Starfrontlet, Sword-billed Hummingbird, Rufous-capped Thornbill, Long-tailed Sylph, Masked Trogon, Grey-breasted Mountain-Toucan*, Yellow-vented and Crimson-mantled Woodpeckers, Olive-backed Woodcreeper, Rufous Spinetail, Spotted Barbtail, Long-tailed Antbird, Undulated Antpitta, Green-and-black and Barred Fruiteaters, Amazonian Umbrellabird, Andean Cock-of-the-Rock, Rufous-headed Pygmy-Tyrant, Golden-faced Tyrannulet, Highland Elaenia, Mottle-cheeked Tyrannulet, Crowned and Rufous-breasted Chat-Tyrants,

Smoky Bush-Tyrant, Rufous-tailed Tyrant, Pale-edged Flycatcher, Sepia-brown Wren, Pale-footed Swallow, Blue-backed and Capped Conebills, Grass-green, Rufous-chested and Olive Tanagers, Bronze-green Euphonia, Flame-faced, Golden-naped and Beryl-spangled Tanagers, Mountain Cacique.

Access

Golden-backed Mountain-Tanager* occurs only well above the road on trails that lead from **Pachachupan**, a few km above the town of Acomayo, south of the Carpish tunnel. Searching for this bird involves a long backpacking trek. However, hardy birders may also see Rufous-browed Hemispingus* and Pardusco here, two more species that rarely descend to the level of the road.

Stop on the Tingo Maria side of the **Carpish tunnel** (km 467) and bird the track on the left-hand side. Roadside birding either side of the tunnel can also be good. Chestnut-bellied Mountain-Tanager and Moustached Flower-piercer occur here. 4.4 km north of the Tingo Maria side of the tunnel stop at the lay-by just below the 'Wilson Ramirez G' shrine on the left. The once brilliant **Paty trail** starts behind this shrine. It leads down, steeply, a few thousand feet to the Paty tea plantation. Although habitat degradation occurs apace, this area may still abound with birds, as it used to. Goodies include Grey-breasted Mountain-Toucan*, Chestnut-crested Cotinga, Band-tailed and Masked Fruiteaters, Inca Flycatcher, White-eared Solitaire and Yellow-scarfed Tanager.

Fasciated Tiger-Heron*, Amazonian Umbrellabird and Andean Cock-of-the-Rock occur at the **Gorge de Río Chinchavito** (km 494), between the tunnel and Tingo Maria. Bird the roadside either side of the 'torrents'.

TINGO MARIA

Map p. 310

This is a very dangerous drug-trafficking region under the control of the Shining Path. Two british birders were murdered near here in 1990. Do not go unless you are sure that it is safe.

Endemics

Masked Fruiteater, Huallaga Tanager.

Specialities

Hooded Tinamou*, Blue-headed Macaw, Oilbird, Gould's Jewelfront, Bluish-fronted Jacamar, Black-streaked Puffbird, Yellow-billed Nunbird, Brown-mandibled Aracari, Bar-breasted Piculet, Crested Foliage-gleaner, Yellow-breasted Antwren, Blackish and Sooty Antbirds, White-browed Purpletuft, Olivaceous Piha, Little Ground-Tyrant, Olivaceous Greenlet, Yellow-crested, Vermilion and Masked Crimson Tanagers, Large-billed Seed-Finch*, Pale-eyed Blackbird.

Others

Brown Tinamou, Plain-breasted Hawk, Wattled Guan*, White-throated Quail-Dove, Military Macaw, White-eyed Parakeet, Blue-winged Parrotlet, Blue-headed Parrot, Lyre-tailed Nightjar, White-bearded and Little Hermits, Grey-breasted Sabrewing, Blue-fronted Lancebill, Long-billed Starthroat, Crested and Golden-headed Quetzals, Lettered and Chestnut-eared Aracaris, Black-mandibled and Cuvier's Toucans, Crimson-bellied Woodpecker, Black-banded Woodcreeper, Black-billed Treehunter, Stipple-throated, Rufous-tailed and Rufous-winged Antwrens, Spot-winged and White-plumed Antbirds, Rufous-breasted Antthrush, Scaled Antpitta, Rufous-vented Tapaculo, Red-ruffed Fruitcrow, Amazonian Umbrellabird, Andean Cock-of-the-Rock, Ochre-bellied Flycatcher, Torrent Tyrannulet, Scale-crested Pygmy-Tyrant, Fulvous-breasted Flatbill, Yellow-breasted, Ornate and Cliff Flycatchers, Black Phoebe, Rufous-tailed Tyrant, White-winged Black-Tyrant, Slaty-capped Shrike-Vireo, Black-billed Thrush, Grey-mantled Wren, Olivaceous Siskin, Neotropical River Warbler, Yellow-browed and Orange-billed Sparrows, Bronze-green Euphonia, Green-and-gold, Golden-eared, Flame-faced, Blue-browed and Blue-necked Tanagers.

Access

The **Cueva de las Lechuzas** in Tingo Maria NP is 8 km north of Tingo Maria. A track which is good for Military and Blue-headed Macaws, as well as Huallaga Tanager, leads through second-growth forest from the entrance to the caves where the Oilbirds are, and beyond.

The highly localised Pale-eyed Blackbird occurs at the **Santa Lucia Marsh** 28 km northwest of Tingo Maria. Bear left on to the Aucayacu road 15 km beyond Tingo Maria. 13 km along here look for a drinking hole on the left. A trail leads right just before this and soon reaches a small, obscure marsh in the midst of pasture. Look for the birds in the trailside bushes and reeds.

DIVISORIA TRAIL

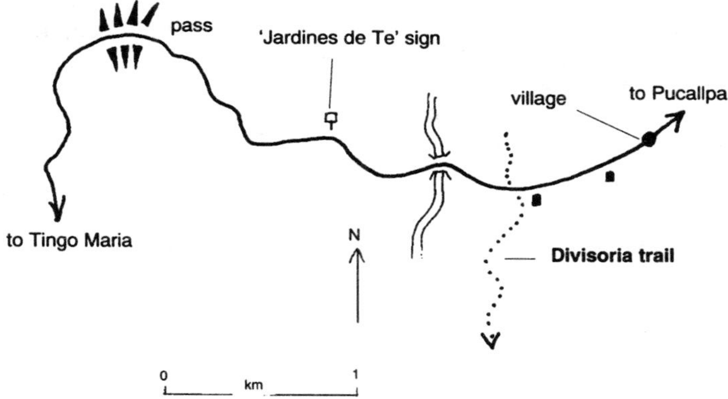

Accommodation: Tingo Maria: Hotel Turistas.

Heading northeast along the central highway *en route* to Pucallpa there are some excellent birding sites. After 28km there are a number of paths through the forest at **Santa Elena**. Look for these paths on the south side of the road. They lead through coffee plantations to fast decreasing remnant forest on the steep slopes beyond. Military Macaw, Gould's

Jewelfront, Black-billed Treehunter, Crested Foliage-gleaner, White-plumed Antbird and Olive Tanager all occur here.

Beyond here the central highway enters the temperate forest of the Cordillera Azul. 38 km northeast of Tingo Maria there is a pass, and just after this is the entrance to the signposted 'Jardines de Te' tea-plantation. 1 km further on, on the right-hand side of the road, is a house where the **Divisoria trail** (Map p. 315) starts. Although subject to rapid habitat degradation, it may still be possible to find Plain-breasted Hawk, Black-streaked Puffbird, Brown-mandibled Aracari, Yellow-breasted Antwren, Blackish Antbird, Rufous-breasted Antthrush, Olivaceous Piha, Grey-mantled Wren and Blue-browed Tanager here.

Amazonian Umbrellabird and Andean Cock-of-the-Rock occur in the **Bosqueron del Padre Abad canyon** 40 km further on towards Aguyatia. Bird either side of the tunnel, especially the trail which leads south, just before the tunnel, over a bridge to a waterfall where the Cock-of-the-Rocks are usually found.

Some good birding is possible in the rainforest at Pucallpa, especially around Lake Yarinacocha.

IQUITOS

Over 500 species have been recorded from the lowland Amazonian rainforest near the city of Iquitos, next to the Amazon in northeast Peru, including some 40 species which do not occur at Tambopata Reserve (see p. 335) and Manu NP (see p. 332) in south Peru. Here, it is possible to follow a roving flock of birds in the canopy (without breaking your neck) for over 400 m, for what is arguably the best canopy walkway in the world, reaching over 30 m (100 ft) above the forest floor, has recently been constructed.

Some rare Amazonian birds occur at Iquitos, including Zigzag Heron*, Nocturnal Curassow, Red-billed Ground-Cuckoo, Black-necked Red-Cotinga and Pearly-breasted Conebill*.

Specialities

Zigzag Heron*, Nocturnal Curassow, Short-tailed Parrot, Red-billed Ground-Cuckoo, Black-throated Brilliant, Brown Nunlet, Bay Hornero, Undulated and Pearly Antshrikes, Black Bushbird, Dugand's and Ash-winged Antwrens, Yellow-browed, Black-and-white, Slate-coloured, White-shouldered, Sooty and Hairy-crested Antbirds, Black-spotted and Reddish-winged Bare-eyes, Ochre-striped Antpitta, Chestnut-belted Gnateater, Black-necked Red-Cotinga, White-browed Purpletuft, Purple-throated* and Plum-throated Cotingas, Black-and-white Tody-Tyrant, Golden-winged Tody-Flycatcher, Lesser Wagtail-Tyrant, Riverside Tyrant, Dusky-chested Flycatcher, Greater Schiffornis, Pearly-breasted Conebill*, Masked Crimson Tanager, Band-tailed Oropendola.

Others

Undulated Tinamou, Horned Screamer, Agami Heron*, Speckled Chachalaca, Sunbittern, Large-billed Tern, Tropical Screech-Owl, Crested and Spectacled Owls, Pavonine Quetzal, Yellow-billed Jacamar, Lanceolated Monklet*, Curve-billed Scythebill, White-bellied Spinetail, Short-billed and Black-tailed Leaftossers, White-plumed and Bicoloured Antbirds, Spangled Cotinga, Amazonian Umbrellabird,

Wire-tailed Manakin, Paradise Tanager.

Access

There are no roads to Iquitos, but there is a good air service. Numerous travel companies operate in the city, offering a whole range of 'jungle' trips, virtually all of which are very expensive. Check with these travel companies which lodge has the new canopy walkway. Currently, the best two lodges for birding are Explorama Lodge at Yanamono, three hours down river from Iquitos, which has an excellent network of trails, and Explornapo Camp at Llachapa, 140 km from Iquitos on the Río Napo. Nocturnal Curassow and Black-necked Red-Cotinga occur here, as well as the rare Pearly-breasted Conebill* on islands in the river.

TRUJILLO–CHICLAYO CIRCUIT

Maps p. 319, p. 320 and p. 321

Adventurous birders willing to try this long, long circuit (over 1,000 km) may be rewarded with a good selection of Tumbesian (birds restricted to southwest Ecuador and northwest Peru) and Maranon (birds restricted to north Peru) endemics, including some of the rarest and least-known birds in South America; such enigmas as White-winged Guan*, Long-whiskered Owlet*, Marvellous Spatuletail*, one of the world's most fabulous hummingbirds, Maranon Crescent-chest*, Peruvian Plantcutter*, Maranon Thrush, and White-capped and Orange-throated* Tanagers.

Endemics

White-winged Guan*, Peruvian Pigeon*, Yellow-faced Parrotlet*, Long-whiskered Owlet*, Spot-throated Hummingbird, Purple-backed Sunbeam*, Neblina Metaltail*, Marvellous Spatuletail*, Black-necked Woodpecker, Coastal Miner, Great Spinetail*, Chestnut-backed Thornbird*, Pale-billed and Ochre-fronted* Antpittas, Maranon Crescent-chest*, Large-footed Tapaculo, Peruvian Plantcutter*, Tumbes Tyrant*, Rufous Flycatcher, Orange-throated Tanager*, Cinereous Finch, Grey-winged*, Buff-bridled and Little* Inca-Finches.

Specialities

Bearded Guan*, Imperial Snipe*, Peruvian Thick-knee, Ochre-bellied Dove*, Red-masked Parakeet*, Pacific Parrotlet, Peruvian Screech-Owl, Grey-chinned Hermit, Amazilia Hummingbird, Rainbow Starfrontlet, Chestnut-breasted Coronet, Purple-throated Sunangel, Emerald-bellied Puffleg, Purple-collared Woodstar, Scarlet-backed and Guayaquil Woodpeckers, Necklaced and Line-cheeked Spinetails, Henna-hooded Foliage-gleaner*, Collared and Chapman's Antshrikes, Grey-headed Antbird*, Elegant Crescent-chest, Chestnut-crested Cotinga, Rufous-winged Tyrannulet, Short-tailed Field-Tyrant, Sooty-crowned and Baird's Flycatchers, Turquoise and White-tailed Jays, Plumbeous-backed and Maranon Thrushes, Long-tailed Mockingbird, Fasciated and Superciliated Wrens, Chestnut-collared Swallow, Bar-winged Wood-Wren, Grey-and-gold and Three-banded Warblers, Black-capped and Tumbes Sparrows, Bay-crowned, White-winged and White-headed Brush-Finches, White-capped, Red-hooded and Silver-backed Tanagers,

The aptly named Marvellous Spatuletail is a rare bird with a limited distribution in north Peru*

Collared Warbling-Finch, Sulphur-throated Finch, Parrot-billed and Chestnut-throated Seedeaters, Masked Saltator*, White-edged Oriole, Peruvian Meadowlark.

Others

Barred Forest-Falcon, Least Seedsnipe, Tawny-throated Dotterel, Andean Lapwing, Andean Gull, Bronzy and Collared Incas, Golden-headed Quetzal, Grey-breasted Mountain-Toucan*, Crimson-bellied Woodpecker, Tyrannine Woodcreeper, Striped Treehunter, Chestnut-crowned Antpitta, Unicoloured Tapaculo, Barred Fruiteater, Black-capped and Mouse-coloured Tyrannulets, Yellow-billed Tit-Tyrant, Tawny-crowned Pygmy-Tyrant, White-winged Black-Tyrant, Masked Water-Tyrant, Spectacled Redstart, Rufous-naped Brush-Finch, Grass-green Tanager, Yellow-throated and Ashy-throated Bush-Tanagers, Rufous-crested Tanager, Blue-winged Mountain-Tanager, Metallic-green Tanager.

Access

The farmland northeast of the Villa del Mar suburb of Trujillo, near the Chan-Chan ruins, and the area around the seaside resort of Juanchaco, 5 km from Villa del Mar, support Short-tailed Field-Tyrant and Black-capped Sparrow.

To begin the circuit head northeast out of Trujillo towards Huamachuco, 190 km east of Trujillo. Purple-backed Sunbeam* occurs at Soquian near Succha, 45 km east of Huamachuco and at Molino, 20 km northwest of Succha (10 km beyond Aricapampa).

Head north from Huamachuco 170 km to Cajamarca, a good place to stay the night, then turn east on to the road to Leimeibamba. West of Balsas, *en route* to Leimeibamba, bird the Acacia woodland and scrub above the farms of 'Hacienda Limon', looking for Chestnut-backed Thornbird* and Grey-winged Inca-Finch*. Peruvian Pigeon occurs

TRUJILLO–CHICLAYO CIRCUIT

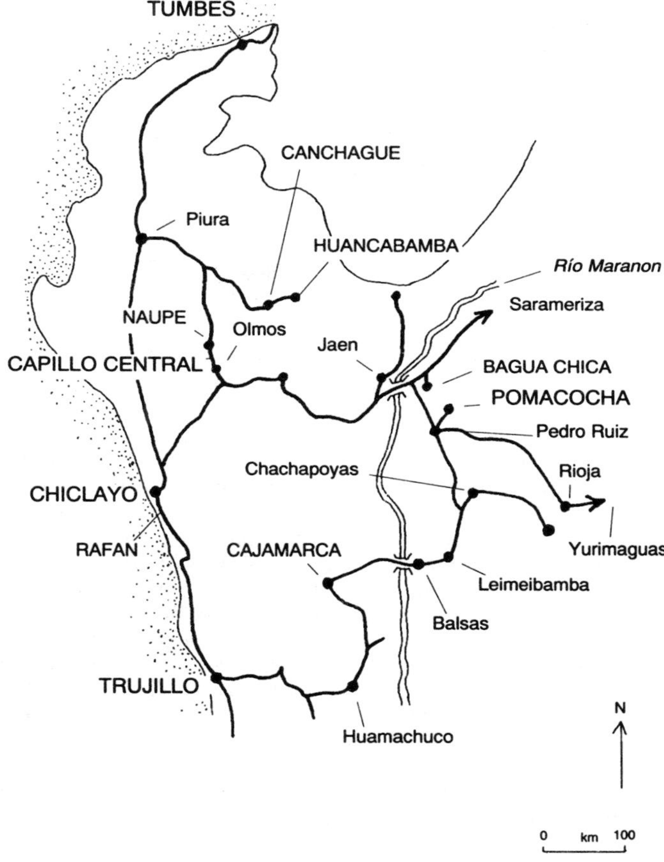

along the river near Balsas while the beautiful Buff-bridled Inca-Finch is found in the dry forest above Balsas. Great Spinetail* occurs in the dense shrubbery on steep slopes (between 2,100 m/6,890 ft and 2,500 m/8,202 ft) near the pass west of Leimeibamba.

Head north from Leimeibamba towards Chachapoyas, ignore the turning to Chachapoyas and continue north to Pedro Ruiz at the junction with the the Jaen–Rioja road. Turn east on to the Pedro Ruiz–Rioja road, which passes through Eastern Cordillera temperate forest 80 km northwest of Rioja. Long-whiskered Owlet*, Pale-billed and Ochre-fronted* Antpittas, and Bar-winged Wood-Wren were all discovered along this road.

POMACOCHA

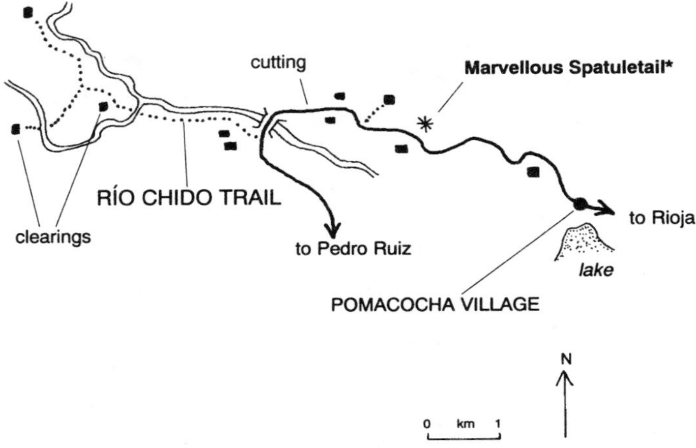

The aptly named Marvellous Spatuletail* is confined to a 100-km stretch of forest along the right bank of the Rio Utcubamba, above Pedro Ruiz. To look for this spectacular localised endemic drive for two hours east from Pedro Ruiz towards the small town of **Pomacocha** (also known as Florida) (Map above). The road crosses a metal bridge 5 km west of Pomacocha. The Rio Chido Trail, which starts here, to the left just before the bridge, and the road between the bridge and Pomacocha, support Marvellous Spatuletail* as well as Chestnut-breasted Coronet, Large-footed Tapaculo, Chestnut-crested Cotinga, Grass-green, White-capped, Red-hooded, Metallic-green (km 320, below the bridge) and Silver-backed Tanagers. Good forest may still remain here, especially along the Río Chido trail.

Accommodation: Pomacocha: Hotel de Turista (A).

Returning to Pedro Ruiz head north towards Jaen. Adventurous birders will consider travelling east from Bagua Chica, 70 km northwest of Pedro Ruiz, on the road to Aramongo and Chiriaco, into the lowland rainforest. The track from the military base at Mesones Muro to the river-post at Urakusa crosses low hills with forested crests where the beautiful Orange-throated Tanager* has been recorded. Maranon Crescent-chest*, Maranon Thrush and Little Inca-Finch* occur in the scrub north and west of Bagua Chica and along the road to Jaen, some 50 km away to the west.

Maranon Thrush also occurs at **Balsahuaycu**, 15 km from Jaen. From Jaen there is a trail on the left just before km post 8. The road between Balsahuaycu and Chamaya (8 km away) is also worth birding, especially east of Chamaya, where Spot-throated Hummingbird and Little Inca-Finch* occur.

Accommodation: Jaen: Hostal Bellow Horizonte (C).

From here head west towards Olmos, 190 km away. Henna-hooded Foliage-gleaner* and Masked Saltator* occur in a ravine 34 km east of Olmos.

White-winged Guan*, Ochre-bellied Dove*, Peruvian Screech-Owl, Spot-throated Hummingbird, Collared Antshrike, Elegant Crescent-chest, Rufous-winged Tyrannulet, White-tailed Jay, Plumbeous-backed Thrush, Three-banded Warbler and Black-capped Sparrow all occur on the trail on the north side of the road 19 km east of Olmos. This trail is over 30 km long and leads through dry forest.

From Olmos head southwest to Chiclayo, to complete the circuit, or turn north from Olmos towards Piura. White-winged Guan*, Tumbes Tyrant* and Tumbes Sparrow occur at **Capillo Central**, 6 km south of Naupe and 50 km north of Olmos (Map below). Bird the dry forest east of the road, reached via a sandy track opposite the church (km post 919). This leads to some houses next to a river (4 km) and on to the San Isidro valley (another 10 km) where the guan was rediscovered in 1977. Ask for guidance at the river.

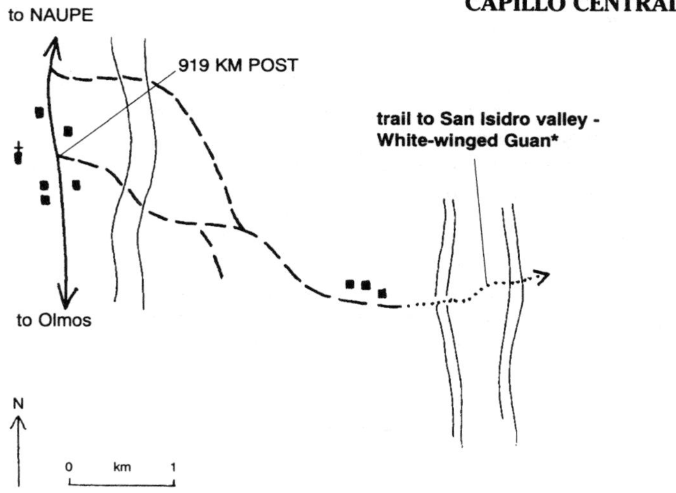

The best birding from **Piura** is along the Huancabamba road above Canchaque, which crosses the Western Cordillera. Before the pass check the roadside scrub near 'Puente Fierro' for Masked Saltator*. A trail leads down from the pass through disturbed forest, where such goodies as Bearded Guan* and Grey-headed Antbird* may still occur.

A few km east of Huancabamba on the track to Sapalache the desert scrub supports Spot-throated Hummingbird and Grey-winged Inca-Finch*. Further along this road, just east of the village of Shapaya, there is an excellent, but arduous, trail up to temperate forest, elfin forest and paramo on the Cerro Chinguela ridge. Imperial Snipe* occurs here.

Beyond the pass the trail descends into superb temperate forest. The many goodies include Neblina Metaltail*, which was only discovered in 1980. Allow six hours to walk from Shapaya to the pass and camp there (probably in heavy rain) or wait until the Huancabamba–Tabaconas–Jaen road is built!

North of Piura the best birding from **Tumbes** is along the road to Puerto Pizarro (where boats to the mangroves can be hired) and anywhere north of town for desert specialities. Bird east of town for dry forest species. Check the river near town for Masked Water-Tyrant.

Back at **Chiclayo**, Peruvian Thick-knee, Tawny-throated Dotterel, Coastal Miner and Sulphur-throated Finch occur at the **Santa Rosa** marshes, 16 km south of Chiclayo. Peruvian Plantcutter* occurs in the acacia wood 1 km north of **Rafan**, a tiny village *en route* to Lagunas, 10 km west of Mocupe, which is 35 km south of Chiclayo.

The basic circuit from Trujillo to Chiclayo is over 1,000 km long, not including the diversion east from Pedro Ruiz along the Rioja road, the diversion north from Pedro Ruiz to Bagua, and the diversion north from Olmos to Piura and Tumbes. Accommodation is sparse and badly situated for the birder. Hence a full-scale camping trip, with personal transport, would be the best way to see the most birds. However, having said that, it is possible to bird this area on public transport.

It is also possible to fly between Trujillo, Chiclayo, Rioja, Cajamarca, Piura and Tumbes.

YUNGAY

The trails around the lakes, 24 km above Yungay (60 km north of Huaraz), offer an opportunity to see a number of species restricted to puna lakes and *Polylepis/Gynoxys* woodland. Peruvian endemics here include White-cheeked Cotinga*, and Tit-like Dacnis* is one of the best specialities.

Endemics

Black-breasted Hillstar, Purple-backed Sunbeam*, Black Metaltail, Rusty-crowned Tit-Spinetail, Canyon Canastero, White-cheeked Cotinga*, Rufous-eared Brush-Finch*, Plain-tailed Warbling-Finch*.

Specialities

Tawny Tit-Spinetail*, Line-cheeked Spinetail, Ash-breasted Tit-Tyrant*, Giant Conebill*, Tit-like Dacnis*.

Others

Andean Condor, Andean Parakeet, Giant Hummingbird, Slender-billed Miner, Stripe-headed Antpitta, Black-billed Shrike-Tyrant, Rufous-webbed Tyrant, Band-tailed Sierra-Finch, Greenish Yellow Finch.

Access

Purple-backed Sunbeam*, Black Metaltail, Canyon Canastero and Plain-tailed Warbling-Finch* occur around the police checkpoint, 17 km from Yungay. Rufous-eared Brush-Finch* and Tit-like Dacnis* occur along the Maria Joseph footpath which leads from the second police checkpoint. Black-breasted Hillstar and White-cheeked Cotinga* occur on the steep slope past the second lake.

Accommodation: Yungay: Hostal Gledel (C).

PARACAS NP AND PISCO **Map p. 324**

The Paracas peninsula and Ballestas Islands lie 300 km south of Lima along the Pan-American highway. Huge numbers of seabirds breed here and their guano has been collected for fertiliser since Inca times. As well as three endemics, Waved Albatross*, Markham's Storm-Petrel*, Peruvian Thick-knee; Swallow-tailed Gull and Peruvian Tern are all possible here.

Endemics

Coastal Miner, Peruvian Seaside Cinclodes, Raimondi's Yellow-Finch.

Specialities

Humboldt Penguin*, Waved Albatross*, Markham's Storm-Petrel*, Peruvian Diving-Petrel*, Blue-footed Booby, Peruvian Thick-knee, Band-tailed and Swallow-tailed Gulls, Peruvian and Inca Terns, Amazilia and Oasis Hummingbirds, Short-tailed Field-Tyrant, Long-tailed Mockingbird, Slender-billed Finch*, Chestnut-throated Seedeater, Peruvian Meadowlark.

Others

Great Grebe, Antarctic Giant Petrel, Southern Fulmar, Cape and White-chinned Petrels, Pink-footed* and Sooty Shearwaters, Wilson's and White-vented Storm-Petrels, Peruvian Booby, Guanay and Red-legged* Cormorants, Peruvian Pelican, White-cheeked Pintail, Chilean Flamingo, Andean Condor, Least Seedsnipe, Blackish Oystercatcher, Puna Plover, Tawny-throated Dotterel, Grey, and Grey-headed Gulls, South American Tern, Chilean Skua, Wren-like Rushbird, Many-coloured Rush-Tyrant, Yellowish Pipit, Grassland Yellow-Finch, Shiny Cowbird.

Other Wildlife

South American Sealion, Southern Fur Seal, Cuy (Guinea Pig),

Access

Boat trips can be arranged in Paracas out to the Ballestas Islands. The best is from the Hotel Paracas. Book in advance. Amazilia and Oasis Hummingbirds occur in the gardens of Hotel Paracas. Markham's Storm-Petrel*, Peruvian Tern and Coastal Miner occur in and alongside the north end of Paracas Bay, accessible from the hotel.

Peruvian Seaside Cinclodes occurs at Lagunillas, 15 km south of Paracas. Punta Arquillo is one of a number of headlands accessible from here which are good for seawatching.

Peruvian Thick-knee and Slender-billed Finch* (in the acacia scrub 7 km east of the junction with the Paracas–Pisco road) occur along the road between the Paracas–Pisco road and the Pan-American highway, just north of Paracas.

Peruvian Tern, Coastal Miner, Many-coloured Rush-Tyrant and Slender-billed Finch* all occur at **Pisco**, north on the coast road from

PARACAS NP AND PISCO

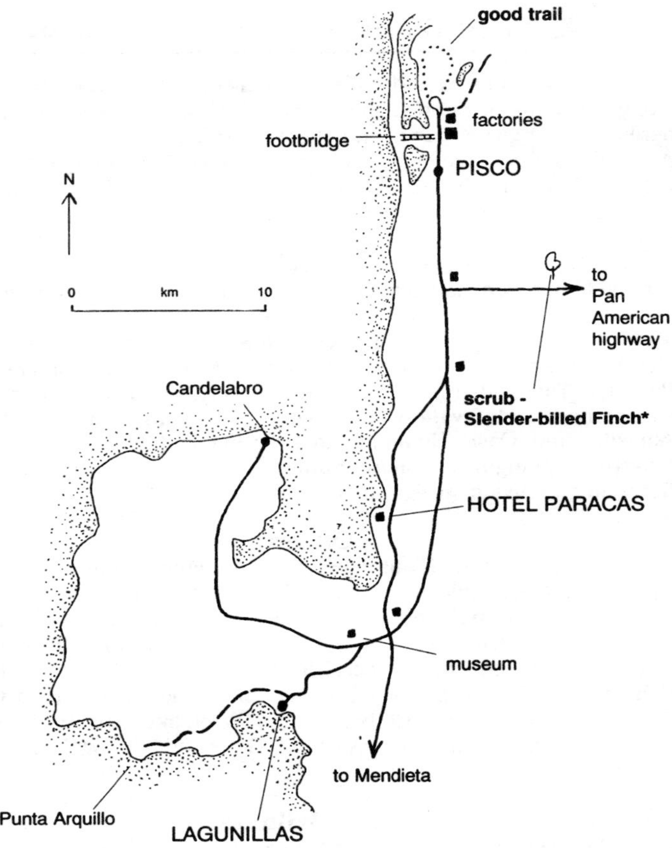

Paracas. Walk the footbridge across the marshes 3 km north of Pisco. The rush-tyrant and Short-tailed Field-Tyrant occur in the damp bushy area just north of the footbridge adjacent to the Río Pisco estuary.

Thick-billed Miner, Cactus Canastero* and Raimondi's Yellow-Finch, all occur on the rocky slopes below (west of) the **Reserva Nacional de Pampa Galeras**, three hours drive east of Nazca (of the famous 'lines'), to the south of Paracas, on the road to Arequipa.

Chestnut-breasted Mountain-Finch occurs in *Polylepis* woodland 29 km west of Chalhuanca, southwest of Abancay, on the Nazca–Abancay road.

ABANCAY

Map below

The small town of Abancay (2,377 m/7,799 ft), *en route* between Paracas and Nazca to Cuzco, lies near the Bosque de Ampay, an area of scrub and remnant temperate forest which supports some very rare and localised endemics.

Endemics

Kalinowski's Tinamou*, Apurimac* and Creamy-crested Spinetails, Rufous-eared Brush-Finch*.

Specialities

Scaled Metaltail, Andean Tyrant, White-browed Conebill.

Others

Spot-winged Pigeon, White-bellied and Sword-billed Hummingbirds, Crimson-mantled Woodpecker, Pearled Treerunner, White-banded Tyrannulet, Andean Tyrant, White-winged Black-Tyrant, Rufous-chested Tanager, Plush-capped Finch, Black-backed Grosbeak, Golden-billed Saltator,

ABANCAY

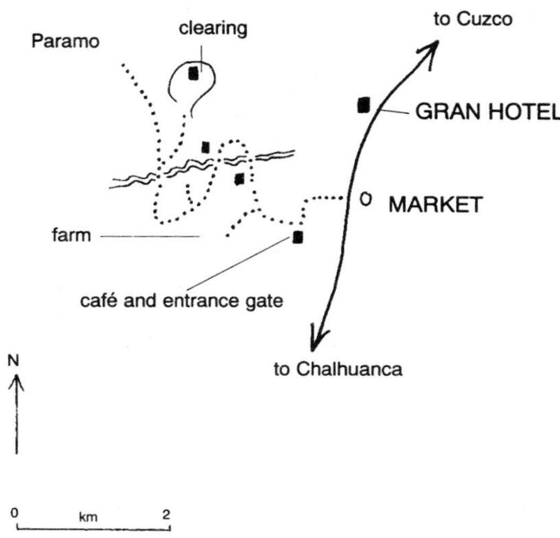

Access

The geographically isolated temperate forest on the slopes of Nevada Ampay above Abancay at **Bosque de Ampay** is the only known locality of Apurimac Spinetail*. A trail starts opposite the market, 1 km beyond the Gran Hotel. Follow the trail up from the Ministero de Agricultura gate, staying right at the first fork, through montane scrub, where Kalinowski's Tinamou*, Creamy-crested Spinetail and Rufous-eared Brush-Finch* occur, for 8 km to the forest, and another 2 km to paramo. Check the bamboo thickets around the clearings for the spinetail, after 8 km.

Accommodation: Gran Hotel (C); Hotel Misti (C).

CUZCO

Map opposite

The attractive, ancient Andean city of Cuzco (3 475 m/11 400 ft), in southeast Peru, is the gateway to the famous Inca ruins of Machu Picchu and a number of superb birding sites including Manu NP where over 1,000 birds have been recorded, the highest total for any NP in the world.

Accommodation: Hotel Espinar; Hostal Suecia (C).

MACHU PICCHU

Maps opposite and p. 328

Even the most single-minded birder should consider visiting the dramatic 'Lost City of the Incas', and not just because it is the best site for the endemic Inca Wren. Machu Picchu is one of the wonders of the world. Other goodies include the endemic Green-and-white Hummingbird, Andean Cock-of-the-Rock, Silver-backed Tanager, and Torrent Duck, which occurs on the Río Urubamba, visible from the train which climbs slowly up to the ruins from Cuzco.

Endemics

Green-and-white Hummingbird, Inca Wren.

Specialities

Fasciated Tiger-Heron*, Drab Hemispingus, Rust-and-yellow and Silver-backed Tanagers, Dusky-Green Oropendola.

Others

Torrent Duck, Andean Guan, Mitred Parakeet, Speckle-faced Parrot, Chestnut-collared and White-tipped Swifts, White-bellied and Speckled Hummingbirds, Green-tailed Trainbearer, White-bellied Woodstar, Ocellated Piculet, Variable Antshrike, Andean Cock-of-the-Rock, Slaty-capped Flycatcher, Ashy-headed Tyrannulet, Highland and Sierran Elaenias, Torrent Tyrannulet, Tufted Tit-Tyrant, Mottle-cheeked Tyrannulet, Cinnamon Flycatcher, Black Phoebe, White-winged Black-Tyrant, Andean Solitaire, Pale-legged Warbler, Chestnut-capped Brush-Finch, Capped Conebill, Yellow-throated, Saffron-crowned, Flame-faced and Beryl-spangled Tanagers, Rusty Flower-piercer, Black-backed Grosbeak, Golden-billed Saltator.

Access

The ruins are only accessible on foot or by train from Cuzco or

CUZCO AREA

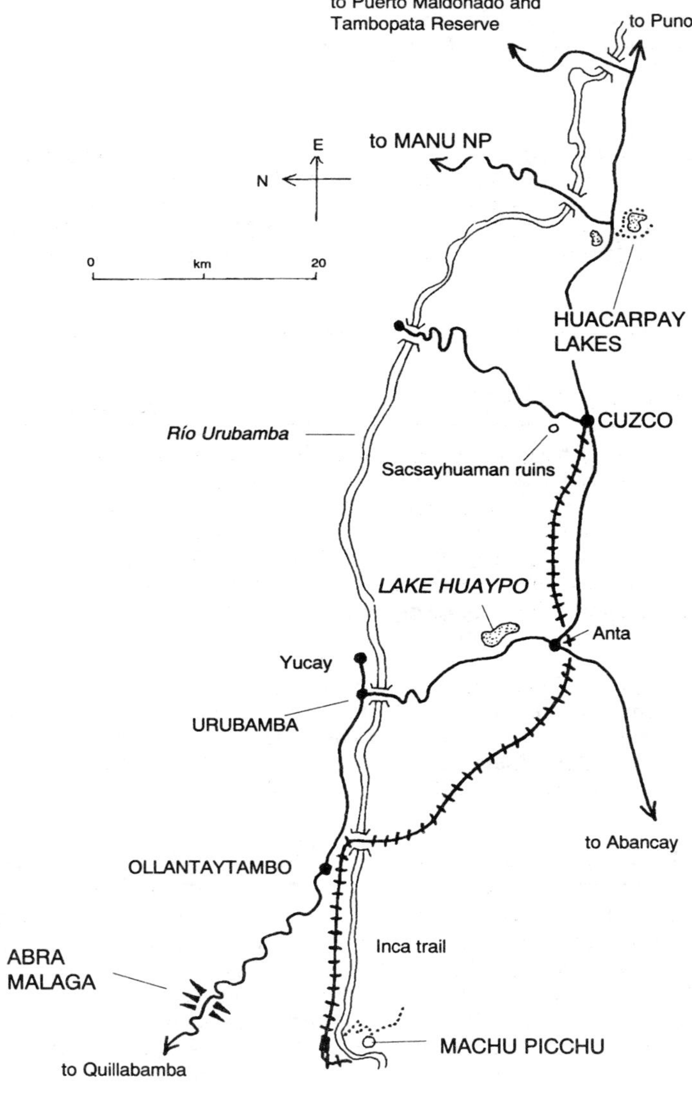
to Puerto Maldonado and
Tambopata Reserve
to Puno
to MANU NP
E
N
0
km
20
HUACARPAY
LAKES
CUZCO
Río Urubamba
Sacsayhuaman ruins
LAKE HUAYPO
Anta
Yucay
URUBAMBA
to Abancay
OLLANTAYTAMBO
Inca trail
ABRA
MALAGA
MACHU PICCHU
to Quillabamba

Ollantaytambo. It is possible to camp near the station, which is 8 km from the ruins. The area around the ruins has been deforested but still supports some birds, notably Inca Wren, which frequents the bamboo thickets around the highest part of the ruins.

Some forest remains along the Río Urubamba beyond Machu Picchu railway station, in the direction of Quillabamba. Green-and-white Hummingbird, White-bellied Woodstar and Andean Cock-of-the-Rock occur here. The best birding is usually past the first major bend in the river below two cliff faces.

MACHU PICCHU

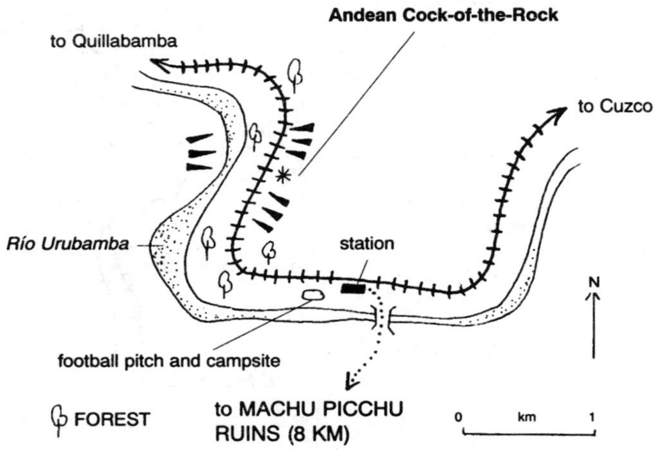

Accommodation: Ruins: Hotel Turista (A). Aguas Calientes (3 km from the railway station): Hotel los Carninantes (C); Hotel Machu Picchu (C); Hotel Municipal (C).

ABRA MALAGA

Map p. 327

Some of the best birding in the Cuzco region may be experienced along the highest part of the road between Ollantaytambo and Quillabamba, one of the few roads in Peru and South America which traverses the dry, scrubby west Andean slope, puna and east Andean slope temperate forest. Peruvian endemics occurring here include Red-and-white Antpitta and Chestnut-breasted Mountain-Finch, whilst the more widespread but scarce Imperial Snipe*, Golden-collared Tanager and Tit-like Dacnis* are also possible.

Endemics

Green-and-White Hummingbird, White-tufted Sunbeam, White-browed Tit-Spinetail*, Puna Thistletail, Marcapata and Creamy-crested Spinetails, Junin Canastero, Red-and-white Antpitta, Unstreaked Tit-

Tyrant, Inca Wren, Parodi's Hemispingus, Chestnut-breasted Mountain-Finch.

Specialities

Imperial Snipe*, Violet-throated Starfrontlet, Scaled Metaltail, Tawny Tit-Spinetail*, Streak-throated and Line-fronted* Canasteros, Ash-breasted Tit-Tyrant*, Ochraceous-breasted Flycatcher, White-browed and Giant* Conebills, Drab and Three-striped Hemispinguses, Golden-collared Tanager, Chestnut-bellied Mountain-Tanager, Tit-like Dacnis*, Moustached Flower-piercer.

Others

Andean Goose, Andean Condor, Plumbeous Rail, Grey-breasted Seedsnipe, Andean Swift, Andean Hillstar, Giant Hummingbird, Shining Sunbeam, Great Sapphirewing, Sword-billed Hummingbird, Sapphire-vented Puffleg, Blue-mantled Thornbill, Bar-bellied Woodpecker, Andean Flicker, Pearled Treerunner, Undulated, Stripe-headed and Rufous Antpittas, Red-crested Cotinga, Tufted Tit-Tyrant, Many-coloured Rush-Tyrant, Slaty-backed and Rufous-breasted Chat-Tyrants, Puna and Plain-capped Ground-Tyrants, Pale-footed Swallow, Paramo Pipit, Thick-billed Siskin, Blue-backed Conebill, Black-capped and Black-eared Hemispinguses, Hooded, Scarlet-bellied and Buff-breasted Mountain-Tanagers, Plush-capped Finch, White-winged Diuca-Finch, Bright-rumped Yellow-Finch, Golden-billed Saltator, Yellow-billed Cacique.

Other Wildlife

Llama, Alpaca.

Access

Turn north 20 km west of Urubamba at Ollantaytambo and head up the road through a series of wicked hairpin bends to the Abra Malaga pass (4,300 m/14,108 ft), and then down towards Quillabamba.

Bird the shrubs about one hour above Ollantaytambo around the **Penas Ruins**, where White-tufted Sunbeam, Creamy-crested Spinetail, Rusty-fronted Canastero and Chestnut-breasted Mountain-Finch all occur, the puna either side of the pass, the *Polylepis* woodland on the west side of the road 1 km before the cafe at the pass, where White-browed Tit-Spinetail* may be found, although probably not for long since the mad axeman has nearly finished the wood off, and the temperate forest down to San Luis. The best area of temperate forest can be found at **Canchaillo**, 18 km beyond the Abra Malaga pass. Bird the forest edge in the direction of Quillabamba, concentrating on bamboo thickets 3 km below Canchaillo. Puna Thistletail, Marcapata Spinetail, Junin Canastero, Unstreaked Tit-Tyrant, Inca Wren and Parodi's Hemispingus all occur here. Many of the species here also occur above and below the **San Luis Restaurant**, 39 km below the Abra Malaga pass. A trail leads into the forest from this restaurant, and is particularly good for Red-and-white Antpitta.

Start out early from Cuzco or Ollantaytambo to reach the east Andean slope in time for the best birding in mid-morning, saving the other sites for the rest of the day.

Accommodation: Ollantaytambo: El Auberge (C).

The glorious endemic Bearded Mountaineer is attracted to tobacco trees which grow in the ravines of the cacti-clad slopes around the **Huacarpay lakes** (Map p. 327), 25 km east of Cuzco, *en route* to Puno and near the junction with the Manu Road. Other species occuring here include White-tufted and Silvery Grebes, Puna Ibis, Cinereous Harrier, Andean Lapwing, Andean Flicker, White-winged Cinclodes, the endemic Rusty-fronted Canastero, Streak-fronted Thornbird, Torrent Tyrannulet, Many-coloured Rush-Tyrant, White-browed Chat-Tyrant, Spot-billed and Puna Ground-Tyrants, Andean Negrito, Brown-bellied Swallow, Black-throated Flower-piercer, and Yellow-winged Blackbird. **Lake Huaypo**, 29 km west of Cuzco *en route* to Urubamba, also supports most of these species.

CUZCO–MANU ROAD **Map p. 332**

This road is one of the few in Peru, and South America, that traverses the east Andean slope, where the temperate and subtropical forests support so many birds, hence it is one of the best birding sites in South America.

Apart from the endemics which include Peruvian Piedtail*, Red-and-white Antpitta and Black-backed Tody-Flycatcher, many of the species possible here are restricted to southeast Peru and west Bolivia. However, of these, only a few, such as White-browed Hermit and Scarlet-breasted Fruiteater*, are easier to find here than in Bolivia. There, two roads pass through superb forest where most of the spectacular specialities, such as Hooded Mountain-Toucan*, occur. Still, it is possible to see such star birds as Blue-banded Toucanet, White-eared Solitaire, Golden-collared Tanager and Moustached Flower-piercer along this road, which leads to Manu NP.

Endemics

Peruvian Piedtail*, Rufous-webbed Brilliant, Coppery Metaltail, Puna Thistletail, Marcapata and Creamy-crested Spinetails, Rufous-fronted Canastero, Red-and-white Antpitta, Black-backed Tody-Flycatcher, Chestnut-breasted Mountain-Finch.

Specialities

White-throated Hawk, Blue-headed Macaw, Speckle-faced Parrot, Andean Pygmy-Owl, Andean Potoo, White-browed Hermit, Violet-throated Starfrontlet, Chestnut-breasted Coronet, Scaled Metaltail, Chestnut-capped and Black-streaked Puffbirds, Versicoloured Barbet, Blue-banded Toucanet, Scribble-tailed Canastero, Crested Foliage-gleaner, Bamboo Antshrike, Striated and Manu Antbirds, White-throated Antpitta, Slaty Gnateater, Scarlet-breasted Fruiteater*, Olivaceous Piha, Black-throated Tody-Tyrant, Bolivian and Rufous-lored Tyrannulets, Unadorned and Ochraceous-breasted Flycatchers, Rufous-bellied Bush-Tyrant, Andean Tyrant, White-collared Jay, White-eared Solitaire, Pale-legged Warbler, White-browed Conebill, Rust-and-yellow, Slaty and Vermilion Tanagers, Chestnut-bellied Mountain-Tanager, Golden-collared Tanager, Moustached Flower-piercer, Dusky-green Oropendola.

Others

Torrent Duck, Semicollared Hawk*, Solitary* and Black-and-chestnut* Eagles, Rufous-breasted Wood-Quail, White-throated Screech-Owl, Band-winged Nightjar, Green Hermit, Many-spotted Hummingbird, Amethyst-throated Sunangel, Greenish Puffleg, Rufous-capped Thornbill, Golden-headed Quetzal, Rufous Motmot, Black-streaked Puffbird, Grey-breasted Mountain-Toucan*, Bar-bellied and Crimson-mantled Woodpeckers, Ash-browed Spinetail, Pearled Treerunner, Stripe-chested and Slaty Antwrens, Rufous-breasted and Barred Antthrushes, Barred Fruiteater, Amazonian Umbrellabird, Andean Cock-of-the-Rock, Mottle-backed Elaenia, Scale-crested Pygmy-Tyrant, Handsome Flycatcher, Rufous-webbed Tyrant, Rufous-breasted Chat-Tyrant, Rufous-naped Ground-Tyrant, Dusky-capped and Lemon-browed Flycatchers, Andean Slaty-Thrush, Chestnut-breasted Wren, Paramo Pipit, Olivaceous Siskin, Two-banded Warbler, Rufous-naped Brush-Finch, Grass-green Tanager, Short-billed Bush-Tanager, Black-capped and Superciliaried Hemispinguses, White-winged and Blue-capped Tanagers, Hooded Mountain-Tanager, Orange-bellied Euphonia, Orange-eared and Blue-browed Tanagers, Black-and-white Seedeater, Lesser Seed-Finch, Masked Flower-piercer, Golden-billed Saltator, Mountain Cacique.

Access

Turn north off the Cuzco–Puno road at Huacarpay lakes and head to Paucartambo. Turn right 25 km beyond here and bird the different habitats from there onwards. The puna and elfin forest (3,400 m/11,155 ft) at the beginning support Puna Thistletail, Paramo Pipit, Golden-collared Tanager and Moustached Flower-piercer. Before reaching Pillihuarta, the road enters temperate forest where Marcapata Spinetail and Drab Hemispingus occur. Just below Pillihuarta in the transition zone between temperate and subtropical forests (2,500 m/8,202 ft); look for Short-billed Bush-Tanager. In the subtropical forests below here there are many specialities including Andean Potoo, Blue-banded Toucanet and Andean Tyrant. At the beginning of the upper tropical forest (1,650 m/5,413 ft) there is an Andean Cock-of-the-Rock lek. Peruvian Piedtail* and Amazonian Umbrellabird occur below here. The road then reaches Patrias. Turn left here for Atalaya and Manu NP. Manu Antbird, Rufous-lored Tyrannulet and Chestnut-breasted Mountain-Finch occur above Atalaya.

Accommodation: camping (food available in Pillihuata and Patrias).

CUZCO–MANU ROAD

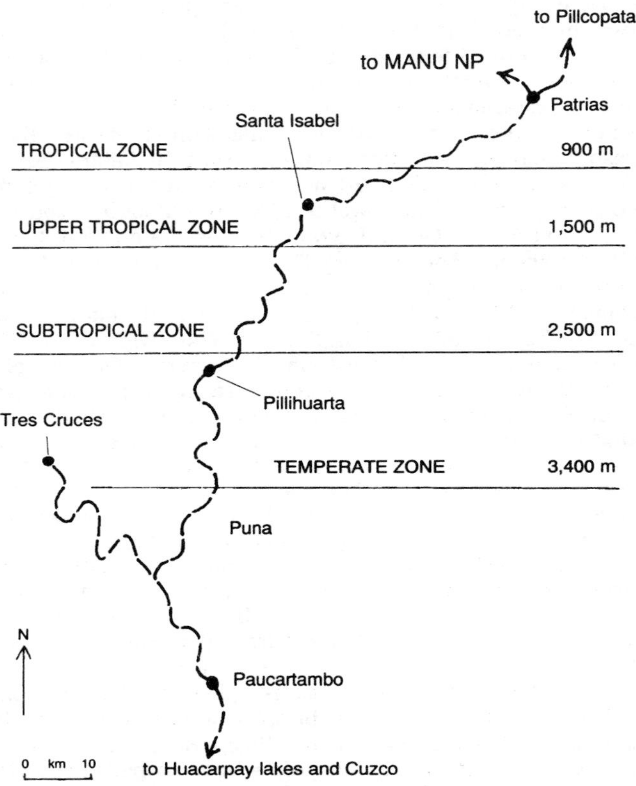

MANU NP

Although a remote and little visited part of Peru for many years, Manu NP is now one of the country's most accessible areas. This is great news for birders, considering over 1,000 species have been recorded here, more than any other park in the world. The reason there are so many birds here is the presence of the world's richest habitat hat-trick: east Andean slope temperate forest, subtropical forest and west Amazonian lowland rainforest.

In September 1986, the late Ted Parker and Scott Robinson set the world day-list record here, recording 331 species in 24 hours. However, do not expect to emulate these two special birders. Ted was famous for his knowledge of bird calls, so many of the species recorded were heard and not seen, as one would expect in Amazonian rainforest. (The record was beaten shortly afterwards, albeit over a much bigger area and by a team equipped with light aircraft, in Kenya, where the total reached was 342.)

The avifauna of the lower part of the reserve is similar to that of the Tambopata Reserve (see p. 335), although Manu NP specialities include Black-capped Tinamou, Koepcke's Hermit*, Chestnut Jacamar, Rufous-capped Nunlet, Fine-barred Piculet, Plain Softtail, Yellow-browed Tody-Flycatcher, and Pale-eyed Blackbird.

It is important to book accommodation at the lodges in the NP well in advance. Book through Barry Walker, Casilla 595, Cuzco (who runs the Cross Keys Pub in Cuzco and organises tours), or Mario Ortiz, Expediciones Manu, PO Box 606, Cuzco (telex 52060 PE).

AMAZONIA LODGE

This old tea-plantation situated at 457 m (1,500 ft), now catering for eco-tourism, lies between subtropical and tropical lowland forests, and therefore offers a rare opportunity to find a number of species which favour this transition zone, as well as many species restricted to both forest types. As well as four endemics, the specialities here include Black-capped Tinamou, Gould's Jewelfront and Hauxwell's Thrush.

Endemics

Koepcke's Hermit*, Scarlet-hooded Barbet*, Fine-barred Piculet, White-cheeked Tody-Tyrant*.

Specialities

Black-capped Tinamou, Fasciated Tiger-Heron*, Stripe-faced and Starred Wood-Quails, Rufous-sided Crake, Blue-headed Macaw, Black-headed Parrot, Pale-rumped Swift, Gould's Jewelfront, Black-throated Brilliant, Chestnut-capped Puffbird, Rufous-capped Nunlet, Dark-breasted Spinetail, Plain Softtail, Crested and Ruddy Foliage-gleaners, Bluish-slate Antshrike, Black, White-lined, Chestnut-tailed, Goeldi's, Sooty and Hairy-crested Antbirds, Amazonian Antpitta, Round-tailed and Fiery-capped Manakins, Johannes' Tody-Tyrant, Rusty-fronted and Yellow-browed Tody-Flycatchers, Violaceous Jay, Lawrence's and Hauxwell's Thrushes, Golden-bellied Warbler, White-winged Shrike-Tanager, Yellow-crested and Opal-crowned Tanagers, Yellow-bellied Dacnis.

Others

Grey and Brown Tinamous, Plumbeous Kite, Speckled Chachalaca, Collared Plover, Pied Lapwing, Grey-fronted Dove, Military and Scarlet Macaws, Hoatzin, Mottled and Black-banded Owls, Grey Potoo, Long-tailed and White-bearded Hermits, Black-eared Fairy, Blue-crowned Trogon, Black-fronted Nunbird, Golden-green Woodpecker, Strong-billed and Lineated Woodcreepers, Pale-legged Hornero, Plain-crowned and Speckled Spinetails, Slender-billed Xenops, Fasciated and Great Antshrikes, Pygmy, Stripe-chested, Grey, Dot-winged and Rufous-rumped Antwrens, Grey, Warbling, Black-throated and Spot-backed Antbirds, Black-faced Antthrush, Thrush-like Antpitta, Mottle-backed Elaenia, Large-headed Flatbill, Long-tailed Tyrant, Short-crested Flycatcher, Thrush-like Wren, Southern Nightingale-Wren, Tawny-faced

Gnatwren, White-thighed Swallow, Olive, Fawn-breasted and Opal-rumped Tanagers.

Other Wildlife

Squirrel, Night and Woolly Monkeys.

Access

Amazonia Lodge is accessible by road from Manu, or by boat via the Río Manu and Río Alto Madre de Dios, from Boca Manu airstrip. Bird the entrance trail to the lodge and around the lodge itself.

MANU LODGE

Over 500 species have been recorded in the vicinity of this lodge, situated on the Cocha Juarez oxbow, making it one of the best sites in South America (see Fig. 6, p. 25). Many species present here also occur in the Tambopata Reserve (see p. 335) except for the two endemics (which include Black-faced Cotinga*), Chestnut Jacamar and Pale-eyed Blackbird.

Endemics

Rufous-fronted Antthrush*, Black-faced Cotinga*.

Specialites

Bartlett's Tinamou, Starred Wood-Quail, Rufous-sided Crake, Pale-winged Trumpeter, Tui Parakeet, Amazonian Parrotlet*, Gould's Jewelfront, Chestnut, White-throated and Bluish-fronted Jacamars, Brown-mandibled and Curl-crested Aracaris, Yellow-ridged Toucan,

The snazzy Pied Lapwing occurs on sandbanks in the rivers of Manu NP

White-throated Woodpecker, Chestnut-winged Hookbill, Plain-winged and Bluish-slate Antshrikes, Sclater's Antwren, Plum-throated Cotinga, Little Ground-Tyrant, Violaceous Jay, White-winged Shrike-Tanager, Yellow-crested, Masked Crimson, Green-and-gold and Opal-crowned Tanagers, Pale-eyed Blackbird.

Others

Cinereous Tinamou, Horned Screamer, Orinoco Goose*, Agami Heron*, Harpy Eagle*, Collared Forest-Falcon, Spix's Guan, Blue-throated Piping-Guan, Razor-billed Curassow, Sungrebe, Sunbittern, Large-billed Tern, Blue-and-yellow, Scarlet and Red-and-green Macaws, Hoatzin, Spectacled Owl, Great Potoo, Ocellated Poorwill, Blue-crowned Trogon, Cuvier's Toucan, Rufous-breasted Piculet, Cinnamon-throated and Spix's Woodcreepers, Rufous-tailed, Rufous-rumped and Olive-backed Foliage-gleaners, Spot-winged and Dusky-throated Antshrikes, Plain-throated Antwren, White-browed, Band-tailed, Silvered and Black-throated Antbirds, Black-spotted Bare-eye, Screaming Piha, Spangled Cotinga, Bare-necked Fruitcrow, Band-tailed, Blue-crowned and Wing-barred Manakins, Golden-crowned Spadebill, Drab Water-Tyrant, Dull-capped Attila, Swainson's Flycatcher, Tawny-crowned Greenlet, Black-capped Donacobius, White-lored Euphonia, Paradise Tanager, Crested Oropendola, Epaulet Oriole.

Other Wildlife

Ten monkeys including Squirrel Monkey, Brown and White-fronted Capuchins, Black Spider-monkey, Red Howler, Emperor Tamarin and Saddle-backed Tamarin, Giant Otter, Jaguar, Capybara, Brazilian Tapir, White and Black Caiman.

Access

This lodge is also accessible by road or river. Bird the trails, oxbows (Pale-eyed Blackbird), salt-licks (macaws) and, a little further up river, the Cocha Salvador trail (Razor-billed Curassow). Black-faced Cotinga* is best looked for from the trail overlooking the oxbow. Giant Otters occur on Otorongo and Salvador lakes.

TAMBOPATA RESERVE

The huge Tambopata Reserve (15,000 km^2) was established to protect lowland Amazonian rainforest at the western edge of the Amazon basin; the most diverse and bird-rich habitat on earth. Although much of the reserve remains inaccessible to all but mini-expeditions, birders can stay at the famous lodge, Explorer's Inn (although this was threatened with closure in 1993). This lodge boasts the biggest birdlist in the world, a mindboggling 587 at the last count (October 1991). However, even the keenest birders will do well to see 300 in a week.

Although the avifauna is similar to the lower reaches of Manu NP (see p. 332), species which occur more reliably at Tambopata include Zigzag Heron*, Long-tailed and Rufous Potoos, Needle-billed Hermit, Golden-collared Toucanet, Peruvian Recurvebill*, Ash-throated Gnateater and Red-headed Manakin.

Naturally, accommodation must be booked well in advance.

Endemics

Scarlet-hooded Barbet*, White-cheeked Tody-Tyrant*.

Specialities

Bartlett's Tinamou, Zigzag Heron*, Slaty-backed Forest-Falcon, Orange-breasted Falcon*, Starred Wood-Quail, Pale-winged Trumpeter, Black-capped Parakeet, Amazonian Parrotlet*, White-bellied Parrot, Least Pygmy-Owl, Long-tailed and Rufous Potoos, Sand-coloured Nighthawk, Ladder-tailed Nightjar, Pale-rumped Swift, White-bearded and Needle-billed Hermits, Gould's Jewelfront, White-throated and Bluish-fronted Jacamars, Semicollared Puffbird, Lemon-throated Barbet, Brown-mandibled and Curl-crested Aracaris, Golden-collared Toucanet, Yellow-ridged Toucan, Bar-breasted Piculet, Point-tailed Palmcreeper, Chestnut-winged Hookbill, Peruvian Recurvebill*, Crested Foliage-gleaner, Bamboo, Plain-winged and Bluish-slate Antshrikes, Sclater's and Ihering's Antwrens, Striated, Black, Manu, White-lined, Chestnut-tailed and Goeldi's Antbirds, Ash-throated Gnateater, Plum-throated Cotinga, Red-headed and Round-tailed Manakins, Sulphur-bellied Tyrant-Manakin, Flammulated Bamboo-Tyrant, Rusty-fronted Tody-Flycatcher, Ringed Antpipit, Dusky-tailed Flatbill, Royal Flycatcher, Little Ground-Tyrant, Violaceous Jay, Lawrence's Thrush, White-winged Shrike-Tanager, Masked Crimson and Green-and-gold Tanagers, Casqued Oropendola.

Others

Great, Cinereous, Little and Undulated Tinamous, Horned Screamer, Capped Heron, King Vulture, Slate-coloured Hawk, Crested* and Harpy* Eagles, Red-throated Caracara, Speckled Chachalaca, Spix's Guan, Blue-throated Piping-Guan, Azure Gallinule, Sungrebe, Sunbittern, Collared Plover, Pied Lapwing, Large-billed and Yellow-billed Terns, Plumbeous Pigeon, Grey-fronted Dove, Ruddy Quail-Dove, Blue-and-yellow, Scarlet, Red-and-green, Chestnut-fronted and Red-bellied Macaws, White-eyed and Dusky-headed Parakeets, Yellow-crowned Parrot, Little Cuckoo, Hoatzin, Spectacled Owl, Great Potoo, Ocellated Poorwill, Pale-tailed Barbthroat, Festive Coquette, White-chinned Sapphire, Pavonine Quetzal, Blue-crowned Trogon, Broad-billed, Rufous and Blue-crowned Motmots, Paradise and Great Jacamars, Swallow-wing, Chestnut-eared Aracari, Cuvier's Toucan, Scaly-breasted, Cream-coloured, Ringed and Red-necked Woodpeckers, Long-billed, Cinnamon-throated and Spix's Woodcreepers, Pale-legged Hornero, Striped Woodhaunter, Cinnamon-rumped, Chestnut-winged, Olive-backed and Chestnut-crowned Foliage-gleaners, Rufous-tailed Xenops, Great, White-shouldered, Amazonian, Spot-winged and Dusky-throated Antshrikes, Pygmy, Plain-throated, White-eyed, Ornate, White-flanked, Slaty, Grey and Dot-winged Antwrens, Blackish, White-browed, Black-faced, Warbling, Silvered, Black-throated and Scale-backed Antbirds, Black-spotted Bare-eye, Rufous-capped Antthrush, Thrush-like Antpitta, White-browed Purpletuft, Screaming Piha, Purple-throated and Bare-necked Fruitcrows, Band-tailed Manakin, Spotted Tody-Flycatcher, Forest and Large Elaenias, Plain Tyrannulet, Short-tailed Pygmy-Tyrant, Ruddy-tailed Flycatcher, Drab Water-Tyrant, Dull-capped Attila, Greyish Mourner, White-winged Becard, Dusky-capped Greenlet, Black-capped Donacobius, Southern Nightingale-Wren, Musician Wren, White-banded

Swallow, Orange-headed Tanager, Red-crowned Ant-Tanager, Paradise Tanager, Yellow-shouldered and Slate-coloured Grosbeaks, Crested and Russet-backed Oropendolas, Solitary Cacique, Troupial.

Other Wildlife

91 mammals including Giant Otter, Tayra, Southern Coati, Red Howler, Red Titi, Squirrel and Night Monkeys, Saddle-backed Tamarin, Giant Armadillo, Giant Anteater, Small-eared Dog, Ocelot, Jaguarundi, Jaguar, White-lipped Peccary and Brocket Deer, as well as Yellow-spotted Sideneck Turtle, Spectacled Caiman, Black Caiman, and over 1,200 butterflies.

Access

Tambopata is accessible via Puerto Maldonado (connected by road and, better still, air to Lima and Cuzco). The basic Explorer's Inn (A) is located on the Río Tambopata, 4 hours (58 km) up river from Puerto Maldonado. Sunbittern, Pied Lapwing and Large-billed Tern occur along the river.

Bird the lodge clearing, 30 km of trails, especially the La Torre, Swamp and High Forest trails, riverbanks and oxbows, especially Cocacocha. There are also canopy walkways well worth birding.

Zigzag Heron* occurs in the swamp off, and Bartlett's Tinamou and Black Antbird on, the main trail which leads to Laguna Cocacocha. Hoatzin and Point-tailed Palmcreeper occur here. There is a Needle-billed Hermit lek near the beginning of the High Forest trail. Plum-throated Cotinga occurs near the beginning of, and Bluish-fronted and White-throated Jacamars by the first bridge, on the La Torre trail which is also good for Chestnut-winged Hookbill. Bamboo Antshrike occurs in the bamboo, appropriately enough, near the beginning of the Katicocha trail. Manu Antbird and White-cheeked Tody-Tyrant* occur in the bamboo near the end of the Heliconia trail. There is a Band-tailed Manakin lek along the Swamp trail. Orange-breasted Falcon*, Starred Wood-Quail, Striated Antbird and White-cheeked Tody-Tyrant* occur on the Bamboo trail. It is possible to get a boat across the Río Tambopata to a bamboo thicket beyond Virgilio's clearing, where Pale-tailed Barbthroat, Bluish-fronted Jacamar, Scarlet-hooded Barbet* and Bamboo Antshrike occur.

Book through Peruvian Safaris S.A., Av. Garcilazo de la Vega 1334, P.O. Box 10088, Lima (1) (telex: 20330 PE PB; fax: 051-14 -328866; tel: 316330/313047). All programmes include reception at Puerto Maldonado airstrip, land and river transport to the Inn, accommodation and three meals per day.

For more information contact The Tambopata Reserve Society (TReeS): U.K.: John Forrest, 64 Belsize Park, London NW3 4EH; USA: W. Widdowson, 5455 Agostino Court, Concord, CA 94521.

AREQUIPA

Map p.339

The attractive city of Arequipa lies at 2,286 m (7,500 ft) on the west Andean slope, below El Misti (5,843 m/19,170 ft) and Chachani (6,096 m/20,000 ft) volcanoes. The city is the gateway to a few excellent birding sites including the Cruz del Condor, one of the best places to see Andean Condors in South America, and the high-altitude Lake Salinas,

At Cruz del Condor near Arequipa it is possible to watch Andean Condors et eye level. They may even be close enough to hear the thermal air rushing through their outstretched primaries

where Andean* and Puna* Flamingos occur.

Access

The snazzy Peruvian Sheartail occurs in the small enclosed garden at the far eastern corner of the **Santa Catalina Monastery** grounds, near the city centre.

Accommodation: Hotel LaFontana; Hotel Excelsior (C).

The road northeast out of Arequipa, to Chivay, goes on to the **Cruz del Condor**, overlooking Colca canyon, one of the best sites in South America for close-up views of Andean Condor (Map opposite). Bird the road up, looking for Diademed Sandpiper-Plover* (at the highest point) as well as Canyon and Cactus* Canasteros. Take the left fork after Sumbay to reach Chivay, birding the boggy area at Pampa de Torca, just after Viscachani, and another boggy area just before the descent to Chivay, where Rufous-bellied Seedsnipe occurs. The right fork after Sumbay (towards Chayoma and Cuzco) leads to a small lake after 5 km where Andean Avocet occurs. The Cruz del Condor is some 40 km beyond Chivay, before the village of Cabanaconde. Andean Condors roost on cliffs in the canyon and occasionally rise up to the lookout point on the morning thermals.

Accommodation: Chivay: Hotel Moderno.

The road west from Arequipa runs to the coast at Mollendo and on to the **Lagunas de Mejia Nature Reserve**. Raimondi's Yellow-Finch

AREQUIPA AREA

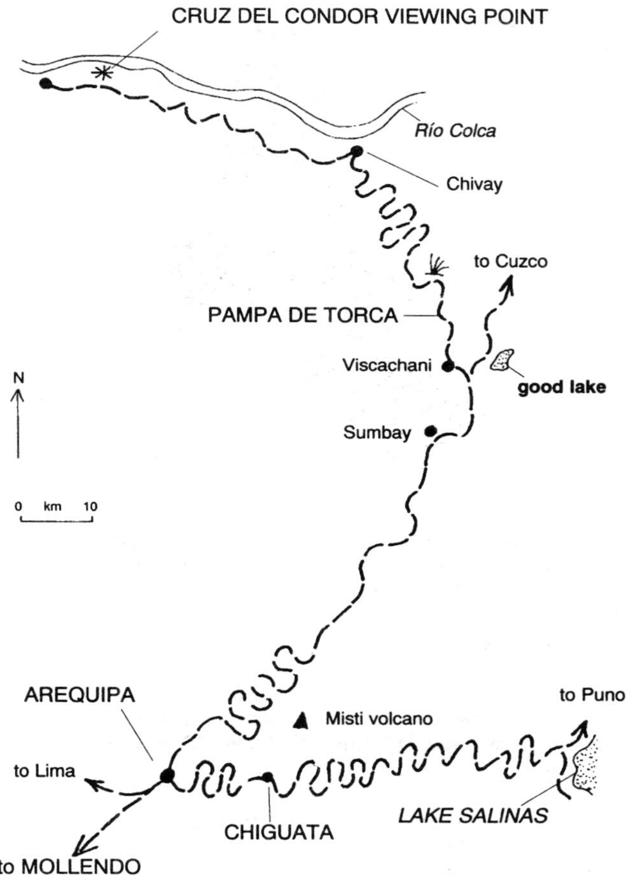

occurs *en route* to Mejia, which is 15 km southeast of Mollendo, itself 70 km southwest of Arequipa. The Nature Reserve (100 ha) is 6 km southeast of Mejia. The coastal lagoons here and around Mejia are the largest permanent water bodies in the 1,500-km stretch of desert that lines the Peruvian coast, hence they are very good for shorebirds. Pied-crested Tit-Tyrant, Short-tailed Field-Tyrant and Slender-billed Finch* occur in the thick scrub and fields just northwest of Mejia.

Peruvian Sheartail and Purple-collared Woodstar occur in the Plaza de Armas in **Moquegua**, 220 km southeast of Arequipa *en route* to the Chilean border. Lesser Rhea (just east), Rufous-bellied Seedsnipe and Mountain Parakeet (both just west) all occur either side of **Puente Humajalso**, 114 km northeast of Moquegua *en route* to Puno, 165 km away. This is high-altitude country so beware of altitude sickness.

AREQUIPA–PUNO ROAD

Map p. 339

This high-altitude road traverses the flanks of Chachani volcano, passing through arid cacti-clad hillsides, then farmland up to *Polylepis* woodland and puna, before reaching Lake Salinas at the giddy height of 4,267 m/14,000 ft. Beware of altitude sickness here and look out for the two rare Andean flamingos, as well as other specialities such as Rufous-bellied Seedsnipe, White-tailed Shrike-Tyrant*, Tamarugo Conebill* and Tit-like Dacnis*.

Endemics

Black Metaltail, Thick-billed Miner, Rufous-crowned Tit-Spinetail, Canyon and Cactus* Canasteros.

Specialities

Andean* and Puna* Flamingoes, Rufous-bellied Seedsnipe, Golden-spotted Ground-Dove, Scaled Metaltail, Oasis Hummingbird, White-throated and Plain-breasted Earthcreepers, Streaked Tit-Spinetail, Streak-throated Canastero, White-tailed Shrike-Tyrant*, Tamarugo* and Giant* Conebills, Tit-like Dacnis*, White-throated Sierra-Finch, Puna Yellowfinch, Black-throated Flower-piercer.

Others

Ornate and Puna Tinamous, Chilean Flamingo, Andean Condor, Red-backed and Puna Hawks, Mountain Caracara, Puna Snipe, Grey-breasted Seedsnipe, Andean Avocet, Puna Plover, Bare-faced and Black-winged Ground-Doves, Giant Hummingbird, Puna and Slender-billed Miners, Straight-billed Earthcreeper, Cordilleran and Creamy-breasted Canasteros, Yellow-billed Tit-Tyrant, D'Orbigny's Chat-Tyrant, Rufous-naped and Puna Ground-Tyrants, Andean Negrito, Andean Swallow, Thick-billed and Black Siskins, Black-hooded and Mourning Sierra-Finches, White-winged Diuca-Finch, Greenish Yellow-Finch, Golden-billed Saltator.

Other Wildlife

Vicuña, Peruvian Huemel (Deer).

Access

Most of the endemics and specialities occur in the three quebradas (small canyons) just before the village of **Chiguata**, 28 km from Arequipa. The woodland above Chiquata (before the silver cross) supports Tamarugo Conebill* and Tit-like Dacnis*. Eventually, the road reaches Lake Salinas, 150 km from Arequipa, which, when the water levels are right, holds thousands of flamingos. Ornate and Puna Tinamous, Rufous-bellied Seedsnipe and White-throated Sierra-Finch all occur sparsely on the slopes north of the lake. Golden-spotted Ground-Dove and Puna Yellow-Finch occur to the northeast of the lake around human settlements.

Short-winged Grebe occurs on Lake Titicaca, either side of the road linking the Hotel Turista to the shore, north of **Puno**, but is more reliable 2 km offshore. Boats can be hired. Other species occuring here include White-tufted and Silvery Grebes, Plumbeous Rail, Andean Lapwing, Bare-faced and Black-winged Ground-Doves, Andean Flicker,

Slender-billed Miner, Wren-like Rushbird, Many-coloured Rush-Tyrant, Mourning Sierra-Finch, and Yellow-winged Blackbird.

Accommodation: Hotel los Incas (C).

Darwin's Nothura and Puna Tinamou occur at **Pucara**, 80 km northeast of Puno, in the grassland 2 km south of the village.

The endemic Green-capped Tanager* occurs at Sandia, a remote town north of Puno near the Bolivian border, and on the west side of Abra de Maruncunca.

ADDITIONAL INFORMATION

Books

An Annotated Checklist of Peruvian Birds, Parker, T., Parker, S. and Plenge, M. 1982. Buteo Books.
The Birds of the Department of Lima, Peru, Koepcke, M., 1992. Harrowood.

PERU ENDEMICS (104)	BEST SITES
Taczanowski's Tinamou*	Central and south: above 2,775 m (9,104 ft)
Kalinowski's Tinamou*	South: Abancay
Puna Grebe*	Central: central highway–Lake Junín
White-winged Guan*	Northwest: Trujillo–Chiclayo circuit
Peruvian Pigeon*	Northwest: Trujillo–Chiclayo circuit
Yellow-faced Parrotlet*	Northwest: Trujillo–Chiclayo circuit
Long-whiskered Owlet*	Northwest: Trujillo–Chiclayo circuit
Koepcke's Hermit*	South: Manu
Spot-throated Hummingbird	Northwest: Trujillo–Chiclayo circuit
Green-and-white Hummingbird	South: Abra Malaga and Machu Picchu
Peruvian Piedtail*	South: Cuzco–Manu road
Rufous-webbed Brilliant	South: Cuzco–Manu road
Black-breasted Hillstar	Central: central highway and Yungay
White-tufted Sunbeam	South: Abra Malaga and Ollantaytambo
Purple-backed Sunbeam*	Northwest and central: Trujillo–Chiclayo circuit and Yungay
Royal Sunangel*	Northwest: rare, Cajamarca and San Martín depts

Bronze-tailed Comet*	Central: central highway
Neblina Metaltail*	Northwest: Trujillo–Chiclayo circuit
Coppery Metaltail	South: Cuzco–Manu road
Fire-throated Metaltail	Central: Huanuco, Junín and Ayacucho depts
Black Metaltail	Central and southwest: central highway, Yungay and Chiguata
Grey-bellied Comet*	Northwest: rare, Maranon valley
Bearded Mountaineer	South: Cuzco
Marvellous Spatuletail*	Northwest: Trujillo–Chiclayo circuit
Scarlet-hooded Barbet*	South: Manu and Tambopata
Yellow-browed Toucanet*	Northwest: Rio Mishollo, La Libertad dept
Speckle-chested Piculet*	Northeast: montane forest
Fine-barred Piculet	South: Manu
Black-necked Woodpecker	Northwest: Trujillo–Chiclayo circuit
Coastal Miner	Northwest and central: Trujillo–Chiclayo circuit and Paracas
Dark-winged Miner	Central: central highway
Thick-billed Miner	Southwest: Pampa Galeras and Chiguata
Striated Earthcreeper	Central: central highway
Peruvian Seaside Cinclodes	Central: Paracas
White-bellied Cinclodes*	Central: central highway
Rusty-crowned Tit-Spinetail	Central and southwest: central highway, Yungay and Chiguata
White-browed Tit-Spinetail*	South: Abra Malaga
Eye-ringed Thistletail	Central: Junín dept
Vilcabamba Thistletail	Central: north end of Vilcabamba mountains
Puna Thistletail	South: Abra Malaga and Cuzco–Manu road
Apurimac Spinetail*	South: Abancay
Maranon Spinetail	Northwest: upper Maranon valley
Russet-bellied Spinetail*	Central: west slope of Ancash dept
Chinchipe Spinetail	Northwest: Maranon and Chinchipe valleys
Baron's Spinetail	Northwest and central: Andes
Marcapata Spinetail	South: Abra Malaga and Cuzco–Manu road
Creamy-crested Spinetail	South: Cuzco sites
Canyon Canastero	Central and southwest: central highway, Yungay, Cruz del Condor and Chiguata
Rusty-fronted Canastero	South: Cuzco sites
Cactus Canaastero*	Southwest: a few sites
Pale-tailed Canastero*	Southwest: Ayacucho and Huancavelica depts

Junin Canastero	Central and south: central highway and Abra Malaga
Great Spinetail*	Northwest: Trujillo–Chiclayo circuit
Chestnut-backed Thornbird*	Northwest: Trujillo–Chiclayo circuit
Russet-mantled Softtail*	Northwest and central: rare
Ash-throated Antwren*	North: San Martín dept
Creamy-bellied Antwren	Central and south: Huanuco to Cuzco depts
Black-tailed Antbird	Northeast: east Loreto dept
White-masked Antbird*	North: known only from one specimen taken in 1938, in north west Loreto
Rufous-fronted Antthrush*	South: Manu
Elusive Antpitta*	East: southeast Loreto dept
Pale-billed Antpitta	Northwest: Trujillo–Chiclayo circuit
Rusty-tinged Antpitta	Northwest: between Maranon and Huallaga rivers
Bay Antpitta	Central: central highway at Carpish pass
Red-and-white Antpitta	South: Abra Malaga and Cuzco–Manu road
Chestnut Antpitta*	North and central: Amazonas and Huanuco depts
Ochre-fronted Antpitta*	Northwest: Trujillo–Chiclayo circuit
Maranon Crescent-chest*	Northwest: Trujillo–Chiclayo circuit
Large-footed Tapaculo	Northwest: Trujillo–Chiclayo circuit
Peruvian Plantcutter*	Northwest: Trujillo–Chiclayo circuit
White-cheeked Cotinga*	Central: central highway and Yungay
Masked Fruiteater	Central: central highway, north from Carpish pass onwards
Black-faced Cotinga*	South: Manu
Cerulean-capped Manakin	Central and south: Andes
Inca Flycatcher	Central: central highway, Carpish pass and Paty trail
White-cheeked Tody-Tyrant*	South: Manu and Tambopata
Black-backed Tody-Flycatcher	South: Cuzco–Manu road
Peruvian Tyrannulet	Central: Huanuco and Junín depts
Unstreaked Tit-Tyrant	South: Abra Malaga
Piura Chat-Tyrant*	Northwest and central: pacific slope
Tumbes Tyrant*	Northwest: Trujillo–Chiclayo circuit
Rufous Flycatcher	Northwest: Trujillo–Chiclayo circuit

Inca Wren	South: Abra Malaga and Machu Picchu
Rusty-bellied Brush-Finch	Central: central highway
Rufous-eared Brush-Finch*	Central and south: Yungay and Abancay
Parodi's Hemispingus	South: Abra Malaga
Rufous-browed Hemispingus*	Central: central highway, Carpish pass
Brown-flanked Tanager	Central: central highway
Huallaga Tanager	Central: central highway, Tingo Maria
Golden-backed Mountain-Tanager*	Central: central highway, above Carpish pass
Green-capped Tanager*	South: around Sandia only, Puno dept
Sira Tanager*	East: on Cerros del Sira only, Huanuco dept
Pardusco	Central: central highway, Carpish pass
Cinereous Finch	Northwest: Trujillo–Chiclayo circuit
Great Inca-Finch	Central: central highway
Rufous-backed Inca-Finch	Central: rare, central highway
Grey-winged Inca-Finch*	Northwest: Trujillo–Chiclayo circuit
Buff-bridled Inca-Finch	Northwest: Trujillo–Chiclayo circuit
Little Inca-Finch*	Northwest: Trujillo–Chiclayo circuit
Plain-tailed Warbling-Finch*	Central: Yungay
Rufous-breasted Warbling-Finch*	Central: central highway
Chestnut-breasted Mountain-Finch	South: Cuzco sites
Raimondi's Yellow-Finch	South: Paracas, Pampas Galeras and Arequipa
Selva Cacique*	East: known only from sighting and specimen, from 1965 at Balta on Río Curanja, southeast Ucalayi

Northwest = Trujillo–Chiclayo circuit.
Northeast = east of Trujillo–Chiclayo circuit.
Central = central highway, Yungay.
South = Cuzco area, Arequipa area.
East = north on Lake Titicaca near Bolivian border.

SURINAME

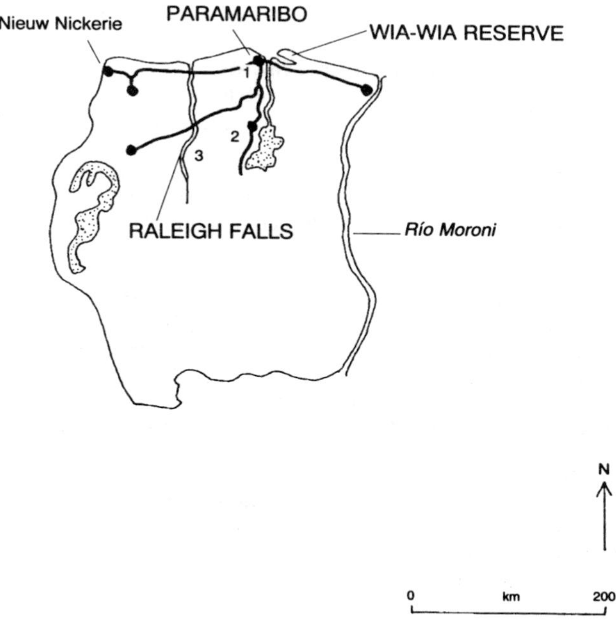

1 Paramaribo
2 Brownsberg
3 Voltberg–Raleigh falls Reserve

INTRODUCTION

Summary

Despite being one of the first South American countries to establish reserves, protecting some very special birds, especially cotingas, Suriname has been racked by terrorism since 1986 and is virtually out of bounds to birders. Even the most ardent birder will find most of the best areas closed and little up to date information on exactly where to watch birds.

Size

Suriname is the second smallest country on the mainland (163,265 km^2). It is slightly larger than England but only a quarter the size of Texas.

Getting Around

There are few good roads and the air and bus networks are poor, but, in normal circumstances, tours can be arranged in Paramaribo through STINASU (the Foundation for Nature Preservation) or the Tourist Office (see Additional Information).

Accommodation and Food

Hotels and resthouses are rare outside Paramaribo so birding Suriname may involve a mini-expedition. The food has heavy Indian and Chinese influences.

Health and Safety

Immunisation against malaria, yellow fever, tetanus and typhoid is recommended. Bilharzia is a threat in water, as are Electric Eels when swimming.

Since 1986 Suriname has been all but closed to travellers owing to what amounts to a civil war. Hence, personal safety is a major problem in and out of Paramaribo. Contact embassy before planning a visit.

Climate and Timing

Although Suriname is a hot and humid tropical country, the northeast trade winds have a cooling effect along the coast. The main wet season is April to August and the main dry season August to November.

Habitats

The swampy coastal plain varies in width from 25 to 80 km. Inland from here the land rises gently and savanna is the major habitat, forming a belt up to 30 km wide. The remote rainforest-cloaked mountains, which cover most of the country, rise from here and extend to the border with Brazil in the south. In the extreme south the forest is broken by some large areas of savanna.

Conservation

The enlightened STINASU established a number of National Parks and reserves many years ago. With the country at war since 1986, some of these may have been damaged, although the very best reserves are rather remote and could be as good for birds as they ever were. All reserves were closed in 1989.

One threatened species occurs in Suriname.

Bird Families

74 of the 92 families which regularly occur in South America are represented in Suriname, including 19 of the 25 Neotropical endemic families, only a few less than the bigger countries, and four of the nine South American endemic families: a screamer, a trumpeter, Hoatzin and a gnateater.

Well-represented families include parrots, antbirds and cotingas.

Bird Species

Just over 600 species have been recorded in Suriname, although the real total may be in excess of 700.

Non-endemic specialities and spectacular species include Scarlet Ibis, Rufous Crab-Hawk, Black Curassow, Racket-tailed Coquette, Guianan Toucanet, Guianan Piculet, Blood-coloured Woodpecker,

White-plumed Antbird, Guianan Red-Cotinga. Spangled and Pompadour Cotingas, Crimson Fruitcrow, Capuchinbird, White Bellbird, Guianan Cock-of-the-Rock, Sharpbill, Guianan Gnatcatcher, Blue-backed Tanager, Finsch's Euphonia and Red-and-black Grosbeak.

Endemics

There are no birds endemic to Suriname.

Expectations

As with the other Guianas, a well-planned two-week trip that included most of the best sites may produce between 350 and 400 species.

PARAMARIBO

The capital of Suriname on the north coast is close to some good forest and marshes which support some excellent birds including the localised Blood-coloured Woodpecker.

Specialities

Guianan Piculet, Blood-coloured Woodpecker, Crimson-hooded Manakin, Cayenne Jay, Finsch's Euphonia.

Others

Azure Gallinule, Paradise Jacamar, Yellow-tufted Woodpecker, Spangled and Pompadour Cotingas, White-bearded Manakin, Cinnamon Attila.

Access

Bird the Botanic Gardens, marshes east of town, the road to the bridge over the Río Saramacca and the road southeast to Kraka.

The **Wia-Wia Reserve** (36,000 ha) on the northeast coast was established to protect breeding sea turtles, but it also supports some interesting birds. Arrange a visit through STINASU.

Accommodation: Matapica: Beach Hut (C).

BROWNSBERG

This small reserve (6,000 ha), three hours drive south of Paramaribo, protects some superb forest. One of the major attractions of this reserve in years gone by was the Capuchinbird lek. A number of other restricted-range species occur here, as well as White-plumed Antbird (one of over twenty antbirds present), White Bellbird and Blue-backed Tanager.

Specialities

Marail Guan, Guianan Toucanet, Rufous-throated Antbird, Capuchinbird, White Bellbird, Guianan Gnatcatcher, Blue-backed Tanager, Red-and-black Grosbeak.

Big, brown and bald, the Capuchinbird sounds even weirder than it looks. The call which resembles a cross between a chain saw and a cow, is likely to be heard before the bird is seen

Others

Dusky Parrot, White-plumed Antbird, Red-billed Pied Tanager, Opal-rumped Tanager.

Other Wildlife

Red-headed Tamarin.

Access

In normal conditions buses travel from Paramaribo to Brownsweg, 14 km downhill from the reserve. STINASU arrange trips and accommodation here. There was a Capuchinbird lek a short distance along the track from the lodge to the waterfall. Listen out for the very distinctive call (described in the caption above).

Accommodation: Central Lodge (A).

VOLTZBERG–RALEIGH FALLS RESERVE

This big reserve (57,000 ha), some 200 km inland from Paramaribo, supports some spectacular birds, notably Black Curassow and Guianan Cock-of-the-Rock.

Specialities

Little Chachalaca, Black Curassow, Guianan Cock-of-the-Rock.

Others

Great Tinamou, Blue-throated Piping-Guan, Grey-winged Trumpeter, Blue-and-yellow, Scarlet, Red-and-green and Chestnut-fronted Macaws, Red-fan Parrot, Blackish Nightjar, Green-tailed Jacamar, Sharpbill, Black-collared Swallow, Green Oropendola.

Other Wildlife

Weeper Capuchin, Red Howler, Black Spider-monkey.

Access

Before the civil war it was possible to fly to Foengoe Island Lodge (40 minutes from Paramaribo), or to drive (3–4 hours) to Bitagron, where the car could be left safely with the Government Geological Service, before taking a two-hour boat trip upriver to the reserve.

Bird Foengoe Island, the trail to Mother's falls and the trail to Voltzberg granite dome where there is a Guianan Cock-of-the-Rock lek (1.5 km from the airstrip).

Accommodation: Foengoe Island Lodge; Lolopasi Lodge.

White-bellied and Capped Seedeaters have been recorded at Sipaliwini in south Suriname.

A SELECTION OF SPECIES WHICH MAY BE SEEN ON A TRIP TO SURINAME

Specialities

Rufous Crab-Hawk, Little Chachalaca, Marail Guan, Black Curassow, Lilac-tailed Parrotlet, Ladder-tailed Nightjar, Green-throated Mango, Tufted and Racket-tailed Coquettes, Green-tailed Goldenthroat, Plain-bellied Emerald, Green and Black-necked Aracaris, Guianan Toucanet, Red-billed Toucan, Guianan Piculet, Blood-coloured, Golden-collared and Waved Woodpeckers, Rufous-bellied, Brown-bellied and Spot-tailed Antwrens, Ferruginous-backed and Rufous-throated Antbirds, Guianan Red-Cotinga, Crimson Fruitcrow, Guianan Cock-of-the-Rock, Crimson-hooded Manakin, Cayenne Jay, Guianan Gnatcatcher, Blue-backed Tanager, Finsch's Euphonia, Red-and-black Grosbeak, Carib Grackle.

Others

Great, Little and Variegated Tinamous, Magnificent Frigatebird, Scarlet Ibis, King Vulture, Double-toothed and Plumbeous Kites, Long-winged Harrier, Ornate Hawk-Eagle, Black and Red-throated Caracaras, Lined Forest-Falcon, Blue-throated Piping-Guan, Marbled Wood-Quail, Grey-necked Wood-Rail, Azure Gallinue, Grey-winged Trumpeter, Yellow-billed Tern, Scaled and Plumbeous Pigeons, Plain-breasted Ground-Dove, Grey-fronted Dove, Ruddy Quail-Dove, Blue-and-yellow, Scarlet, Red-and-green and Chestnut-fronted Macaws, Brown-throated and Painted Parakeets, Green-rumped Parrotlet, Golden-winged Parakeet, Black-headed, Dusky, Orange-winged, Mealy and Red-fan Parrots, Little and Striped Cuckoos, Spectacled Owl, Grey Potoo, White-tailed and Blackish Nightjars, Band-rumped, Short-tailed and Lesser Swallow-tailed Swifts, Long-tailed, Straight-billed, Reddish and Little Hermits, Grey-breasted Sabrewing, White-necked Jacobin, Blue-tailed Emerald, White-chinned Sapphire, Black-eared Fairy, White-tailed Trogon, Green-and-rufous and American Pygmy Kingfishers, Yellow-billed, Green-tailed and Paradise Jacamars, Pied, Spotted and Collared Puffbirds, Swallow-wing, Channel-billed Toucan, Golden-spangled Piculet, Yellow-tufted, Yellow-throated, Golden-green, Chestnut and Red-necked Woodpeckers, Wedge-billed, Barred, Chestnut-rumped, Buff-throated and Lineated Woodcreepers, Plain-crowned Spinetail, Rufous-rumped and Ruddy Foliage-gleaners, Black-crested, Mouse-coloured, Eastern Slaty, Amazonian, Dusky-throated and Cinereous Antshrikes, Long-winged, Grey and Dot-winged Antwrens, Grey, Dusky, Black-headed, White-plumed and Scale-backed Antbirds, Screaming Piha, Spangled and Pompadour Cotingas, Purple-throated Fruitcrow, Capuchinbird, White Bellbird, Sharpbill, Golden-headed and White-bearded Manakins, McConnell's Flycatcher, Spotted and Painted Tody-Flycatchers, Slender-footed Tyrannulet, Plain-crested Elaenia, Helmeted Pygmy-Tyrant, Cinnamon Attila, Rusty-margined Flycatcher, Cinereous Becard, Black-tailed Tityra, Ashy-headed Greenlet, Black-capped Donacobius, Coraya and Buff-breasted Wrens, White-banded and Black-collared Swallows, Bicoloured Conebill, Red-billed Pied Tanager, Fulvous Shrike-Tanager, Fulvous-crested Tanager, Golden-sided Euphonia, Turquoise and Opal-rumped Tanagers, White-bellied and Capped Seedeaters, Yellow-green Grosbeak, Green Oropendola, Yellow-rumped Cacique, Moriche and Yellow Orioles, Yellow-hooded Blackbird, Giant Cowbird.

ADDITIONAL INFORMATION

Addresses

STINASU (Foundation for Nature Preservation), PO Box 436, Jongbawstraat 14, Paramaribo (tel: 75845/71856).
Tourist Office, PO Box 656, Waterkant 8, Paramaribo (tel: 71163/78421).

Books

The Birds of Surinam, Haverschmidt, F., 1968. Oliver and Boyd.
Birds of Surinam: An Annotated Checklist, Donahue, P. and Pierson, J., 1982. South Harpswell.
A Guide to the Birds of Venezuela, Meyer de Schauensee, R. and Phelps, N., 1978. Princeton UP.

The **Dutch Birding Travel Reports Service** (see Useful Addresses, p. 409) holds a number of reports on Suriname, in Dutch, for the period 1975–1982.

SURINAME ENDEMICS

None.

TRINIDAD AND TOBAGO

TRINIDAD

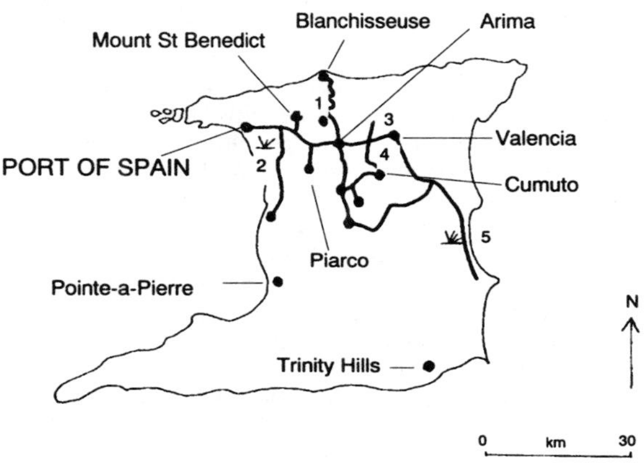

1 Asa Wright Nature Centre
2 Caroni Swamp
3 Cerro del Aripo
4 Wallerfield
5 Nariva Swamp

INTRODUCTION

Summary

The tiny islands of Trinidad and Tobago support representatives from half of the Neotropical endemic families, but relatively few species. Together with the modern infrastructure and a superb field guide, the 'family reprsentatives' make the islands an ideal destination for the first time visitor to South America who fears being overwhelmed on the mainland. Although some accommodation and permits need to be arranged before visiting, this extra effort may be rewarded with some 200 species in two weeks, including Scarlet Ibis and Oilbird.

Size

At 5,130 km^2 this is a tiny archipelago, 4% the size of England and 1% the size of Texas. Trinidad is just 80 km by 60 km and Tobago 80 km by 12 km.

Getting Around

Most buses are confined to the busier routes, but there are many taxis, including those on set routes which are cheaper, and cars may be hired. There are frequent flight and ferry services between the islands.

Accommodation and Food

Accommodation is a bit thin on the ground and expensive in the high season (mid-December to mid-April). It is best to book in advance. Food is varied and fine.

Health and Safety

Immunisation against hepatitis, malaria, typhoid and yellow fever is recommended. Both islands are very friendly, although robberies have occured at Port-of-Spain sewage works and in the capital itself at night.

Climate and Timing

Although rain is possible at any time most falls between May and December so January to March is the best time for tourists and birders, since Red-billed Tropicbirds breed on islands offshore from Tobago from December to February.

Habitats

There are a wide range of habitats, from coastal islets (mainly off Tobago), mangrove swamps and marshes to savanna and tropical forest. Since much of the lowlands are cultivated, most of the forest is restricted to the highlands. These include the northern range (rising to 915 m/3,020 ft), central range and southeast range (Trinity Hills) on Trinidad, and the central spine of Tobago (rising to 576 m/1,890 ft).

Conservation

Few birds seem to be threatened on the islands, although the only endemic, Trinidad Piping-Guan*, is rarely seen, possibly owing to hunting.

Two threatened species, of which one is endemic, occur on Trinidad and Tobago.

Bird Families

66 of the 92 families which regularly occur in South America are represented, an amazing cross-section for such a small archipelago. 12 of the 25 Neotropical endemic families are present, a better total than all the other archipelagos, as well as mainland Chile and Uruguay. However, only one of the nine South American endemic families is present, the Oilbird.

Well-represented families include hummingbirds and 'tanagers'. Other families with one or two representatives include motmots, jacamars, toucans, woodcreepers, antbirds and manakins.

Bird Species

430 species have been recorded on the islands, just ten less than the total for Chile, which is nearly 150 times larger! Approximately 250 species breed.

Non-endemic specialities and spectacular species include Red-billed Tropicbird, Red-footed Booby, Scarlet Ibis, Rufous-vented Chachalaca, Lilac-tailed Parrotlet, White-tailed Sabrewing*, Green-throated Mango, Ruby-topaz Hummingbird, Tufted Coquette, White-chested Emerald, Blue-crowned Motmot, Rufous-tailed Jacamar, Channel-billed Toucan, White-bellied Antbird, Bearded Bellbird, Blue-backed Manakin, Venezuelan Flycatcher, Bare-eyed Thrush, Long-billed Gnatwren, Caribbean Martin and Sooty Grassquit.

Endemics

The sole endemic, Trinidad Piping-Guan* is rare and restricted to the remote reaches of the Northern Range, around Aripo, Cumaca and Matura, and the Trinity Hills, in the southeast, on Trinidad.

Expectations

A two-week trip that includes both islands is likely to produce around 200 species including most of the goodies.

TRINIDAD

ASA WRIGHT NATURE CENTRE

This former plantation (170 ha) is the most popular base from which to bird all the sites on Trinidad. The lodge, complete with feeders which attract many birds including ten species of hummer, overlooks the Arima valley in the mountains of the northern range. The centre is surrounded by forest, and Oilbirds are present in caves in the grounds.

Specialities

Lilac-tailed Parrotlet, Oilbird, Tufted Coquette, White-chested Emerald, Bearded Bellbird, Golden-fronted Greenlet, Bare-eyed Thrush, Trinidad Euphonia.

Others

Grey-headed and Double-toothed Kites, White Hawk, Ornate Hawk-Eagle, Scaled Pigeon, Gray-fronted Dove, Ruddy Quail-Dove, Blue-headed Parrot, Tropical Screech-Owl, Spectacled Owl, Chestnut-collared, Band-rumped and Grey-rumped Swifts, Rufous-breasted, Green and Little Hermits, White-necked Jacobin, Blue-chinned Sapphire, Copper-rumped Hummingbird, Long-billed Starthroat, White-tailed, Collared and Violaceous Trogons, Blue-crowned Motmot, Channel-billed Toucan, Golden-olive, Chestnut and Lineated Woodpeckers, Plain-brown and Buff-throated Woodcreepers, Grey-throated Leaftosser, Streaked Xenops, Great Antshrike, Plain Antvireo, White-flanked Antwren, Black-faced Antthrush, Golden-headed and White-bearded Manakins, Ochre-bellied and Slaty-capped Flycatchers, Southern Beardless-Tyrannulet, Yellow-breasted and Olive-sided Flycatchers, Tropical Pewee, White-winged Becard, Black-tailed Tityra, Rufous-browed Peppershrike, Cocoa and White-necked Thrushes, Rufous-breasted Wren, Long-billed Gnatwren, White-shouldered and White-lined Tanagers, Red-crowned Ant-Tanager, Silver-beaked Tanager, Violaceous Euphonia, Turquoise and Bay-headed Tanagers, Green, Purple and Red-legged Honeycreepers, Greyish Saltator, Crested Oropendola.

Other Wildlife

Agouti, Red-tailed Squirrel, Opossum, Tarantula.

Access

To reach the Nature Centre take the Arima bypass off the Churchill-Roosevelt highway and carry straight on across Eastern Main Road towards Blanchisseuse. The turning to the Nature Centre is to the west, 11 km north of Arima. Guides are available for walks, including one to the caves where approximately 130 Oilbirds usually breed.

Accommodation: Asa Wright Nature Centre (A) (tel: 809 667 4655; fax: 667 0493); Mount St Benedict PAX Guesthouse (A), half the price of Asa Wright (tel: 662 4084; fax: 622 7263).

There are a number of good birding sites along the **Arima–Blanchisseuse road**, which crosses the northern range. The forest this precarious road passes through receives up to 380 cm of rain per annum, although it is relatively dry from December to March. In addition to most of the species listed above, Little Tinamou, Stripe-breasted Spinetail, White-bellied Antbird, Euler's Flycatcher, Yellow-legged Thrush, Tropical Parula, Speckled Tanager and Sooty Grassquit occur along here.

Once across Eastern Main Road from the Arima bypass, pass Temple village to the east and cross a concrete bridge, then take the track east to a sand quarry. Turn north at the quarry on to the road up to the Simla Research Station. Trinidad Euphonia occurs here. Turn east off the Arima–Blanchisseuse road 3 km before the Asa Wright turn-off for **Lalaja Trace**, a good site for forest skulkers such as Little Tinamou, White-flanked Antwren and White-bellied Antbird, as well as Trinidad Euphonia and Blue Dacnis.

North of the Nature Centre at the top of the Arima–Blanchisseuse road turn right on to the TSTT road which is good for Channel-billed Toucan, or walk left opposite this turning along a track through scrub to good forest where Red-legged Honeycreeper occurs.

Rufous Nightjar may be seen at dawn up from the car park at the Rehabilitation Centre near **Mount St Benedict Guesthouse** (250 m/820 ft) (Map below). Turn north off Eastern Main Road on to St John Road at Tunapuna, 9.5 km west of Arima (13 km east of Port of Spain)

MOUNT ST. BENEDICT

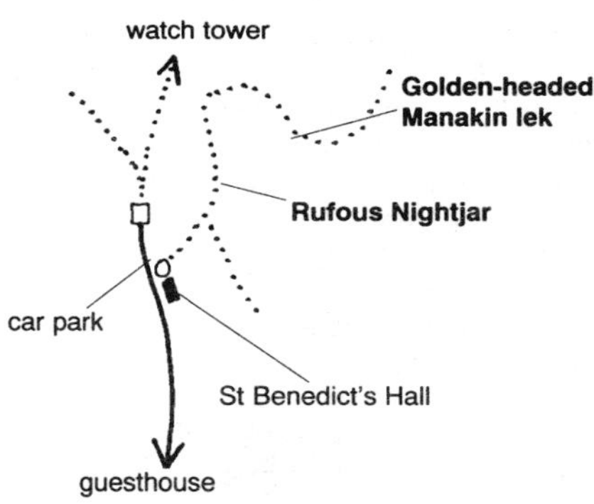

to reach here. There is also a Golden-headed Manakin lek here, and Little Tinamou, Tropical Screech-Owl, and Mottled Owl also occur.

CARONI SWAMP

Map below

Watching hundreds of Scarlet Ibises gliding into the mangroves of Caroni Swamp at sundown is one of the world's ornithological high-

CARONI SWAMP

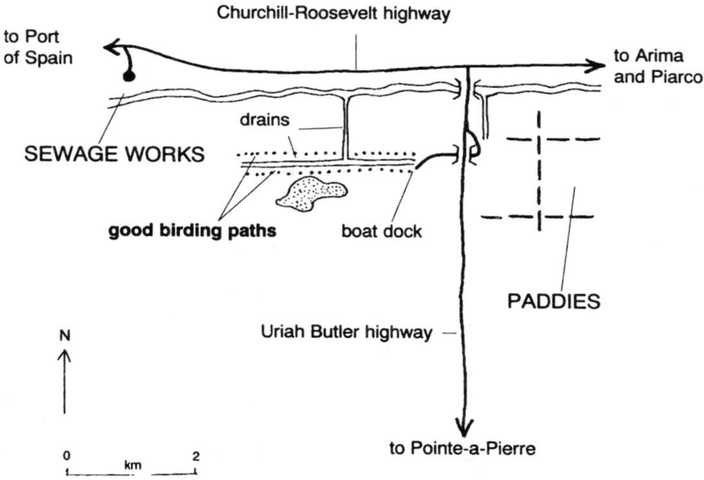

lights. The ibis ceased breeding on Trinidad in the early 1970s but large numbers still spend the non-breeding season here, mainly September to January. Most birds leave for Venezuela by March or April.

Specialities

Scarlet Ibis, Green-throated Mango.

Others

Masked Duck, Boat-billed Heron, Striped-backed Bittern, Limpkin, Mangrove Cuckoo, Gray Potoo, American Pygmy Kingfisher, Straight-billed Woodcreeper, Black-crested Antshrike, Red-capped Cardinal, Bicoloured Conebill.

Other Wildlife

Two-toed Silky Anteater, Spectacled Caiman, Tree Boa.

Access

To see the most Scarlet Ibises it is best to arrange a boat and guide (at the Asa Wright Nature Centre) which takes you deep into the mangroves and marshes, and to the island where the Scarlet Ibises roost.

Otherwise, it is possible to visit without a guide to see Green-throated Mango, which occurs along the paths through the mangroves from the boat dock. Turn south off the Churchill-Roosevelt highway on to Uriah Butler highway, 11 km east of Port-of-Spain, then take the left exit which then crosses the road back west to the boat dock. Stripe-backed Bittern occurs in the paddies east of the Uriah Butler highway junction. Walk or drive the tracks between the paddies.

Scarlet Ibis also occurs at **Port of Spain Sewage Works** (See Caroni Swamp map p. 355). This site is accessible from the Beetham highway, just east of the Port of Spain 'flyover', itself just east of the capital. Turn south on to a 1-km track here to reach the pools, which are also a good site for shorebirds such as Collared Plover, as well as Greater Ani, Pied Water-Tyrant, White-headed Marsh-Tyrant and Grey Kingbird.

The **Piarco Water Treatment Works**, opposite Piarco airport, 16 km southeast of Port-of-Spain, are also good for waterbirds (permission to visit required from Water and Sewage Authority). Turn west on leaving the airport, then immediately turn south to a gate. Collared Plover occurs at the airport itself. More wetlands can be found at Trincity Water Works, viewable from the road 3 km west of the Piarco crossroads. Long-winged Harrier and Yellow-billed Tern occur here.

Turn north off the Eastern Main Road, 4.5 km east of Arima, to reach the **Arima Agricultural Research Station**. Ask for permission to enter or contact the director in advance (by phone). This is one of the few areas where the only endemic bird of the island, Trinidad Piping-Guan*, has been seen recently, albeit just once in many years. Scour the northern edge of the site. Collared Plover, Yellow-chinned Spinetail, Pied Water-Tyrant, White-headed Marsh-Tyrant, Fork-tailed Flycatcher and Blue-black Grassquit also occur here.

CERRO DEL ARIPO

Trinidad Piping-Guan* occurs (rarely) around Cerro del Aripo, the highest peak in Trinidad (915 m/3,020 ft).

Endemics

Trinidad Piping-Guan*.

Specialities

Bearded Bellbird, Orange-billed Nightingale-Thrush.

Others

Bat Falcon, Chestnut-collared Swift, Brown Violet-ear, White-tailed and Collared Trogons, Grey-throated Leaftosser, Plain Antvireo, Black-faced Antthrush, Short-tailed Pygmy-Tyrant, Bright-rumped Attila, Black-tailed Tityra, Yellow-legged Thrush, Red-crowned Ant-Tanager, Blue-capped Tanager.

Access

Turn north off Eastern Main Road, 6.5 km east of Arima, opposite the turning south to Wallerfield and Cumuto. Bird the road from the turn-off. At the 6.5 km post continue straight on then right after 0.5 km onto Ruiz Trace, a steep road. The guan is best looked for as far as possible along this road.

WALLERFIELD

This abandoned US airbase near Arima represents Trinidad's only remaining 'savanna'.

Others

Grey-headed and Pearl Kites, Savanna Hawk, Yellow-headed Caracara, Southern Lapwing, Ruddy Ground-Dove, Striped Cuckoo, Fork-tailed Palm-Swift, Ruby-topaz Hummingbird, Blue-crowned Motmot, Black-crested Antshrike, Bran-coloured and Sulphury Flycatchers, Masked Yellowthroat, Yellow-rumped Cacique, Yellow-hooded and Red-breasted Blackbirds, Shiny and Giant Cowbirds.

Access

Turn south off Eastern Main Road, 6.5 km east of Arima on to the Cumuto road. After crossing the Aripo river turn west at a large derelict building on to the 'Circle B track'. Black-crested Antshrike occurs along here. Otherwise there is a maze of tracks and roads in this area worth exploring.

Large-billed Tern and Pied Water-Tyrant occur at Arena Dam, reached by turning south some 4 km west of Cumuto (permission to visit required from the Water and Sewage Authority).

NARIVA SWAMP

This large freshwater marsh on the east coast, 40 km southeast of Arima, is the best site on Trinidad for Pinnated Bittern, Azure Gallinule and Red-bellied Macaw.

Specialities

Pinnated Bittern, Azure Gallinule.

Others

Least Bittern, Pearl and Plumbeous Kites, Black-collared and Savanna Hawks, Limpkin, Plain-breasted Ground-Dove, Red-bellied Macaw, American Pygmy Kingfisher, Streak-headed Woodcreeper, Black-crested Antshrike, Silvered Antbird, Pied Water-Tyrant, White-headed Marsh-Tyrant, Fork-tailed Flycatcher, Dickcissel, Moriche Oriole, Yellow-hooded Blackbird.

Other Wildlife

Red Howler, Manatee.

Access

Bird the coastal road which skirts the east side of the marsh. Silvered Antbird occurs along Bush Bush Creek, approximately halfway along this road, 20 km south of Lower Manzanilla. Red-bellied Macaws usually fly in to roost in the Moriche palms (after 5 pm), 3 km south of here.

TOBAGO

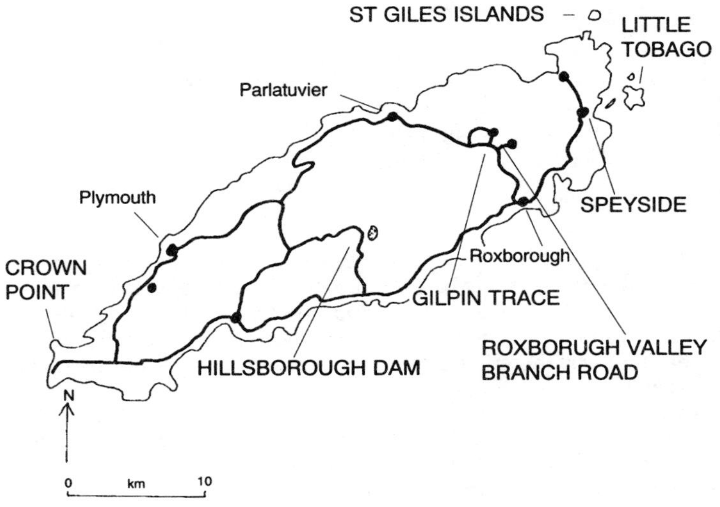

This tiny island (300 km^2), 40 km northeast of Trinidad, is almost what everyone imagines a truly tropical island to be, complete with breeding Red-billed Tropicbirds. Twelve 'landbirds' breed here which do not breed on Trinidad. They are: Rufous-vented Chachalaca, Striped Owl, White-tailed Sabrewing*, Red-crowned Woodpecker, Olivaceous Woodcreeper, White-fringed Antwren, Blue-backed Manakin, Venezuelan Flycatcher, Caribbean Martin, Scrub Greenlet, Variable Seedeater, and Black-faced Grassquit.

Accommodation: Crown Point Beach Hotel (A); Arthur's-on-Sea (A). Charlotteville: Almandoz Beachhouse (A); Man o' War Bay Cottages (A). Speyside: Blue Waters Inn (A).

Specialities

Red-billed Tropicbird, Red-footed Booby, Rufous-vented Chachalaca, Brown Noddy, Green-rumped Parrotlet, White-tailed Sabrewing*, Stripe-breasted Spinetail, Venezuelan Flycatcher, Bare-eyed Thrush, Caribbean Martin, Black-faced Grassquit, Carib Grackle.

Others

Audubon's Shearwater, Magnificent Frigatebird, Brown Booby, White-cheeked Pintail, Wattled Jacana, Collared Plover, Southern Lapwing, Royal, Bridled and Sooty Terns, Pale-vented Pigeon, Orange-winged Parrot, Striped Owl, White-tailed Nightjar, Short-tailed Swift, Rufous-breasted Hermit, White-necked Jacobin, Ruby-topaz and Copper-rumped Hummingbirds, Collared Trogon, Blue-crowned Motmot, Rufous-tailed Jacamar, Red-crowned, Red-rumped and Golden-olive

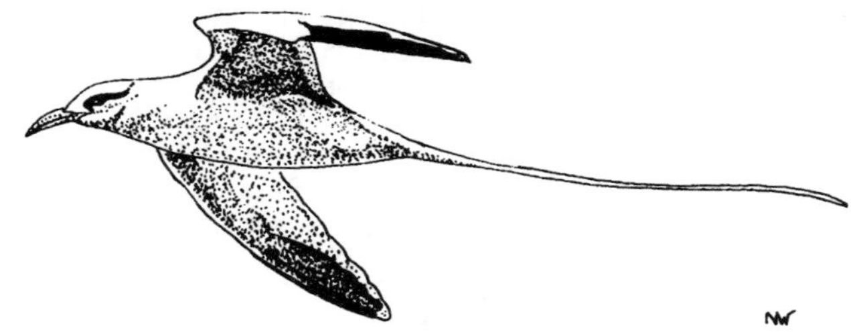

The elegant Red-billed Tropicbird may be the highlight of a boat trip to Little Tobago or St Giles Islands, off Tobago

Woodpeckers, Olivaceous and Buff-throated Woodcreepers, Barred Antshrike, White-fringed Antwren, Blue-backed Manakin, Yellow-bellied Elaenia, Yellow-breasted, Fuscous and Brown-crested Flycatchers, Grey Kingbird, Streaked Flycatcher, Scrub Greenlet, Yellow-legged and White-necked Thrushes, Tropical Mockingbird, Rufous-breasted Wren, White-lined Tanager, Variable and Ruddy-breasted Seedeaters, Red-legged Honeycreeper, Crested Oropendola, Giant Cowbird.

Other Wildlife

Leatherback and Hawksbill Turtles, Caiman, Queen and French Angelfish, Blue Tang, Doctorfish, Grouper, French Grunt, Rainbow, Spotlight and Lipstick Parrotfish, Sergeant Major, Squirrelfish, Triggerfish, Trumpetfish, Stingray.

Birding Sites

At the western end of Tobago, Bon Accord lagoon supports White-cheeked Pintail and is great for snorkelling.

Crown Point is a good site for Ruby-topaz Hummingbird. White-tailed Nightjar occurs at the Kilgwyn swamp, east of Crown Point airport. Grafton Estate is also worth a look. Turn right just before the Grafton Estate Hotel off the road north from Crown Point to Plymouth on the north coast.

At the centre of the island is **Hillsborough Dam** (permission to visit required from Water and Sewage Authority) on the Mount St George–Castara road. Blue-backed Manakin occurs along the road beyond the dam. Forest birding is best along the **Roxborough–Parlatuvier (Bloody Bay) road**. White-tailed Sabrewing* and Blue-backed Manakin occur along the Roxborough valley branch road, which runs east 5 km north of Roxborough. The sabrewing occurs at the fork 400 m along here, and there is a Blue-backed Manakin lek opposite a large stand of bamboo, a 100 m or so along the left fork from here. Gilpin Trace, the road east 9 km north of Roxborough, is another excellent area. This is one of the few sites which escaped the worst of the 1963 hurricane and is a good site for White-tailed Sabrewing*, as well as White-fringed Antwren, Blue-backed Manakin, Yellow-legged Thrush and Red-legged Honeycreeper.

At the eastern end of Tobago, **Speyside** is a good site, and the hilly **Little Tobago Island** (113 ha), off the northeast tip, is a must. It supports breeding Red-billed Tropicbird, Bridled and Sooty Terns and Brown Noddy; Blue-crowned Motmot is common. Boats leave from the Blue Waters Hotel, Speyside; permission to visit the island must be obtained from the Department of Agriculture, Forestry and Fisheries.

Red-footed Booby is rare on Little Tobago Island but breeds, along with all the species recorded from Little Tobago, on **St Giles Islands**, also off the island's northeast tip. Landing is not allowed but the islands can be viewed from boats hired at Speyside or Charlotteville.

Snorkelling Sites

At Arnos Bay, ask permission to swim where the reef is adjacent to the beach. At Pirate's Bay, Charlotteville it is a short swim to the reef on the left-hand side of the bay. Blue Waters, Speyside is the best reef but there is a 100-m swim to it.

ADDITIONAL INFORMATION

Addresses

Water and Sewage Authority, Farm Road, Valsaya, St Joseph.
Department of Agriculture, Forestry and Fisheries, Windward Road, Mount St George, or phone the Botanic Station at Scarborough on 809 639 3265.

Books

A Guide to the Birds of Trinidad and Tobago, ffrench, R., 1992. Helm.
Birds of Trinidad and Tobago, ffrench, R., 1986. Macmillan. (Photographic guide)
A Birder's Guide to Trinidad and Tobago, Murphy, W., 1987. Peregrine Enterprises.

TRINIDAD AND TOBAGO ENDEMICS (1)	**BEST SITES**
Trinidad Piping-Guan*	Very rare: Arima Agricultural Research Station and Cerro del Aripo, Trinidad

URUGUAY

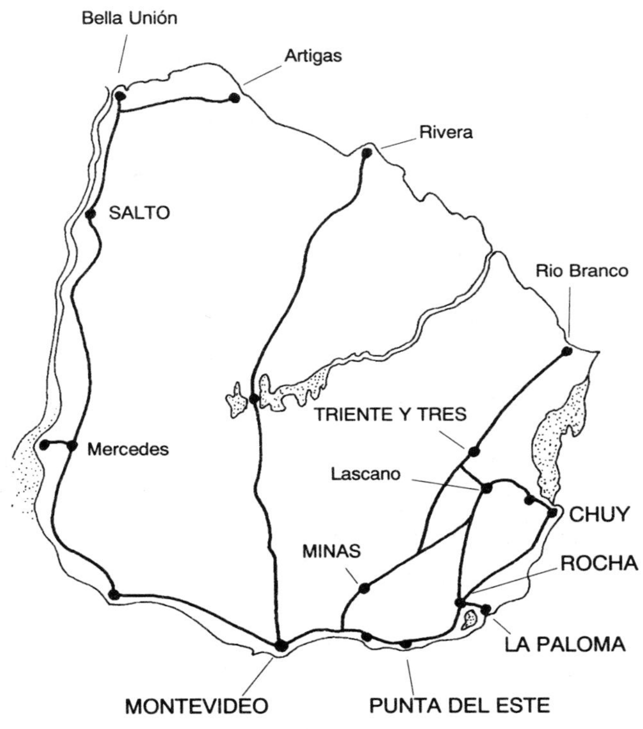
Bella Unión
Artigas
Rivera
SALTO
Rio Branco
TRIENTE Y TRES
Mercedes
Lascano
CHUY
MINAS
ROCHA
LA PALOMA
MONTEVIDEO
PUNTA DEL ESTE

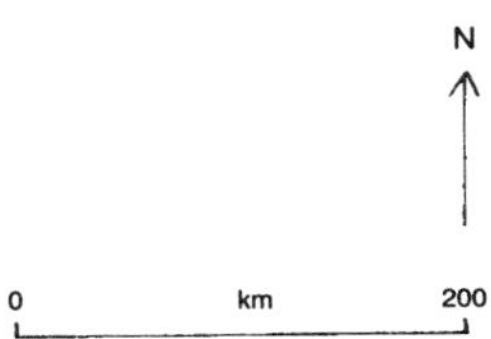
N
0
km
200

INTRODUCTION

Summary

This small, relatively expensive country has been neglected by most birders since it supports few of South America's more spectacular species. However, it has an excellent infrastructure and field guide, and presents an exciting prospect for the pioneer, especially if interested in rails, furnariids and tyrants, some of which are very rare and localised in South America.

Size

At 176,215 km^2 Uruguay is one of the smallest South American countries, only slightly larger than England, only a quarter the size of Texas, and one fifth the size of Venezuela.

Getting Around

Uruguay has a modern infrastructure utilised by an extensive bus service. Hire cars are available in Montevideo. There is also a cheap internal air network.

Accommodation and Food

There are few cheap hotels in Uruguay, although some youth hostels provide a useful back-up for the budget-birder. Spicy soups and beef dishes dominate the menus.

Health and Safety

Immunisation against polio, typhoid and hepatitis is recommended. Uruguay is a friendly country with few political problems, and personal safety is not a major cause for concern.

Climate and Timing

Uruguay's climate is very variable. Generally, the winter (April to September) is wet and windy and the summer (mainly December to March) is drier. This is therefore the best time to visit, although it can be very hot along the coast.

Habitats

This is primarily a land of pampas grassland and remnant woodland lying on and between low rolling hills which rise to 610 m (2,001 ft). However, the south and, especially, east coasts are lined with lagoons and marshes. The Río Uruguay runs along the western border with Argentina.

Conservation

Much of Uruguay lies within the pampas zone, and the lagoons and marshes along southeast coast, an area known as the Banados del Este, form one of South America's most important wetlands.

11 threatened species occur in Uruguay.

Bird Families

68 of the 92 families which regularly occur in South America are represented, just one fewer than Paraguay. These include ten of the 25 Neotropical endemic families and four (equal to the total for Venezuela) of the nine South American endemic families. Trumpeters,

Magellanic Plover, Hoatzin, Oilbird and gnateaters are all absent, as well as Sunbittern, motmots, jacamars, puffbirds, toucans and manakins.

Well-represented families include waterfowl, rails, furnariids, tyrants, pipits and icterids.

Bird Species

Uruguay's list of 404 species is poor by South American standards. Fewer birds occur here than in the Trinidad and Tobago archipelago,a fraction of Uruguay's size. However, despite being a quarter the size of Chile, it supports only 40 fewer species.

Non-endemic specialities and spectacular species include Greater Rhea*, Crowned Eagle*, Speckled Rail*, Red-and-white and Dot-winged* Crakes, Red-legged Seriema, American Painted-Snipe, Snowy Sheathbill, Olrog's Gull*, Snowy-crowned Tern, Rufous Hornero, Hudson's Canastero, Grey-bellied and Sulphur-bearded Spinetails, Curve-billed and Straight-billed* Reedhaunters, Brown Cachalote, White-tipped Plantcutter, Black-and-white Monjita*, Blue-billed Black-Tyrant, Strange-tailed Tyrant*, Diademed and Chestnut-backed Tanagers, Marsh Seedeater*, and Saffron-cowled Blackbird*.

In addition to these a large number of seabird species have been recorded, including two penguins, four albatrosses and 17 shearwaters.

It is possible to see the rare Black-and-white Monjita along the Uruguay coast circuit*

Endemics

Uruguay has no endemic species. However, a few species, such as Marsh Seedeater*, are virtually confined to the country's boundaries.

Expectations

With such a small list, trip totals will not compare favourably with other countries. A week along the southeast coast may produce over 100 species.

MONTEVIDEO–CHUY–MINAS CIRCUIT

It is possible to see over 100 species on this circuit, which is approximately 800 km long. The route runs 330 km east from Montevideo to Chuy on the Brazilian border, and back west inland via Lascano and Minas. It includes the Banados del Este (300,000 ha) to the east, the most important wetland area in Uruguay where some 10,000 Black-necked Swans and 1,000 Coscoroba Swans have been recorded, as well as Speckled Rail*, Straight-billed Reedhaunter*, Black-and-white Monjita* and Saffron-cowled Blackbird*.

Specialities

Spotted Nothura, Ringed Teal, Crowned Eagle*, Speckled Rail*, Giant Wood-Rail, Olrog's Gull*, Snowy-crowned Tern, White-spotted Woodpecker, Olive Spinetail, Straight-billed Reedhaunter*, Black-and-white Monjita*, Chocolate-vented Tyrant (April–Sep), Red-crested Cardinal, Diademed Tanager, Lesser Grass-Finch*, Rusty-collared and Marsh* Seedeaters, Glaucous-blue Grosbeak, Brown-and-yellow Marshbird, Saffron-cowled* and Scarlet-headed Blackbirds.

Others

Greater Rhea*, White-tufted and Great Grebes, Magnificent Frigatebird, Southern Screamer, Fulvous and White-faced Whistling-Ducks, Black-necked and Coscoroba Swans, Brazilian Teal, Chiloe Wigeon, Speckled Teal, Yellow-billed Pintail, Silver Teal, Rosy-billed Pochard, Whistling and Cocoi Herons, Bare-faced, White-faced and Plumbeous Ibises, Roseate Spoonbill, Wood and Maguari Storks, Snail Kite, Long-winged Harrier, Chimango Caracara, Grey-necked Wood-Rail, Plumbeous Rail, Spot-flanked Gallinule, White-winged Coot, Limpkin, Wattled Jacana, South American Snipe, Southern Lapwing, Brown-hooded Gull, Yellow-billed Tern, Black Skimmer, Spot-winged Pigeon, Eared Dove, Ruddy and Picui Ground-Doves, Grey-fronted Dove, Monk Parakeet, Dark-billed and Guira Cuckoos, Glittering-bellied Emerald, Rufous-throated and Gilded Sapphires, White-throated Hummingbird, Green-barred Woodpecker, Campo Flicker, Rufous Hornero, Chicli Spinetail, Firewood-gatherer, White-tipped Plantcutter, Southern Beardless-Tyrannulet, Small-billed Elaenia, White-crested Tyrannulet, Bran-coloured Flycatcher, Grey and White Monjitas, Spectacled, Yellow-browed and Cattle Tyrants, Swainson's and Fork-tailed Flycatchers, White-winged Becard, Rufous-bellied and Creamy-bellied Thrushes, Chalk-browed Mockingbird, Masked Gnatcatcher, White-rumped Swallow, Hooded Siskin, Tropical Parula, Masked Yellowthroat, Blue-and-yellow Tanager, Black-and-rufous Warbling-Finch, Grassland Yellow-Finch, Great Pampa-Finch, Blue-black Grassquit, Double-collared Seedeater, Yellow-winged, Chestnut-capped and White-browed Blackbirds, Bay-winged and Shiny Cowbirds.

Access

Olrog's Gull* occurs on the Río Soca estuary 5 km east of **Atlántida**, 45 km east of Montevideo, the closest 'natural' habitat to the capital. Bird the riverside marsh. Head east 94 km from here to **Punta del Este** (139 km east of Montevideo). Coscoroba Swan occurs on the lagoons north of town. The 72-km drive east to **Rocha** (211 km from Montevideo) through scrub, lagoons and rolling grassland is excellent for birds. Giant

Wood-Rail, Rufous-throated Sapphire, White-spotted Woodpecker, Olive Spinetail, White-tipped Plantcutter, Diademed Tanager, Marsh Seedeater* and Glaucous-blue Grosbeak have all been recorded around Rocha. Great Grebe, Magnificent Frigatebird and Snowy-crowned Tern occur at **La Paloma**, 28 km south of Rocha.

From Rocha head 95 km further east to the **Santa Teresa NP**, where Chocolate-vented Tyrant occurs (at least in December), and then a further 24 km to **Chuy** (330 km east of Montevideo) on the Brazilian border. Many waterbirds including Ringed Teal occur in the marshes around here, and Crowned Eagle*, White-spotted Woodpecker, Diademed Tanager and Lesser Grass-Finch* occur around **San Miguel Fortress**, approximately 25 km northwest of Chuy, *en route* inland to Lascano. The road from Chuy to **Lascano**, where Scarlet-headed Blackbird occurs, passes through wet grasslands and Palm savanna. From Lascano head southwest to the town of **Minas**, where the parks with waterfalls and woodland may be worth a look. From here it is 90 km back to Montevideo.

The **Quebrada de los Cuervos NP**, a land of rocky hills and streams 325 km northeast of Montevideo, is probably a good area for birds. Turn west 20 km north of Treinta y Tres (286 km northeast of Montevideo) along the 19-km track through the NP.

The long road northwest from Montevideo to **Salto**, on the Río Uruguay, is good for pampas species including Greater Rhea*, which also occurs near Termas del Arapey. Turn east 61 km north of Salto on to the road to Termas del Arapey, 35 km away. This is a good area for a number of pampas species.

The ranch country around Artigas, 100 km east of Bella Unión, in far northwest Uruguay and on the border with Brazil, is a good area for the rare Marsh Seedeater*. Other species here could include those listed under the Uruguaiana Site, just across the border in Brazil (see p. 117).

For details of possible seabirds on the Montevideo–Falkland Islands crossing see p. 261.

ADDITIONAL INFORMATION

Books

Birds of Argentina and Uruguay: A Field Guide, Narosky, T. and Yzurieta, D., 1989.
Lista de Referencia y Bibliografia de las Aves Uruguayes, Cuello, J., 1985.
Las Aves del Uruguay, Gore, M. and Gepp, A., 1978. Montevideo, Musca. Hnos.

URUGUAY ENDEMICS

None.

VENEZUELA

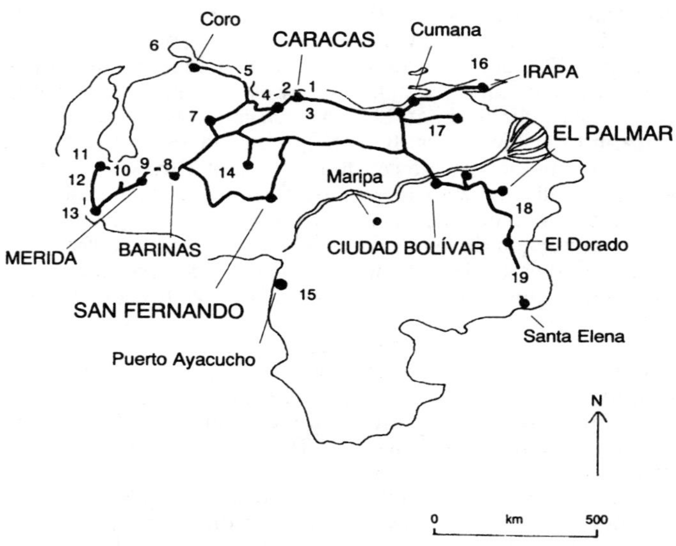

1 Caracas
2 Colonia Tovar
3 Guatapo NP
4 Henri Pittier NP
5 Chichiviriche
6 Paraguaná Peninsula
7 Barquisimeto
8 Santo Domingo Valley
9 Pico Humboldt Trail
10 La Azulita Road
11 Encontrados Road
12 San Juan de Colón
13 Paramo de Tama
14 The Llanos
15 Amazonas
16 Cerro de Humo
17 Cueva de los Guacharos NP
18 Imataca FR and El Dorado
19 The Escalera

INTRODUCTION

Summary

For those who have no fear of being overwhelmed, Venezuela ranks alongside Ecuador as the best introduction to South American birding. The modern infrastructure, excellent field guide and friendly atmosphere combine to aid the birder who can expect a list high in quantity and quality. The southeast, where no less than fifteen cotingas are present and Harpy Eagle* is a realistic possibility, is one of the most exciting birding areas on the continent.

Size

At 912,050 km^2 Venezuela is seven times the size of England but only a little larger than Texas. Although it is three times the size of Ecuador, most of the best birding sites can be visited on a three-week trip.

Getting Around

Venezuela is a 'modern' country, but away from the major towns many of the roads are in poor condition. However, a four-wheel-drive with high clearance is not needed for most sites and buses go virtually everywhere. Should some distances between sites seem daunting there is an excellent, cheap internal air network, which also reaches remote destinations such as Amazonas.

Accommodation and Food

All types of accommodation can be found in the major cities but, with the exception of the Llanos and the Merida Andes, where most hotels, lodges and ranches are expensive, accommodation is relatively basic. Finding good food is not a problem in Venezuela, and the beer is also excellent.

Health and Safety

Immunisation against yellow fever, typhoid and malaria is recommended, and beware of bilharzia and dengue fever.

Although this is one of South America's most friendly countries it is wise to be cautious in built-up areas and some remote settlements at night, such as 'km 88' in Bolívar.

Climate and Timing

Rain is possible at any time in this tropical country but the main wet season is between April and November. It is driest between December and March so, overall, this is the best time to visit. However, June–July and October–November are also good times.

Habitats

There are three main mountain ranges in Venezuela. The Cordillera de la Costa runs along the northern coast; it supports arid scrub on the low northern slopes near the coast, subtropical forest, cloud (temperate) forest at its crest, and deciduous woodland on the drier southern slopes. The Andes reach their northeastern limit in southwest Venezuela. Their slopes support subtropical forests and paramo. The Guianan highlands and their distinctive avifauna extend into southeast Venezuela where the flat-topped mountains known as 'tepuis' dominate the scenery made famous in Conan Doyle's *Lost World*.

The swampy Maracaibo basin lies to the west of the Andes, whilst semi-desert covers much of the land to the north of these mountains.

The seasonally-wet grasslands known as the 'Llanos' cover much of the land south of the coastal mountains, east of the Andes, and north of the Orinoco river, which flows west to east across the middle of the country. South of the Orinoco river, in the state of Amazonas, the Llanos give way to one of the largest tracts of untouched lowland rainforest left in South America, stretching as far as the eye can see, even from a plane, to the Brazilian border.

Conservation

Venezuela boasts an impressive 35 National Parks as well as plenty of private reserves which depend, to some extent, on ecotourism for their income. However, some of this country's rich ecosystems are not protected, and at least one of them, the famous Llanos, is under threat from various hydrological schemes aiming to turn these vast wetlands into pasture land.

20 threatened species, of which 11 are endemic, occur in Venezuela. A further six endemics are near-threatened.

Bird Families

79 of the 92 families which reguarly occur in South America are represented, including 20 of the 25 Neotropical endemic families, a diversity to rival Argentina and Ecuador. However, only four of the nine South American endemic families are represented, at least two less than most of the other mainland countries. There are no rheas, seriemas, Magellanic Plover, seedsnipes or gnateaters.

Well-represented families include ibises, cracids, hummingbirds, jacamars, toucans, antbirds, antpittas, cotingas, manakins and 'tanagers'.

Bird Species

Approximately 1,360 species have been recorded in Venezuela, nearly 200 fewer than in Ecuador, which is a third the size.

Non-endemic specialities and spectacular species include Northern Screamer*, Scarlet Ibis, Harpy Eagle*, Yellow-knobbed Curassow, Sungrebe, Sunbittern, Grey-winged Trumpeter, Hoatzin, Oilbird, White-tipped Quetzal, Guianan Red-Cotinga, Dusky Purpletuft, Purple-breasted and Spangled Cotingas, Capuchinbird, Guianan and Andean Cock-of-the-Rocks, Rosy Thrush-Tanager and Blue-backed Tanager.

Endemics

Venezuela's impressive total of 41 endemic species is the fourth highest country total in South America, well behind Brazil and Peru, but well above Argentina and Ecuador.

Over ten of these species unique to Venezuela are extremely rare and localised. For example, Saffron-breasted Redstart is known only from Cerro Guaiguinima, a remote mountain in central Bolívar. The twenty or so endemics likely to be seen on an extensive trip to Venezuela include some superb birds, not least Scissor-tailed Hummingbird*, Merida Sunangel, Ochre-browed Thistletail, Handsome Fruiteater and White-fronted Redstart.

The long list of near-endemics, which are hard to get to, let alone see in neighbouring countries, include, in the north and west, Pygmy Swift, Bearded Helmetcrest, White-tipped Quetzal and Vermilion Cardinal, and in the southeast, Peacock Coquette, Red-banded Fruiteater, Rose-collared Piha, Scarlet-horned Manakin and Greater Flower-piercer.

Expectations

Some amazing species totals are possible on trips to Venezuela. Well prepared experienced birders may record 500 species in just three weeks in the northwest, and over 600 species over the same period of time, if the sites in Bolívar are also visited.

CARACAS

Venezuela's capital, on the north-central coast, supports some interesting birds, including two endemics, and is close to some excellent sites, not least Henri Pittier NP, which is just two hours away, once you have found your way out of this city's bamboozling road system.

Endemics

Violet-chested Hummingbird, Black-throated Spinetail.

Specialities

Band-tailed Guan, Crested Spinetail, Bare-eyed Thrush, Fulvous-headed Tanager.

Others

Brown Pelican, Plumbeous Kite, Grey-necked Wood-Rail, Brown-throated and White-eared Parakeets, Green-rumped Parrotlet, White-collared Swift, Pale-bellied Hermit, Black-throated Mango, Bronzy Inca, Collared Trogon, Rufous-tailed Jacamar, Scaled Piculet, Barred Antshrike, White-fringed Antwren, Jet and White-bellied Antbirds, Slate-crowned Antpitta, Brown-rumped Tapaculo, Southern Beardless-Tyrannulet, Mountain Elaenia, Flavescent Flycatcher, Chestnut-crowned Becard, Stripe-backed Wren, Tropical Gnatcatcher, Black-striped Sparrow, Chestnut-capped Brush-Finch, Black-faced Grassquit, Buff-throated and Greyish Saltators, Yellow-billed Cacique, Yellow-backed Oriole, Oriole Blackbird, Carib Grackle.

Access

The **Parque del Este** is a good place to go birding if time is really short. Grey-necked Wood-Rail and Oriole Blackbird occur here, but little else of note. Many of the species listed above occur in the forested hills behind **Macuto**, a beach resort a few km east of Caracas International airport. Band-tailed Guan, Violet-chested Hummingbird and Fulvous-headed Tanager occur in **Avila NP**, reached via cable-car from Avenida Perimetral de Mariperez or via the Sabas Nieves trail, which starts 100 m west of the Tarzilandia restaurant in Decima Transversal Street, Altamira.

Accommodation: Caracas: Hotel Campo Alegre; Hotel Avila (B). Macuto: Hotel Melia Caribe (A):

Stripe-tailed Yellow-Finch, Wedge-tailed Grass-Finch and Ruddy-breasted Seedeater occur in the grassy areas alongside the road between La Victoria and Colonia Tovar, west of the city.

COLONIA TOVAR

The pretty and popular German hill-resort of Colonia Tovar, situated at 1,890 m (6,201 ft) in the coastal mountains, is only 50 km west of Caracas via El Junquito. The subtropical forest here supports White-tipped Quetzal, Golden-breasted Fruiteater and Plush-capped Finch, but it can take two hours to get here from the capital, so a visit to Henri Pittier NP (see p. 371), also two hours from Caracas, may be a wiser choice if time does not allow both sites to be visited.

Endemics

Black-throated Spinetail.

Specialities

Rufous-shafted Woodstar, White-tipped Quetzal, Groove-billed Toucanet, Crested Spinetail, Brown-rumped Tapaculo, Golden-breasted Fruiteater, Venezuelan Tyrannulet, Ochre-breasted Brush-Finch.

Others

Lazuline Sabrewing, Green and Sparkling Violet-ears, Violet-fronted Brilliant, Long-tailed Sylph, Spot-crowned Woodcreeper, Streaked Tuftedcheek, Green-and-black Fruiteater, Mountain Elaenia, Marble-faced Bristle-Tyrant, Yellow-bellied Chat-Tyrant, Black-and-white Becard, Andean Solitaire, Black-crested Warbler, Chestnut-capped Brush-Finch, Capped Conebill, Blue-winged Mountain-Tanager, Blue-naped Chlorophonia, Black-capped Tanager, Plush-capped Finch, Rusty and Bluish Flower-piercers.

Access

Birding Colonia Tovar involves a certain amount of patience since trails are hard to find. The track, on the left 7 km before Colonia Tovar and *en route* from La Victoria, can be good (Plush-capped Finch occurs here), and any other side-tracks or trails are worth investigating.

Accommodation: many to choose from. When open, the Hotel Drei Tannan, just east of town, has excellent gardens which attract many hummers.

GUATAPO NP

Guatapo NP is also two hours from Caracas, to the southeast, via Santa Theresa. The subtropical forest here is particularly good for raptors (ten are possible in a day) including Solitary Eagle* and Ornate Hawk-Eagle. It is also the best site in Venezuela for Military Macaw.

Specialities

Military Macaw, White-eared Parakeet, Sooty-capped Hermit, Venezuelan Tyrannulet, Golden-fronted Greenlet, Yellow Oriole.

Others

Plumbeous Kite, Solitary Eagle*, Black-and-white* and Ornate Hawk-Eagles, Grey-necked Wood-Rail, Mottled Owl, Rufous-breasted Hermit, White-necked Jacobin, Violet-headed Hummingbird, Golden-tailed Sapphire, Rufous-tailed Jacamar, Lineated Woodpecker, Rufous-winged Antwren, Wire-tailed, Golden-headed and Lance-tailed Manakins, Brown-capped Tyrannulet, Forest Elaenia, Marble-faced Bristle-Tyrant, Rufous-and-white Wren.

Access

The Santa Theresa–Altagracia road passes through the park. Bird the roadside, the HQ where a feeding station attracts tanagers, and the trail around the lagoon on the west side of the road a few km south of the HQ. For Military Macaw, the areas south of the HQ and outside the NP on the road to Altagracia are the best.

Accommodation: Altagracia: Hotel Amazor (C).

HENRI PITTIER NP

Map p. 373

This large NP (107,800 ha), which straddles the coastal mountains, boasts one of the biggest bird lists for any South American NP; over 500. Just over 100 km west of Caracas it is *the* place to head for once through customs at the airport. Running more or less parallel to the dry coast, where arid scrub and woodlands are present, the best habitat here is the superb cloud (temperate) forest between 914 m (3,000 ft) and 1,829 m (6,000 ft). It is possible to see over 100 species in this NP in one day including mixed-feeding flocks of up to 30 species, Helmeted Curassow*, White-tipped Quetzal, and the aptly named Handsome Fruiteater, one of ten endemics present.

Endemics

Venezuelan Wood-Quail*, Red-eared Parakeet, Green-tailed Emerald, Violet-chested Hummingbird, Black-throated Spinetail, Guttulated Foliage-gleaner, Scallop-breasted Antpitta*, Handsome Fruiteater, Venezuelan Bristle-Tyrant*, Rufous-cheeked Tanager.

Specialities

Fasciated Tiger-Heron*, Band-tailed Guan, Helmeted Curassow*, Scarlet-fronted Parakeet, Green-rumped and Lilac-tailed Parrotlets, Pale-bellied and Sooty-capped Hermits, Buffy Hummingbird, Rufous-shafted Woodstar, White-tipped Quetzal, Russet-throated and Moustached Puffbirds, Groove-billed Toucanet, Scaled Piculet, Crested Spinetail, Streak-capped Treehunter, Black-backed Antshrike, Scalloped Antthrush, Plain-backed Antpitta, Bearded Bellbird, Rufous-lored Tyrannulet, Black-hooded Thrush, Stripe-backed Wren, Golden-winged Sparrow, Ochre-breasted Brush-Finch, Fulvous-headed Tanager, Rosy Thrush-Tanager, Pileated Finch, Dull-coloured Grassquit.

Others

Magnificent Frigatebird, Brown Booby, White Hawk, Ornate Hawk-Eagle, Rufous-vented Chachalaca, Lined Quail-Dove, Brown-throated Parakeet, Black-and-white Owl, Rufous Nightjar, White-tipped Swift, Blue-chinned and Golden-tailed Sapphires, Copper-rumped Hummingbird, White-vented Plumeleteer, Violet-fronted Brilliant, Bronzy Inca, Long-tailed Sylph, Wedge-billed Hummingbird, Rufous-tailed Jacamar, Crimson-crested Woodpecker, Montane and Buff-fronted Foliage-gleaners, Grey-throated Leaftosser, Sharp-tailed Streamcreeper, White-streaked Antshrike, Plain Antvireo, Slaty and Rufous-winged Antwrens, White-bellied Antbird, Black-faced Antthrush, Rusty-breasted Antpitta, Golden-breasted Fruiteater, Wire-tailed and Lance-tailed Manakins, Pearly-vented Tody-Tyrant, Cinnamon and Pale-edged Flycatchers, Black-and-white and Black-capped Becards, Golden-fronted Greenlet, Orange-billed Nightingale-Thrush, White-necked Thrush, Whiskered, Rufous-breasted and Rufous-and-white Wrens, Grey-breasted Wood-Wren, Flavescent Warbler, Black-striped Sparrow, White-eared Conebill, White-winged and Blue-capped

Rancho Grande, in Henri Pittier NP, is the most reliable site for the dazzling White-tipped Quetzal, which occurs only in north Venezuela and north Colombia

Tanagers, Blue-winged Mountain-Tanager, Thick-billed Euphonia, Golden, Bay-headed, Beryl-spangled and Black-capped Tanagers, Green Honeycreeper, Swallow-Tanager, Black-faced Grassquit, Bluish Flower-piercer, Crested Oropendola, Yellow-rumped Cacique.

Other Wildlife

Red Howler Monkey, Three-toed Sloth.

Access

The NP is just 10 km north of Maracay, 110 km west of Caracas. Two roads cross the mountains and pass through the NP before reaching the coast. One of the best areas for birding is **Rancho Grande**, which is 20 km from Maracay (a 45 minute, 28 km drive from the Hotel Maracay) along the **Ocumare road**. Venezuelan Bristle-Tyrant occurs in the grounds around the building here, and Helmeted Curassow*, Venezuelan Wood-Quail*, Guttulated Foliage-gleaner and Plain-backed Antpitta occur along the Pico Guacamayo trail which starts behind the building. Violet-chested Hummingbird and Handsome Fruiteater occur around Rancho Grande, including the roadside forest either side of the entrance and at the Portachuelo pass (1,130 m/3,707 ft), which is towards Ocumare. Handsome Fruiteater even comes to food put out by the Rancho Grande staff. Further along the road towards Ocumare, Bearded Bellbird occurs around km 36. Fasciated Tiger-Heron*, Pale-bellied Hermit and Lance-tailed Manakin occur along the **Turiamo road**. Take the left fork 10 km before Ocumare at the military check-point. The manakin occurs along the track to the right 2.5 km along here, just before a village; the tiger-heron frequents the river.

HENRI PITTIER NP

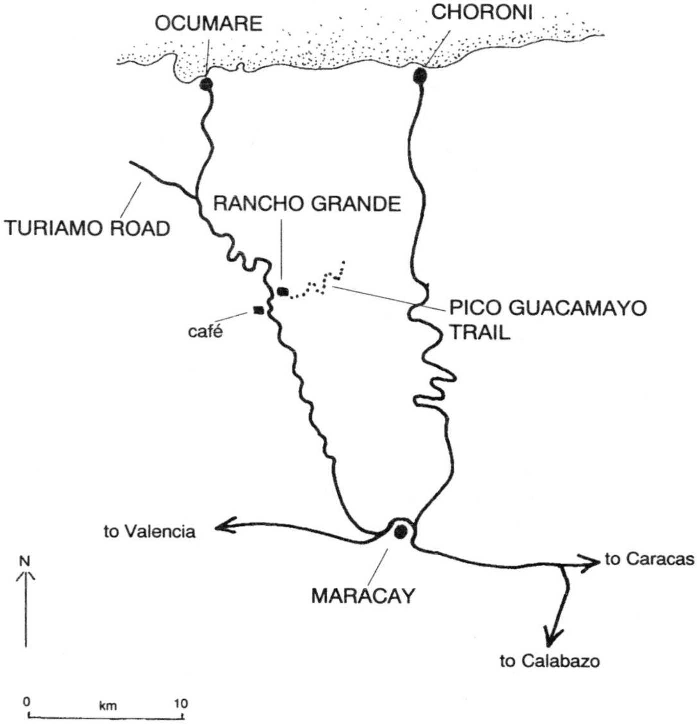

The **Choroni road** ascends to 1,830 m (6,001 ft) before descending into forest with large areas of bamboo, then dry scrub, where Black-backed Antshrike occurs, and, finally, desert scrub along the coast where the rare Rosy Thrush-Tanager is found. The blinding Wire-tailed Manakin occurs behind the Museo Cadafe, on the right, shortly before Choroni and just north of the La Grande Posado café. Cross the stream behind the museum and bird the trail to the left. Black-and-white Owl also occurs here. 1–2 km north of the La Grande Posado café there is an abandoned building on the right-hand side of the road. White-bellied Antbird and Rosy Thrush-Tanager occur along the track to the side of this building.

Accommodation: Hotel Maracay (B), in the Las Delicias (eastern) suburbs of Maracay, is an excellent base. Its large grounds and nearby hillsides support Tropical Screech-Owl, Grey Potoo, Rufous Nightjar, Russet-throated Puffbird, Stripe-backed Wren, Flavescent Warbler, Golden-winged and Black-striped Sparrows. Hotel Pipo (B). Hotel Vladimir (C).

For permission to visit and, possibly, stay at Rancho Grande contact the Audubon Society (VAS) Venezuela, (address p. 399).

CHICHIVIRICHE

This coastal town, some 300 km west of Caracas, is a good base from which to explore nearby Morrocoy NP, where coral cays and mangroves support a wide selection of waterbirds including Scarlet Ibis.

The xerophytic vegetation around Chichiviriche supports the rare endemic Short-tailed Tody-Flycatcher* and a few specialities restricted to northern Venezuela and northeast Colombia.

Endemics

Short-tailed Tody-Flycatcher*.

Specialities

Bare-eyed Pigeon, Yellow-shouldered Parrot*, Buffy Hummingbird, Russet-throated Puffbird, White-whiskered Spinetail, Black-backed Antshrike, Slender-billed Tyrannulet, Black-chested Jay, Golden-winged and Tocuyo Sparrows, Glaucous Tanager, Vermilion Cardinal.

Others

Magnificent Frigatebird, Brown Booby, Greater Flamingo, Reddish Egret, Stripe-backed Bittern, Scarlet Ibis, King Vulture, Pearl and Snail Kites, Limpkin, Ruby-topaz Hummingbird, Pale-legged Hornero, Fuscous Flycatcher, Black-capped Donacobius, Bicoloured Conebill.

Access

Bird **Morrocoy NP**, accessible from Tucacas, the lagoons around Chichiviriche itself, and the Tacarigua (Paez) dam, 10 km north of Chichiviriche, *en route* to San Juan. Turn west immediately after crossing the Río Tocuyo, then south after 7 km. Stripe-backed Bittern occurs here. Black-chested Jay and Glaucous Tanager occur along the San José de la Costa road.

Chichiviriche also lies on the edge of the Cuare Wildlife Refuge, one of only six sites where the extremely rare endemic Plain-flanked Rail* has been recorded, and currently threatened by hotel developments.

Accommodation: Chichiviriche: Hotel Nautico.

PARAGUANÁ PENINSULA

The dry Paraguaná peninsula supports a number of xerophytic specialities which are restricted to north Venezuela and northeast Colombia, as well as the rare endemic Short-tailed Tody-Flycatcher*.

Endemics

Short-tailed Tody-Flycatcher*.

Specialities

Buffy Hummingbird, Russet-throated Puffbird, Tocuyo Sparrow, Black-faced Grassquit, Vermilion Cardinal.

Others

Crested Bobwhite, Straight-billed Woodcreeper, Pale-breasted Spinetail, Black-crested Antshrike, White-fringed Antwren, Northern Scrub-Flycatcher, Troupial.

Access

Most of the species listed above occur on Cerro Santa Ana, a hill in the centre of the peninsula. Head for the village of Moruy from Santa Ana. Ignore the left turn to the actual village and continue straight on towards the La Hija restaurant. Take the track to the right just before the restaurant to a parking area where a trail up Cerro Santa Ana begins.

Short-tailed Tody-Flycatcher* has been recorded along the track south (to El Cayude) from the Coro–Punto Fijo road.

Coro, 177 km northwest of Tucacas, would make a good base for birding this peninsula.

BARQUISIMETO

Venezuela's fourth largest city lies next to semi-desert where a number of specialities can be found, including the scarce Orinocan Saltator.

Specialities

Bare-eyed Pigeon, Buffy Hummingbird, White-whiskered Spinetail, Slender-billed Tyrannulet, Vermilion Cardinal, Trinidad Euphonia, Pileated Finch, Black-faced Grassquit, Orinocan Saltator.

Others

Pearl Kite, Straight-billed Woodcreeper, Black-crested Antshrike, Southern Beardless Tyrannulet, Northern Scrub-Flycatcher, Pale-eyed Pygmy-Tyrant, Bicoloured Wren, Troupial.

Access

Some exploration is required to find good habitat but **El Pandito** has been productive in the past. Turn north 1 km west of the Hotel Paris, just west of Barquisimeto on the road to Carora. Bird the area north of the village of El Pandito, 5 km from the main road.

The endemic Rusty-flanked Crake* as well as a number of specialities including Band-tailed Guan, Helmeted Curassow*, Bronzy Inca and Golden-breasted Fruiteater, occur in **Yacambu NP**, near the Andean town of Sanare (the best base), southeast of Barquisimeto. This NP contains some of the best forest remaining in the Venezuelan Andes. A 40-km stretch of road runs through this forest. The crake occurs at the Laguna El Blanquito, 1 km past the park HQ. The low arid hills north of Sanare support a number of species restricted to north Venezuela and northeast Colombia, including Russet-throated Puffbird, Slender-billed Tyrannulet, Tocuyo Sparrow and Glaucous Tanager, as well as Rosy Thrush-Tanager.

MÉRIDA ANDES

Mérida lies in the centre of the Venezuelan Andes, 674 km southwest of Caracas. The avifauna of this, the northeastern extremity of the Andes, is composed of species at the edge of ranges centred further south, birds which otherwise occur only in the largely inaccessible regions of east Colombia, and birds which are totally endemic to the Venezuelan Andes.

SANTO DOMINGO VALLEY

Map p. 378

The 84-km road that traverses the Santo Domingo valley between Barinas at the western edge of the Llanos and Apartaderos north of Mérida, offers the birder a great opportunity to see some very localised species, including seven endemics, the unique Bearded Helmetcrest and Pale-headed Jacamar. The only known site in Venezuela for Andean Cock-of-the-Rock is also in this valley.

Endemics

Rose-headed Parakeet*, Mérida Sunangel, Ochre-browed Thistletail, Guttulated Foliage-gleaner, Paramo Wren, White-fronted Redstart, Mérida Flowerpiercer.

Specialities

Andean Snipe, Blue-fronted Parrotlet, Orange-throated Sunangel, Bearded Helmetcrest, Pale-headed Jacamar, Moustached Puffbird, Paramo Pipit, Crimson-backed and Orange-eared Tanagers.

Others

Torrent Duck, Black-chested Buzzard-Eagle, Black-and-Chestnut Eagle*, Lined Quail-Dove, Rufous-breasted and Sooty-capped Hermits, Sword-billed Humingbird, Booted Racket-tail, Long-tailed Sylph, Long-billed Starthroat, Crested and Golden-headed Quetzals, Red-headed Barbet, Buff-throated Woodcreeper, Azara's Spinetail, Streak-backed Canastero, Montane Foliage-gleaner, Chestnut-crowned Antpitta, Brown-rumped and Ocellated Tapaculos, Golden-breasted Fruiteater, Red-ruffed Fruitcrow, Andean Cock-of-the-Rock, Wire-tailed, Golden-winged and White-bearded Manakins, Sepia-capped Flycatcher, Black-headed Tody-Flycatcher, Golden-faced and Torrent Tyrannulets, Scale-crested Pygmy-Tyrant, Cliff Flycatcher, Slaty-backed and Brown-backed Chat-Tyrants, Streak-throated Bush-Tyrant, White-capped Dipper, Spotted Nightingale-Thrush, Yellow-legged and Chestnut-bellied Thrushes, Rufous-breasted and Rufous-and-White Wrens, Pectoral Sparrow, Chestnut-capped Brush-Finch, Blue-backed Conebill, Guira Tanager, Red-crowned Ant-Tanager, Lacrimose and Buff-breasted Mountain-Tanagers, Golden-rumped Euphonia, Chestnut-breasted Chlorophonia, Saffron-crowned Tanager, Plumbeous Sierra-Finch, Rusty, White-sided, Bluish and Masked Flower-piercers, Scarlet-rumped Cacique, Yellow-backed Oriole.

Access

It is possible to fly to Barinas from Caracas, a distance of over 500 km by a road which is rough and slow in places. Barinas lies next to the Llanos (see p. 384).

The following sites are discussed in km numbers northwest from Barinas *en route* to Apartaderos; Pale-headed Jacamar, Wire-tailed Manakin and Rufous-breasted Wren occur along the trail to the north just past **Río Barragan**, at km 28, a few km beyond the village of

The bizarre Bearded Helmetcrest, a rare hummingbird which feeds mainly on insects, occurs on the paramo slopes below Pico Aguila, near the top of the Santo Domingo valley

Barinitas. Recent clearance work may have seriously depleted the habitat here. Moustached Puffbird and Rufous-and-white Wren occur along the road north to **Altamira** at km 35. The undriveable La Soledad track, to the south on a hairpin bend at km 39, passes through degraded forest, but is worth a look if time allows. Rose-headed Parakeet*, Crested Quetzal, Guttulated Foliage-gleaner, Red-ruffed Fruitcrow, Andean Cock-of-the-Rock, Spotted Nightingale-Thrush, and Orange-eared Tanager occur in the ravine just past the quarry, 3 km along the **San Isidro track** (1,829 m/6,000 ft) to the south, on another hairpin bend at km 49. Owing to quarry operations, this site is changing constantly. In March 1993 it was necessary to traverse the quarry face on foot to reach the track which used to pass, albeit precariously, across the top. It may be impossible, or very dangerous, to cross the quarry and reach the best habitat on a working day, although an approach from below the quarry may be possible for the persistent birder.

Torrent Duck occurs on the rivers just south of Santo Domingo at km 63. Chestnut-crowned Antpitta occurs in the ravines adjacent to the grounds of the Hotel Morruco 1 km above Santo Domingo at km 67. There is a short track to the right, just before the hotel entrance, worth birding. The roadside up from here to **Laguna Mucubaji** (3,658 m/12,000 ft), at km 81, is worth exploring for Ocellated Tapaculo, mountain-tanagers and flower-piercers including the endemic Mérida Flower-piercer. The grounds of the **Hotel Los Frailes**, just below the Laguna, are good for birding. Turn right at Apartaderos for **Pico Aguila** (4,115 m/13,501 ft). Black-chested Buzzard-Eagle, Bearded Helmetcrest, Ochre-browed Thistletail, Streak-backed Canastero, Paramo Wren, Paramo Pipit and Plumbeous Sierra-Finch all occur around here, especially beyond the high point of the road (marked by a statue) around the small lagoons to the east of and below the road, by an old wall. Although the helmetcrest feeds mainly on insects, it does seem to prefer

SANTO DOMINGO VALLEY

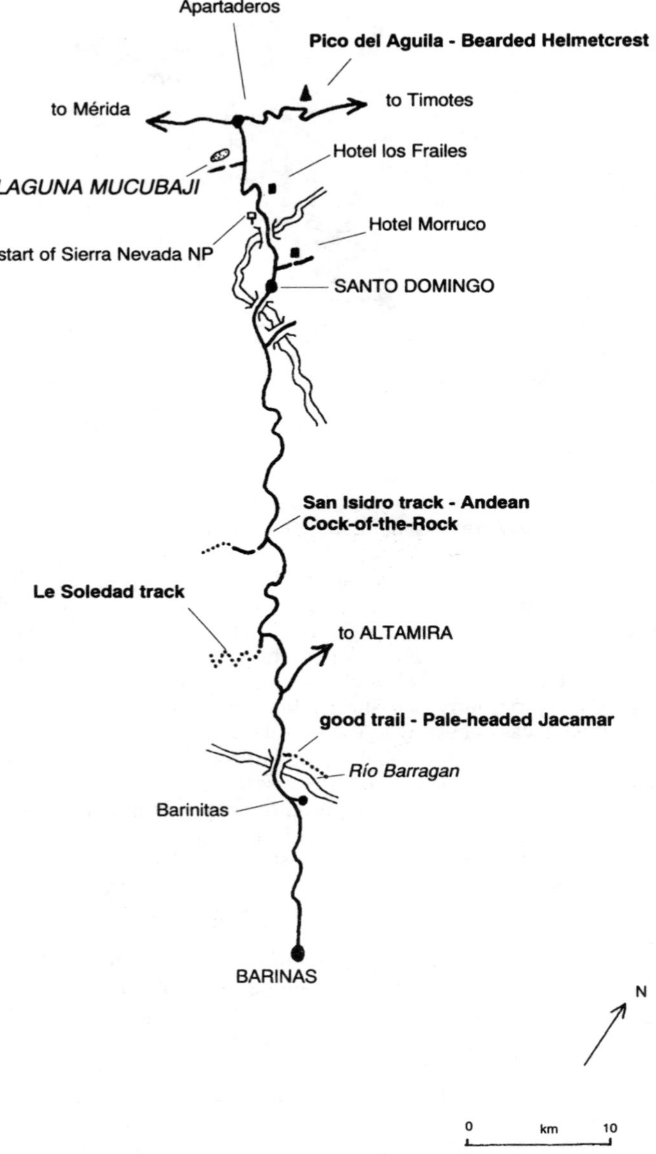

areas where the yellow-flowered *Frailjon* is in bloom.

Accommodation: Barinas: Hotel Bristol (B). Santo Domingo: Hotel Morruco (A). Above Santo Domingo: Hotel Los Frailes (A). Apartaderos: Hotel Parque Turistico (B). Mérida: Hotel Belensate (B).

Turn south at Apartaderos to reach the Pico Humboldt trail and Mérida.

PICO HUMBOLDT TRAIL

Situated within the Sierra Nevada NP the Pico Humboldt trail, 18 km north of Mérida, passes through excellent subtropical and temperate forests, where the possible endemics include Grey-naped Antpitta, two hemispinguses and Mérida Flower-piercer.

Endemics

Rose-headed Parakeet*, Mérida Sunangel, Grey-naped Antpitta, White-fronted Redstart, Grey-capped and Slaty-backed* Hemispinguses, Mérida Flower-piercer.

Specialities

Golden-bellied Starfrontlet, Purple-backed Thornbill, Grey-throated Warbler*, Moustached Brush-Finch.

Although widespread throughout the Andes, the 'three whistles' of the Chestnut-crowned Antpitta is a particularly familiar sound on the Pico Humboldt trail. However, hearing this striking bird is one thing, seeing it is another

Others

Andean Guan, White-necked Jacobin, Collared Inca, Golden-headed Quetzal, Masked Trogon, Emerald Toucanet, Crimson-mantled Woodpecker, Black-banded Woodcreeper, Pearled Treerunner, Streaked Tuftedcheek, Undulated, Chestnut-crowned and Slate-crowned Antpittas, Unicoloured Tapaculo, Red-crested Cotinga, Green-and-black and Barred Fruiteaters, Red-ruffed Fruitcrow, Black-capped and White-throated Tyrannulets, Yellow-bellied and Slaty-backed Chat-Tyrants, Black-collared Jay, Black-crested, Russet-crowned and Three-striped Warblers, Slaty Brush-Finch, Blue-backed Conebill, Superciliared and Oleaginous Hemispinguses, Fawn-breasted Tanager, Black-capped Tanager, Plush-capped Finch, White-sided and Masked Flower-piercers, Yellow-billed Cacique.

Access

To reach the Pico Humboldt trail head northeast out of Mérida 11 km to Tabay. In Tabay take the road from the north of the plaza, to the left of the main through road, up to the NP HQ, some 7 km away. The trail begins here. Although the lower part of the trail is excellent, the very best birds occur higher up, between the first refugio and Laguna Coromoto. Grey-capped and Slaty-backed* (in bamboo) Hemispinguses occur in this area and Mérida Flower-piercer around the laguna.

Accommodation: Tabay: Hotel Les Cumbres (B) (good birding).

LA AZULITA ROAD

The subtropical forest along the road to La Azulita, especially within the Universty of Andes Reserve, 60 km southwest of Mérida, supports such specialities as Wattled Guan*, and is excellent for antpittas. Otherwise, the birds occuring here are very similar to those on the lower reaches of the Pico Humboldt trail.

Endemics

Rose-headed Parakeet*, Grey-capped Hemispingus.

Specialities

Wattled Guan*, Rufous-banded Owl, Black-mandibled Toucan, Golden-breasted Fruiteater.

Others

White-rumped Hawk, Andean Guan, Ruddy Pigeon, Collared Inca, Booted Racket-tail, Long-tailed Sylph, Crested Quetzal, Masked Trogon, Emerald Toucanet, Slate-crowned Antpitta, Andean Tapaculo, Red-ruffed Fruitcrow, Mountain Wren, Saffron-crowned and Black-capped Tanagers.

Access

To reach the La Azulita road, turn west 40 km south of Mérida (3 km north of Jaji) on the road to Tovar. The University of Andes Forest Reserve is on the left 59 km from Mérida. It is an hour's steady walking from the entrance gate to the best forest, where there are many trails.

This is a good site for Wattled Guan*, Rose-headed Parakeet*, Rufous-banded Owl, and Grey-capped Hemispingus. The main road down from the reserve entrance *en route* to La Azulita is also worth prolonged birding.

Accommodation: Jaji: a delightful little town with a couple of small hotels.

Birders with time to spare may wish to take the cable-car from Mérida to **Pico Espejo** (4,877 m/16,000 ft) and bird between the second and first stations on the way down, or the grounds of Hotels Belensate, or the road up from the **Hotel Valle Grande** (9 km north of the city), where Sword-billed Hummingbird occurs.

Pygmy Swift occurs around the palms behind the Hotel Gran Sasso on the main road through **El Vigia**, opposite the La Grita restaurant.

ENCONTRADOS ROAD

This road, between San Carlos de Zulia and Encontrados, passes through wetlands at the southern edge of the Maracaibo basin, some 50 km northwest of El Vigia. It is one of the few sites where the rare Northern Screamer* and Saffron-headed Parrot occur.

Specialities

Northern Screamer*, Saffron-headed Parrot, Shining-green Hummingbird, Russet-throated Puffbird, Crimson-backed Tanager, Grey Seedeater.

Others

Horned Screamer, Capped Heron, Scarlet Ibis, Lesser Yellow-headed and King Vultures, Savanna and Black-collared Hawks, Brown-throated Parakeet, Green-rumped Parrotlet, White-chinned Sapphire, Red-crowned Woodpecker, White-fringed Antwren, Slate-headed Tody-Flycatcher, Mouse-coloured Tyrannulet, Tawny-crowned Pygmy-Tyrant, Panama Flycatcher, Cinereous Becard, Ruddy-breasted Seedeater, Yellow Oriole.

Access

20 km east of Encontrados is a good area with both screamers occurring near the canalised river here.

SAN JUAN DE COLÓN

The roadside coffee plantations north of San Juan de Colón, near the Colombian border, support some species which are difficult to see elsewhere in Venezuela, such as Black-mandibled Toucan and White-eared Conebill, as well as the recently rediscovered endemic Tachira Emerald*.

Endemics

Tachira Emerald*.

Specialities

Green-rumped Parrotlet, Pygmy Swift, Black-mandibled Toucan, Bare-eyed Thrush, White-eared Conebill, Crimson-backed Tanager, Grey Seedeater.

Others

Little Tinamou, Blue Ground-Dove, Striped Cuckoo, Short-tailed Swift, White-necked Jacobin, Rufous-tailed Hummingbird, Red-rumped Woodpecker, White-bearded Manakin, Sooty-headed and Golden-faced Tyrannulets, Yellow-bellied Elaenia, Yellow-legged and Black-billed Thrushes, White-lined and Blue-necked Tanagers, Lesser Seed-Finch, Crested Oropendola.

Access

Bird the roadside, which is best up to 11 km north of San Juan de Colón. There are also some side-tracks worth exploring along the road north to La Fria.

PARAMO DE TAMA

Map opposite

This remote site in extreme southwest Venezuela, near the Colombian border, supports some species, such as Black-billed Mountain-Toucan, with ranges which do not extend to the Venezuelan Andes owing to the Tachira gap, which separates these mountains from the rest of the Andes. Most of the birds here, however, may be seen more easily in Ecuador (and Colombia).

Endemics

Venezuelan Wood-Quail*, Grey-naped Antpitta, White-fronted Redstart.

Specialities

Plain-breasted Hawk, Band-tailed Guan, Rusty-faced* and Speckle-faced Parrots, Andean Potoo, Golden-bellied and Blue-throated Starfrontlets, Glowing Puffleg, Moustached Puffbird, Black-billed Mountain-Toucan*, Golden-fronted Redstart, Pale-naped and Ochre-breasted Brush-Finches, Black-capped and Black-eared Hemispinguses, Slaty Finch.

Others

Andean Guan, White-throated Screech-Owl, Rufous-banded Owl, Amethyst-throated Sunangel, Booted Racket-tail, Rufous Spinetail, Rusty-winged Barbtail, Undulated, Chestnut-crowned, Rufous and Slate-crowned Antpittas, Unicoloured, Brown-rumped and Ocellated Tapaculos, Red-crested Cotinga, Barred Fruiteater, Rufous-breasted Flycatcher, Black-collared Jay, Capped Conebill, Grey-hooded Bush-Tanager, Hooded, Black-chested and Lacrimose Mountain-Tanagers, Golden-crowned Tanager, Chestnut-breasted Chlorophonia, Glossy Flower-piercer.

Access

Head from San Cristobal to Las Delicias and Betania, preferably with a four-wheel-drive vehicle, although buses seem to make it without too many problems. Ocellated Tapaculo occurs alongside the Las

PARAMO DE TAMA

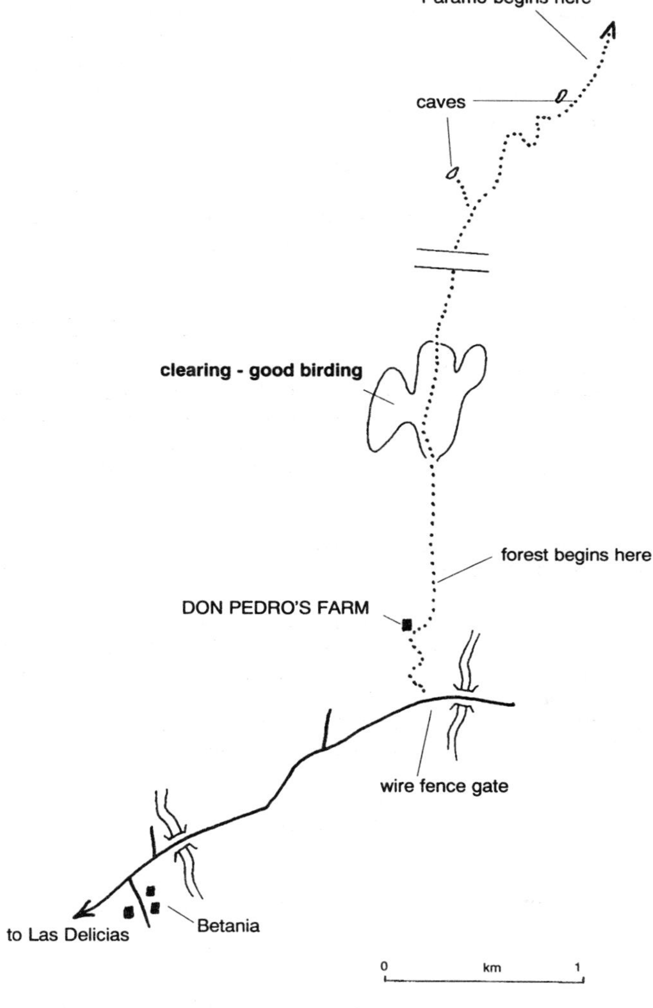

Delicias–Betania road. Once in Betania, ask to speak to Don Pedro, who usually gives permission to walk through his farm, 2 km from Betania, to the forest beyond. The first large clearing above the farm is a particularly good spot.

THE LLANOS

Maps opposite and p. 387

This huge expanse of seasonally-flooded savanna is one of the best birding areas in the world, with 150 species possible in a day, a total that could include 15 herons (including Agami*), seven ibises (including Scarlet), 20 raptors (including Slender-billed Kite), the very localised Yellow-knobbed Curassow, Sungrebe, Sunbittern, Double-striped Thick-knee, Scarlet Macaw, Hoatzin, Great Potoo, all five South American kingfishers, and the endemic White-bearded Flycatcher. In other words, a phenomenal selection of some of South America's finest birds.

Furthermore, many waterbirds occur in astounding numbers, especially in the lower, wetter Llanos, in the south (December to April).

Endemics

White-bearded Flycatcher.

Specialities

Agami Heron*, Yellow-knobbed Curassow, Pale-headed Jacamar, Russet-throated Puffbird, White-naped Xenopsaris, Orinocan Saltator.

Others

Horned Screamer, White-faced and Black-bellied Whistling-Ducks, Orinoco Goose*, Muscovy and Comb Ducks, Brazilian Teal, Whistling, Capped and Boat-billed Herons, Rufescent Tiger-Heron, Pinnated Bittern, Scarlet, Bare-faced, Sharp-tailed, Buff-necked and Green Ibises, Maguari Stork, Jabiru, Lesser Yellow-headed and King Vultures, Snail and Slender-billed Kites, Crane, Savanna, Black-collared and Grey-lined Hawks, Aplomado Falcon, Grey-necked Wood-Rail, Azure Gallinule, Sungrebe, Sunbittern, Limpkin, Wattled Jacana, Double-striped Thick-knee, Collared Plover, Pied and Southern Lapwings, Large-billed and Yellow-billed Terns, Plain-breasted Ground-Dove, Scarlet and Chestnut-fronted Macaws, Dwarf Cuckoo, Hoatzin, Spectacled and Striped Owls, Great and Grey Potoos, Band-tailed and Nacunda Nighthawks, White-tailed and Little Nightjars, Ruby-Topaz Hummingbird, Glittering-throated Emerald, Green-and-rufous and American Pygmy Kingfishers, Golden-green and Spot-breasted Woodpeckers, Pale-breasted, Rusty-backed and Yellow-chinned Spinetails, Common Thornbird, Lance-tailed Manakin, Pale-tipped Tyrannulet, Yellow-breasted Flycatcher, Riverside Tyrant, Pied Water-Tyrant, White-headed Marsh-Tyrant, Cattle Tyrant, Fork-tailed Flycatcher, Grey Kingbird, Bicoloured Wren, Yellowish Pipit, Red-capped Cardinal, Hooded and Burnished-buff Tanagers, Orange-fronted Yellow-Finch, Saffron Finch, Troupial, Oriole and Yellow-hooded Blackbirds.

Other Wildlife

Capybara, Red Howler, Wedge-capped Capuchin, Giant and Lesser Anteaters, Crab-eating Raccoon, Armadillo, White-tailed Deer, Collared Peccary, Tayra, Crab-eating Fox, Ocelot, Puma, Jaguar, Jaguarundi, Vampire and Fishing Bats, Spectacled Caiman, Orinoco River Dolphin, Anaconda, Killer Bee!

Access

It is possible to see most of these species on the 530-km Calabozo–San Fernando–Mantecal–Bruzual–Barinas road which passes through the east and south Llanos, but there are few places to stay. Anywhere along this road could produce pulsating birding.

Calabozo is 272 km south of Caracas. The 80-km stretch between here and San Fernando is particularly good. **Blohm's Ranch**, some 35 km south of Calabozo, is private, and permission should be sought before birding here. Ask for Thomas Blohm. To reach here head south from Calabozo to a huge TV transmitter on the east side of the road. Wires cross the road 4 km south of here. Just south of here tracks go east 3 km to a wooded area where Yellow-knobbed Curassow, Scarlet Macaw and Orinocan Saltator occur, and west to the ranch where Sunbittern occurs. Hoatzin is found in the woodland 3 km south of these tracks, on the west side of the road.

THE LLANOS

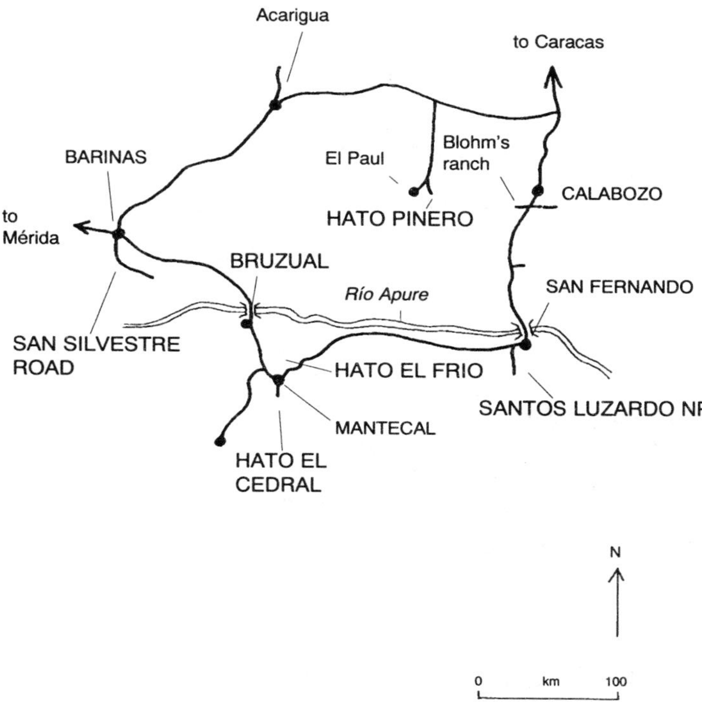

South of San Fernando, the road to Puerto Paez passes through **Santos Luzardo NP** where Orinoco Goose*, Sharp-tailed Ibis, Double-striped Thick-knee, Pied Lapwing, Nacunda Nighthawk, White-tailed Nightjar, Pale-tipped Tyrannulet and Yellowish Pipit all occur. The road west from San Fernando to Mantecal passes Hato El Frio and Hato El Cedral ranches (see below). Turn north after Mantecal and head to Bruzual. Pied Lapwing, Large-billed Tern, Dwarf Cuckoo and Orinoco River Dolphin occur on the Río Apure at the suspension bridge at **Bruzual**.

The roadside ponds, a few km to the north, are also worth stopping at. The pond to the east of the road can be worked from the village alongside it. (Map opposite). From here it is 120 km northwest to **Barinas**, which is at the base of the Santo Domingo valley in the Mérida Andes (see p. 376). Capped Heron, Double-striped Thick-knee, Dwarf Cuckoo, Hoatzin and Nacunda Nighthawk all occur along the San Silvestre road, which runs from the junction next to the Hotel Cacique, in Barinas, to San Silvestre.

Accommodation: San Fernando: Hotel La Fuente; Hotel Plaza. Calabozo: Motel Tiuna (B).

There are also a number of expensive ranches which cater for birders in the Llanos:

Hato Pinero: over 350 species have been recorded on this huge (80,940 ha) working ranch, which lies in the higher, drier Llanos and supports speciality species such as Agami Heron* and Yellow-knobbed Curassow. It is possible to fly (very expensively) in from Caracas or to drive, a journey of about five hours from Caracas. Turn south approximately halfway between Acarigua and El Sombrero towards El Baul. The turning to Hato Pinero is 15 km north of El Baul. Take a hit-list and show it immediately to your appointed guide in order to make sure they know you are a lean, mean birding machine!

Accommodation: Ranch: (A+). Very expensive, but price includes food, boat and truck trips, a guide, as many free drinks as you like, and all those birds! Book well in advance through the Venezuelan Audubon Society (VAS).

A stalking Sunbittern, in a family of its own, can be one of the highlights of a boat trip at Hato Pinero. Such a sight more than justifies the expense of staying at the ranch

BRUZUAL

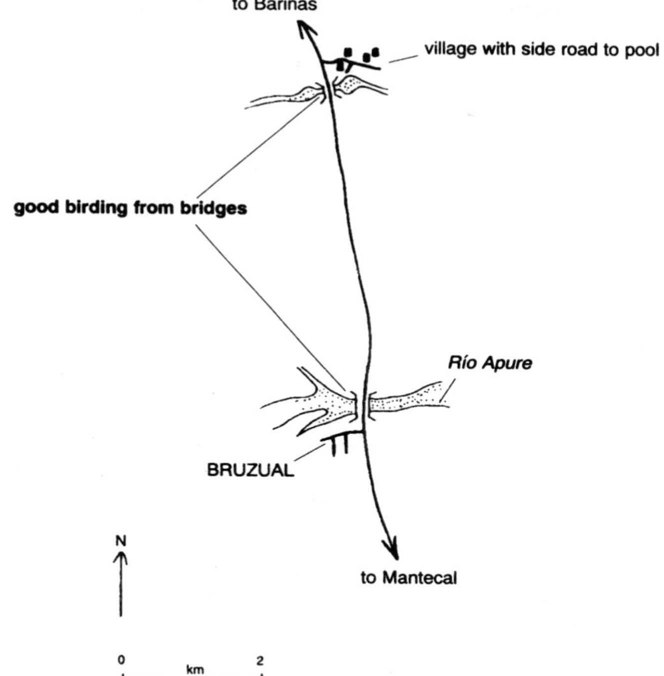

Hato el Frio: This large ranch, east of Mantecal (turn north at km post 351), in the low Llanos, is not as good as El Cedral and Hato Pinero, although Orinoco Goose*, Lesser Yellow-headed Vulture and Collared Plover occur here, and there is a Scarlet Ibis roost.

Accommodation: Ranch (A). Basic but still expensive. Book through the Venezuelan Audubon Society (VAS). Take food. Cost includes guided tours.

Hato El Cedral: This ranch (42,898 ha), arguably the best of the bunch, is situated in the lower, wetter Llanos, just south of Mantecal, and supports a number of species rarely seen at Hato Pinero including Orinoco Goose*, Pinnated Bittern, Azure Gallinule, Giant Snipe, Riverside Tyrant, White-naped Xenopsaris and Orinocan Saltator. Hato El Cedral is also the best ranch for sheer numbers of birds, especially between December and April.

Accommodation: Ranch (A).

AMAZONAS

The state of Amazonas in south Venezuela is one of South America's true wildernesses. Here, untouched lowland rainforest still stretches as far as the eye can see, even out of an aeroplane. And, hidden amongst the forest, there are three remote lodges catering for ecotourism.

Whilst the avifauna of this region includes many species associated with western Amazonia, and the species recorded are similar to those at sites such as La Selva in east Ecuador (see p. 230), and Iquitos (see p. 316), Tambopata (see p. 335) and Manu NP (see p. 332) in Peru, the forests here also support species restricted to the forests of the upper Río Negro drainage and the foothills of the tepuis, in south Veneuela, southwest Guyana and north Brazil. These specialities include Rufous-winged Ground-Cuckoo, Guianan Cock-of-the-Rock and Yellow-crested Manakin.

Puerto Ayacucho, approximately 600 km due south of Caracas, is the gateway to the lodges here. Both Puerto Ayacucho and the lodges are accessible by air.

Endemics

Orinoco Softtail*.

Specialities

Zigzag Heron*, Crestless, Razor-billed and Black Curassows, Grey-winged Trumpeter, Scarlet-shouldered Parrotlet, Rufous-winged Ground-Cuckoo, Long-tailed Potoo, Blackish Nightjar, Brown-banded Puffbird, Ruddy Spinetail, Black-throated and White-shouldered Antshrikes, Cherrie's Antwren, Guianan Cock-of-the-Rock, Yellow-crested Manakin, Amazonian Black-Tyrant, Pale-bellied Mourner, Brown-headed Greenlet, Short-billed Honeycreeper.

Others

Great, Cinereous and Undulated Tinamous, Capped and Agami* Herons, Tiny Hawk, Lined and Collared Forest-Falcons, Little Chachalaca, Spix's Guan, Blue-throated Piping-Guan, Marbled Wood-Quail, Sungrebe, Sunbittern, Blue-and-yellow, Scarlet and Red-and-green Macaws, Black-headed and Orange-cheeked Parrots, Great and Grey Potoos, Least and Band-tailed Nighthawks, Dusky-throated Hermit, Black-eared Fairy, Blue-crowned Motmot, Brown, Yellow-billed, Green-tailed, Paradise and Great Jacamars, Spotted Puffbird, Rusty-breasted Nunlet, Swallow-wing, Green, Ivory-billed and Many-banded Aracaris, Scaly-breasted, Chestnut, Cream-coloured and Ringed Woodpeckers, Long-billed and Lineated Woodcreepers, Point-tailed Palmcreeper, Cinnamon-rumped Foliage-gleaner, Blackish-grey, Amazonian and Spot-winged Antshrikes, White-browed, Black-chinned, White-bellied, White-plumed and Rufous-throated Antbirds, Rufous-capped Antthrush, White-browed Purpletuft, Spangled and Pompadour Cotingas, Bare-necked Fruitcrow, Amazonian Umbrellabird, Black Manakin, Dwarf Tyrant-Manakin, Rufous-tailed Flatbill, White-crested Spadebill, Drab Water-Tyrant, Greyish Mourner, Sulphury and Dusky-chested Flycatchers, Violaceous Jay, Slaty-capped Shrike-Vireo, Buff-cheeked Greenlet, White-banded and Black-collared Swallows, Pectoral Sparrow, Black-faced, Yellow-backed and Flame-crested Tanagers, Rufous-bellied Euphonia, Turquoise, Paradise, Masked and Opal-rumped Tanagers, Orange-fronted Yellow-Finch, Olive Oropendola, Moriche Oriole.

Other Wildlife

This, especially Junglaven, is an exceptional area for butterflies as well as Brown-bearded Saki, Giant Otter, Tayra, River Dolphin, and Jaguar.

Access

The best birding site in Amazonas is the area around the rustic **Junglaven Lodge**, which is a one-hour charter flight from Puerto Ayacucho (160 km). Bird the 12-km track through superb forest from the airstrip to the lodge, around the lodge, Camani creek (by boat) and the Río Ventuari (by boat). Goodies include Crestless, Razor-billed and Black Curassows, Grey-winged Trumpeter, Ringed Woodpecker, Black-throated Antshrike, Cherrie's Antwren, Amazonian Umbrellabird, Yellow-crested Manakin, Amazonian Black-Tyrant, Pale-bellied Mourner and Brown-headed Greenlet.

Camturama Amazonas Lodge, on the banks of the Orinoco, 20 km from Puerto Ayacucho airport, is just as good, if not better. The savanna, gallery forest, lowland rainforest and palm swamps here support Point-tailed Palmcreeper, White-browed Purpletuft, Spangled and Pompadour Cotingas, Amazonian Umbrellabird and Guianan Cock-of-the-Rock. Bird the lodge clearing, roads and trails, where up to 200 species are possible in a week.

The expensive lodge at **Camani** is also an hour's flight from Puerto Ayacucho. Notable species here include Black Curassow and Rufous-winged Ground-Cuckoo.

It is possible to fly from Puerto Ayacucho, south to **San Carlos**, which is near the **Serrania de la Neblina NP**. From San Carlos it is possible to take a cargo boat up the Cano Casiquiare. Possible goodies include Grey-legged Tinamou, Tawny-tufted Toucanet, Azure-naped Jay and Scaled Flower-piercer.

It is also possible to fly to **Esmeralda**, which lies near the base of Cerro Duida, where the endemic Duida Grass-Finch occurs.

CERRO DE HUMO

Map p. 390

The isolated mountains of the Paria peninsula in northeast Venezuela support no less than five endemic species: Scissor-tailed Hummingbird*, White-throated Barbtail*, Yellow-faced Redstart*, Grey-headed Warbler*, and Venezuelan Flower-piercer*, which are all rare, owing to deforestation. However, all but the warbler occur in remnant cloud forest above 1,000 m (3,281 ft) on Cerro de Humo, alias Smoke Mountain, near Irapa.

This is also a good site to look for the localised White-tailed Sabrewing*, Tufted Coquette and Handsome Fruiteater.

Endemics

Scissor-tailed Hummingbird*, White-throated Barbtail*, Handsome Fruiteater, Yellow-faced Redstart*, Venezuelan Flower-piercer*.

Specialities

Band-tailed Guan, White-eared Parakeet, Tepui and Lilac-tailed

CERRO DE HUMO

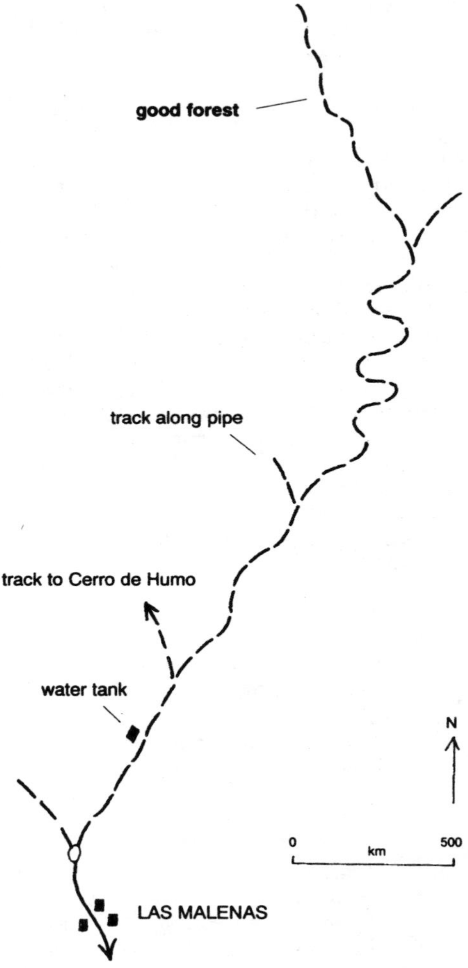

Parrotlets, White-tailed Sabrewing*, Tufted Coquette, Copper-rumped Hummingbird, Groove-billed Toucanet, Stripe-breasted Spinetail, Plain-backed Antpitta, Bearded Bellbird, Venezuelan Tyrannulet, Bare-eyed Thrush, Ochre-breasted Brush-Finch, Trinidad Euphonia, Orange-crowned Oriole.

Others

Little and Red-legged Tinamous, White Hawk, Ornate Hawk-Eagle, Grey-fronted Dove, Orange-winged Parrot, Green Hermit, Fork-tailed Woodnymph, Golden-tailed Sapphire, Copper-rumped Hummingbird, Golden-olive Woodpecker, Streak-headed Woodcreeper, Slaty Antwren, White-bellied Antbird, Black-faced Antthrush, Slate-crowned Antpitta, Bright-rumped Attila, Chestnut-crowned Becard, Rufous-and-white Wren, Yellow-bellied Siskin, Three-striped Warbler, Oleaginous Hemispingus, Blue-capped, Speckled, Bay-headed and Black-headed Tanagers.

Other Wildlife

Land-crab, orchids.

Access

Head east from Carúpano (which is accessible by air from Caracas) to Irapa and on to Río Grande de Arriras. From there head up 10 km to the village of **Las Malenas** (also called La Melena), where visitor accommodation is planned. From the parking area above the village take the right fork on to the track past the water tank, ignore the left fork which is a track leading to Cerro de Humo, and continue on the main track to the top of the ridge where the mixed coffee plantations and cloud forest begins. White-throated Barbtail* feeds like a mouse amongst fallen logs, Scissor-tailed Hummingbird* and Venezuelan Flowerpiercer* are attracted to red tubular flowers in the lower canopy, and Yellow-faced Redstart* occasionally accompanies tanager flocks at all levels.

BALNEARIO SABACUAL

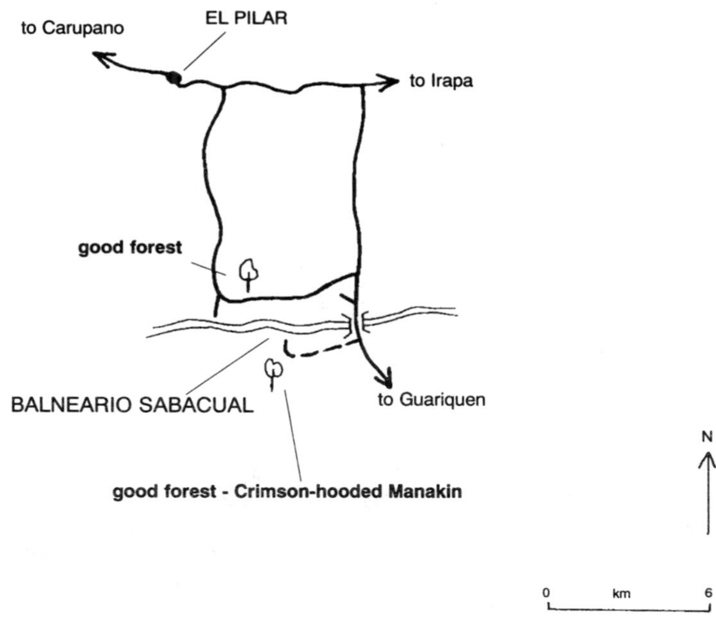

Accommodation: a visitors' 'lodge' is planned. Otherwise, Irapa would make a good base.

After one record in 30 years, the gravely endangered Grey-headed Warbler* was recorded on Cerro Negro in 1993.

The localised Crimson-hooded Manakin occurs near the Carupano–Irapa road, at **Balneario Sabacual**, which is signposted east of El Pilar town (Map p. 391). Turn south towards Guariquen from the main El Pilar–Irapa road, then west after 10 km, just past a bridge. White-bellied Piculet and Velvet-fronted Grackle also occur in the remnant forest here.

CUEVA DE LOS GUACHAROS NP

The biggest cave (10 km long) in Venezuela supports up to 15,000 Oilbirds, and these unique birds make a spectacular sight as they leave the cave at dusk to feed. April to May is the best time, although there are few left between August and January.

Specialities

Oilbird, Groove-billed Toucanet, Scaled Piculet.

Others

Bat Falcon, Scarlet-fronted Parakeet, Band-rumped and Grey-rumped Swifts, Sooty-capped Hermit, White-vented Plumeleteer, Golden-olive Woodpecker, Red-billed Scythebill, Rufous-breasted and Rufous-and-white Wrens, Pectoral Sparrow, Burnished-buff Tanager.

Access

The NP HQ is just off the Cumana–Caripe road, 12 km from Caripe.

Accommodation: Caripe: Hotel Guacharo.

Bridled Tern, Bare-eyed Pigeon, Buffy and Ruby-topaz Hummingbirds, Bicoloured Conebill and Vermilion Cardinal occur on the **Isla de Margarita**, accessible by ferry from Cumana. Bird the mangrove-fringed lagoon lying next to the island ferry terminal. White-cheeked Pintail, Scarlet Ibis and Bicoloured Conebill occur around **Laguna de Unare**, just off the Caracas–Cumana road, west of Barcelona.

BOLÍVAR

The state of Bolívar in southeast Venezuela is one of the top birding areas in South America. Here is the Imataca Forest Reserve, the most reliable site on the continent for Harpy Eagle*, and the road south from El Dorado to Brazil, which quite simply, should be called 'Paradise Road'. The lowland 'guianan' rainforest and 'tepui' foothill forest alongside this road is full of jacamars, toucans, antbirds (including White-plumed), brilliant cotingas (15 species) and tepui endemics, many of which are hard to see elsewhere.

The cities of Ciudad Bolívar (640 km from Caracas) and Ciudad

Guayana are the gateways to Bolívar, and both are accessible by road and air from Caracas.

IMATACA FOREST RESERVE AND EL DORADO

Maps p. 395 and p. 398

The selectively-logged Imataca Forest Reserve (also known as Rio Grande) near El Palmar is the most reliable site in South America to see the monkey-eating Harpy Eagle*, one of the world's most impressive raptors. Many other spectacular specialities occur here and along the trails which lead off the road south from El Dorado, including Black-faced Hawk, Rufous-winged Ground-Cuckoo, Guianan Toucanet, White-plumed Antbird, Guianan Red-Cotinga, Dusky Purpletuft, Purple-breasted Cotinga, Capuchinbird, and Rose-breasted Chat.

Some of these species and many of those listed below also occur around Manaus in Brazil (see p. 143).

Specialities

Black-faced Hawk, Crested* and Harpy* Eagles, Marail Guan, Black Curassow, Grey-winged Trumpeter, Rufous-winged Ground-Cuckoo, Long-tailed Potoo, Racket-tailed Coquette, Crimson Topaz, Guianan Toucanet, Ruddy Spinetail, Cinnamon-rumped Foliage-gleaner, Black-throated Antshrike, Spotted Antpitta, Guianan Red-Cotinga, Dusky Purpletuft, Purple-breasted Cotinga, White and Bearded Bellbirds, Capuchinbird, Painted Tody-Flycatcher, White-naped Xenopsaris, Rose-breasted Chat, Red-and-black Grosbeak, Orinocan Saltator.

Others

Variegated Tinamou, Capped Heron, King Vulture, Double-toothed and Plumbeous Kites, Ornate Hawk-Eagle, Red-throated Caracara, Little Chachalaca, Spix's Guan, Blue-throated Piping-Guan, Grey-fronted Dove, Red-and-green Macaw, Painted Parakeet, Dusky-billed Parrotlet, Golden-winged Parakeet, Black-headed, Caica, Mealy and Red-fan Parrots, Black-bellied Cuckoo, Tawny-bellied Screech-Owl, Black-banded, Crested and Spectacled Owls, Least Pygmy-Owl, Great Potoo, Blackish Nightjar, Band-rumped, Grey-rumped, Short-tailed and Lesser Swallow-tailed Swifts, Pale-tailed Barbthroat, Long-tailed, Straight-billed, Sooty-capped and Reddish Hermits, Blue-fronted Lancebill, Grey-breasted Sabrewing, Blue-tailed Emerald, Rufous-throated and White-chinned Sapphires, Black-eared Fairy, Black-tailed Trogon, Brown, Yellow-billed, Green-tailed, Paradise and Great Jacamars, Pied Puffbird, Chestnut-tipped Toucanet, Green and Black-necked Aracaris, Channel-billed and Red-billed Toucans, Golden-spangled Piculet, Yellow-tufted, Golden-collared, Waved, Cream-coloured and Red-necked Woodpeckers, Cinnamon-throated, Red-billed and Chestnut-rumped Woodcreepers, Red-billed Scythebill, Point-tailed Palmcreeper, Black-tailed Leaftosser, Mouse-coloured, Dusky-throated and Cinereous Antshrikes, Rufous-bellied, Brown-bellied, Stipple-throated, Long-winged, Grey and Spot-tailed Antwrens, Grey, Dusky, White-browed, Warbling, Spot-winged, Ferruginous-backed, White-plumed, Rufous-throated, Wing-banded and Scale-backed Antbirds, Rufous-capped Antthrush, Thrush-like Antpitta, Screaming Piha, Spangled and Pompadour Cotingas, Purple-throated Fruitcrow, Tiny Tyrant-Manakin,

*Imataca Forest Reserve is the most reliable site in South America for one of the world's most impressive birds of prey, the huge Harpy Eagle**

Ringed Antpipit, White-lored Tyrannulet, Helmeted Pygmy-Tyrant, Rufous-tailed Flatbill, Cinnamon-crested and White-crested Spadebills, Ruddy-tailed and Cliff Flycatchers, Smoke-coloured Pewee, Glossy-backed Becard, Cayenne Jay, Lemon-chested and Buff-cheeked Greenlets, Black-billed and Cocoa Thrushes, Bicoloured, Coraya and Musician Wrens, Long-billed Gnatwren, Black-collared Swallow, Pectoral Sparrow, Magpie, Flame-crested, Fulvous-crested and Red-shouldered Tanagers, Purple-throated and Golden-sided Euphonias, Turquoise, Paradise, Yellow-bellied, Spotted and Opal-rumped Tanagers, Black-faced Dacnis, Lesser Seed-Finch, Green Oropendola, Red-rumped Cacique.

Other Wildlife

Jaguar, Giant Otter, Red-rumped Agouti, White-faced Saki, Ocelot.

Access

Black-collared Swallow occurs on the Río Caroni in the Parque Cachamay adjacent to Puerto Ordaz, just west of Ciudad Guayana. Bird from the Embalse Macagua dam, accessible from San Felix.

The private ranch (22,000 ha) at **Caurama**, southwest of Ciudad Bolívar, near Maripa, is sometimes open to birders. It lies between the easternmost edges of the Llanos and the westernmost lowland rainforest; over 400 species have been recorded and 350 are possible in a week. The **Caura Forest Reserve** is 30 km southeast of Maripa. Turn south in response to 'Los Trincheras' sign 15 km east of Maripa. Good forest starts 15 km along here. Goodies include Scarlet Ibis, Black-and-white Hawk-Eagle*, Blue-throated Piping-Guan, Black Curassow, Grey-winged Trumpeter, Painted Parakeet, Red-fan Parrot, Black-bellied Cuckoo, Rufous-winged Ground-Cuckoo, Long-tailed Potoo, an incredible ten nightjars, Yellow-billed Jacamar, Green and Black-necked Aracaris, Golden-spangled Piculet, Cream-coloured and Ringed

IMATACA FOREST RESERVE

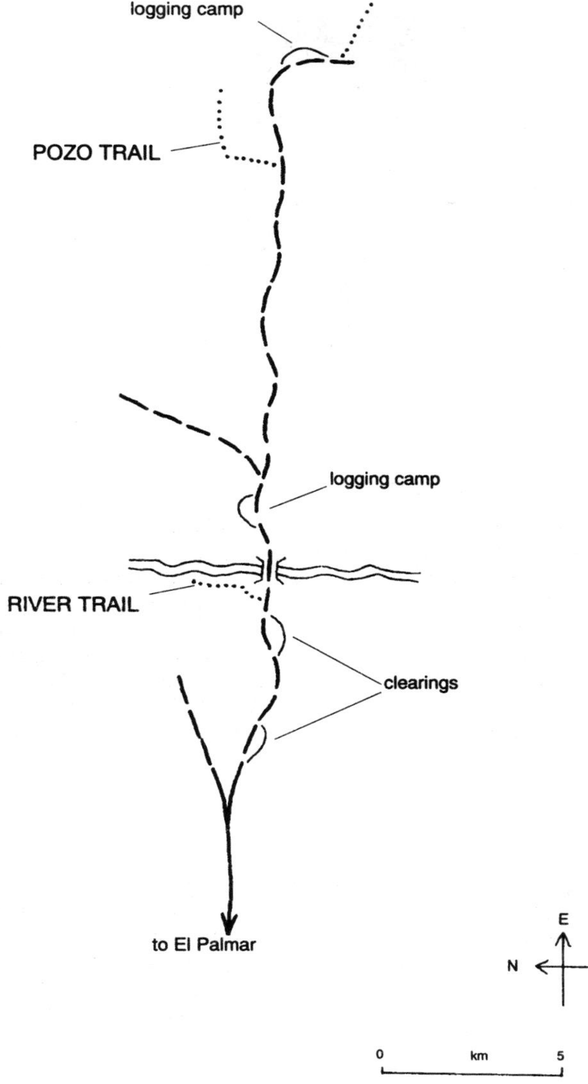

Woodpeckers, White-plumed Antbird, Spangled Cotinga, Bare-necked Fruitcrow, Helmeted Pygmy-Tyrant, Cayenne Jay and Rose-breasted Chat. The roadside palms 20 km west of Maripa support Point-tailed Palmcreeper. Bearded Tachuri* has also been seen in grasslands around Maripa.

To reach the Imataca Forest Reserve from Ciudad Bolívar or Ciudad

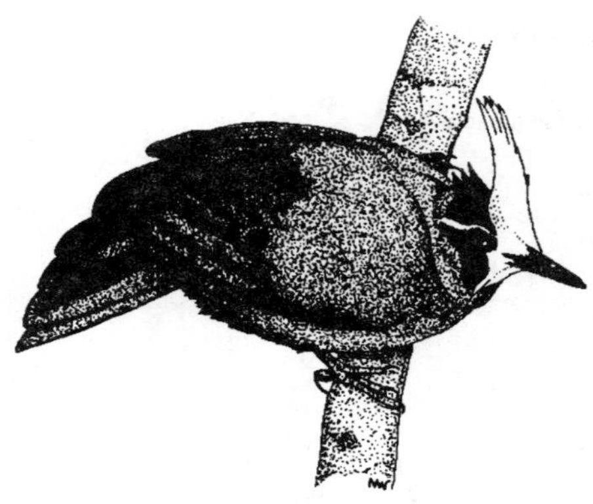

If you hear an antbird flock whilst walking the trails in Bolívar, tuck your trousers into your socks and walk into the forest in search of the ants and their attendant birds. It may be worth a few bites to see the fabulous White-plumed Antbird

Guayana head east and/or south to Upata. Orinocan Saltator occurs in the scrub southeast of **Upata**. Take the track into the scrub 5 km beyond Hotel Andrea.

From Upata head 19 km south to Villa Lola, then east 30 km to El Palmar. Based in El Palmar, ask the owner of the Parador Taguapire hotel (the best base) if it is possible to arrange a guide to see Harpy Eagle*. From this hotel it is 25 km east to the **Imataca Forest Reserve** (Map p. 395). Bird the 5-km stretch of track before the river, the river trail on the west bank heading north (good for Grey-winged Trumpeter), the main track to the first logging camp, and side-tracks leaving this. Imataca specialities include Black-faced Hawk, Harpy Eagle*, Little Chachalaca, Black Curassow, Grey-winged Trumpeter, Least Pygmy-Owl, Long-tailed Potoo, Racket-tailed Coquette, Cinnamon-throated Woodcreeper, Ruddy Spinetail, Rufous-capped Antthrush, Thrush-like Antpitta, Black-tailed Leaftosser, Rufous-bellied and Brown-bellied Antwrens, Ferruginous-backed Antbird, Helmeted Pygmy-Tyrant and Rose-breasted Chat.

Accommodation: El Palmar: Parador Taguapire (B); Campamento Wasaina

En route south from Villa Lola stop at the Río Cuyuni bridge, 6 km south of the turning to the town of El Dorado. Black-collared Swallow occurs here and in the late afternoon hundreds of 'parrots' fly over the river to their roosting sites. South of the Río Cuyuni bridge (km 0) there are a number of trails either side of the road (Map p. 398). There are no maps of the individual trails since they are constantly changing, owing mainly to fallen trees. It is important to remember that new trails and tracks are also appearing now the area is 'opening up' and these will no doubt repay some exploration. The following km numbers are taken from the Río Cuyuni bridge (km 0):

KM 16-20: There is a trail leading east from just south of the village, just before km 17; Marail Guan, Yellow-billed Jacamar, Guianan Toucanet, and Spangled Cotinga have been recorded along here. Marbled Wood-Quail occurs along the track leading west just south of here, and Red-and-black Grosbeak occurs along the wide track leading east just before km 20.

KMS 67 (leading east), **70** (leading west), **71** (leading east), **73** (the **Guyana trail** leading east), and **74** (leading east): Goodies recorded along these trails include Crested Eagle*, Little Chachalaca, Marail Guan, Blackish Nightjar, Yellow-billed Jacamar, Guianan Toucanet, Mouse-coloured Antshrike, Brown-bellied Antwren, White-plumed Antbird, Guianan Red-Cotinga, Purple-breasted Cotinga, Capuchinbird and Red-and-black Grosbeak.

KM 83: Henry Cleeve's Guesthouse, a fine base for this region (A), is situated on the west side of the road here. Crimson Topaz and Guianan Red-Cotinga have been recorded in the garden, and more cotingas in the surrounding forest.

KM 85/between 85 and Las Claritas: This expanding 'frontier settlement' currently stretches to KM 88, also known as San Isidro, where there are cafés, a petrol station and accommodation. The track west, opposite the bank, leads to a Capuchinbird lek after 5 km.

Accommodation: Hotel La Pilonera (B); Anaconda Lodge (B).

KM 92: This trail, on the west side of the road, also leads to a Capuchinbird lek. Fifty metres before km 92 check the small pool on the east side of the road near dusk for Crimson Topaz.

THE ESCALERA

Map p. 398

Beyond km 98 'Paradise Road' suddenly begins to ascend, passing through superb forest full of birds, many of them restricted to the 'tepuis' (table mountains) of southern Venezuela, west Guyana and north Brazil. This is a birder's 'El Dorado', where the star birds include four Venezuelan endemics, Peacock Coquette, the gorgeous Red-banded Fruiteater, Rose-collared Piha, Guianan Cock-of-the-Rock, Scarlet-horned Manakin, Ruddy Tody-Flycatcher, Rufous-brown Solitaire*, Blue-backed Tanager, and Greater Flower-piercer, many of which are no where near as accessible elsewhere in South America.

Endemics

Tepui Goldenthroat, Streak-backed Antshrike, Great Elaenia, Black-fronted Tyrannulet.

Specialities

Ocellated Crake*, Giant Snipe, Fiery-shouldered Parakeet, Tepui Parrotlet, Blue-cheeked Parrot*, Tepui Swift, Rufous-breasted Sabrewing, Peacock Coquette, Velvet-browed Brilliant, MacConnell's Spinetail, Roraiman Barbtail, White-throated Foliage-gleaner, Roraiman Antwren, Tepui Antpitta, Red-banded Fruiteater, Rose-collared Piha, Guianan Cock-of-the-Rock, Sharpbill, Scarlet-horned, Tepui and Olive Manakins, Ruddy Tody-Flycatcher, Chapman's Bristle-Tyrant,

EL DORADO - ESCALERA ROAD

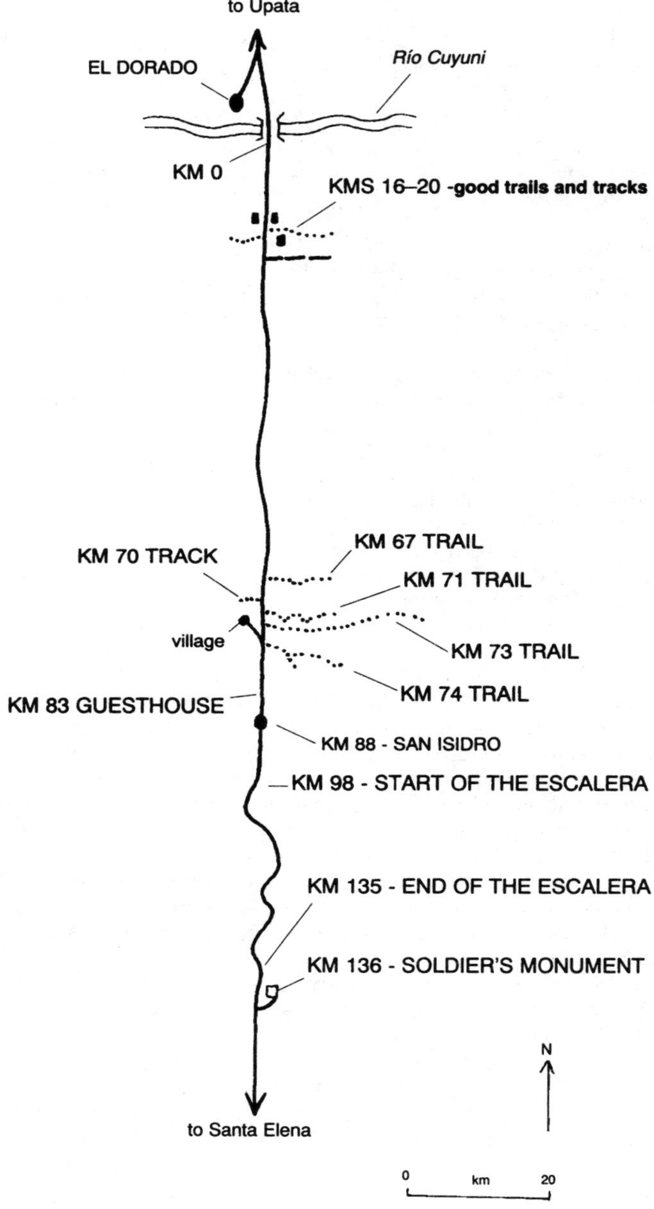

The loud, far-carrying 'bing-bing' bell-like call of the White Bellbird rings out over the Escalera

Tepui Greenlet, Rufous-brown Solitaire*, Black-hooded Thrush, Flutist Wren, Tepui Redstart, Tepui Brush-Finch, Olive-backed and Blue-backed Tanagers, Greater Flower-piercer, Golden-tufted Grackle.

Access

Bird the roadside. Although most species are present at various altitudes along this road previously productive spots have included km 100 (Blue-cheeked Parrot*), km 111–112 (Roraiman Barbtail, Guianan Cock-of-the-Rock, Sharpbill, Rufous-brown Solitaire* and Blue-backed Tanager), km 117 (Peacock Coquette and Olive Manakin), km 121–123 (Greater Flower-piercer), km 123 (White-throated Foliage-gleaner, Ruddy Tody-Flycatcher and Flutist Wren), km 125 (Red-banded Fruiteater), and km 131–134 (MacConnell's Spinetail, Red-banded Fruiteater, Rose-collared Piha, Scarlet-horned and Tepui Manakins, and Rufous-brown Solitaire*).

Beyond km 135 lies the Gran Sabana, savanna country. Tepui Goldenthroat occurs in the scrub around the monument at km 136. Ocellated Crake*, Giant Snipe (in marsh 200 m west of the road at km 142), Great Elaenia (km 139) and Tawny-headed Swallow occur beyond here.

Red-shouldered Macaw and Peacock Coquette have been recorded at km 290. At Santa Elena, km 348, there are more birding sites. Blue-fronted Lancebill, White-shouldered Antshrike, Pygmy Antwren and Dusky Purpletuft occur at **El Pauji**, 85 west of Santa Elena.

ADDITIONAL INFORMATION

Addresses

Venezuelan Audubon Society (VAS), Aparto No. 80450, Caracas 1080-A, Venezuela. The Caracas office is located on the ground floor of Paseo Las Mercedes shopping complex.

Books

Birding in Venezuela, Goodwin, M. L., 1990. VAS.
A Guide to the Birds of Venezuela, Meyer de Schauensee, R. and Phelps, W., 1978. Princeton UP.

VENEZUELAN ENDEMICS (41)	BEST SITES
Tepui Tinamou	South: rare, on remote tepuis
Venezuelan Wood-Quail*	North and southwest: Henri Pittier NP and Paramo de Tama
Rusty-flanked Crake*	North: Yacambu NP near Barquisimeto
Plain-flanked Rail*	North: very rare, Chichiviriche
Red-eared Parakeet	North: Henri Pittier NP
Rose-headed Parakeet*	Southwest: Mérida Andes
Roraiman Nightjar*	South: rare, on remote tepuis
Green-tailed Emerald	North: Henri Pittier NP
Tepui Goldenthroat	Southeast: the Escalera
Tachira Emerald*	Southwest: previously known only from one 1936 specimen from Tachira, this bird was rediscovered in 1992 in coffee plantations near San Juan de Colón
Scissor-tailed Hummingbird*	Northeast: Cerro de Humo
Violet-chested Hummingbird	North: Caracas and Henri Pittier NP
Mérida Sunangel	Southwest: Mérida Andes
Ochre-browed Thistletail	Southwest: Pico Aguila
Black-throated Spinetail	North: Colonia Tovar and Henri Pittier NP
Orinoco Softtail*	South: Amazonas
White-throated Barbtail*	Northeast: Cerro de Humo
Guttulated Foliage-gleaner	North and southwest: Henri Pittier NP and Santo Domingo valley
Streak-backed Antshrike	Southeast: the Escalera
Great Antpitta*	North and southwest: rare, mountains
Tachira Antpitta*	Southwest: not seen since type specimen was collected in south west Tachira
Grey-naped Antpitta	Southwest: Humboldt trail and Paramo de Tama
Scallop-breasted Antpitta*	North: Henri Pittier NP
Handsome Fruiteater	North and northeast: Henri Pittier NP and Cerro de Humo
Short-tailed Tody-Flycatcher*	North: Chichiviriche and Paraguaná peninsula
Great Elaenia	Southeast: the Escalera
Venezuelan Bristle-Tyrant*	North: Henri Pittier NP
Black-fronted Tyrannulet	Southeast: the Escalera

White-bearded Flycatcher	Central: the Llanos
Paramo Wren	Southwest: Pico Aguila, Mérida Andes
Yellow-faced Redstart*	Northeast: Cerro de Humo
White-faced Redstart	South: rare, on three remote tepuis in south (Guany, Yavi and Paraque)
Saffron-breasted Redstart	Southeast: rare, on one tepuis (Guaiquinima)
White-fronted Redstart	Southwest: Mérida Andes and Paramo de Tama
Grey-headed Warbler*	Northeast: rare, Cerro Negro (pre viously recorded on Cerro Turumiquire)
Grey-capped Hemispingus	Southwest: Mérida Andes
Slaty-backed Hemispingus*	Southwest: Humboldt trail
Rufous-cheeked Tanager	North: Henri Pittier NP
Duida Grass-Finch	South: rare, on one tepuis(Duida)
Venezuelan Flowerpiercer*	Northeast: Cerro de Humo
Mérida Flowerpiercer	Southwest: Mérida Andes

North = Caracas east to Guatapo, west to Colombia and south to Barquisimeto.
Northeast = Sucre.
Central = the Llanos.
Southwest = Mérida Andes.
South = Amazonas.
Southeast = Bolívar.

ANTARCTICA

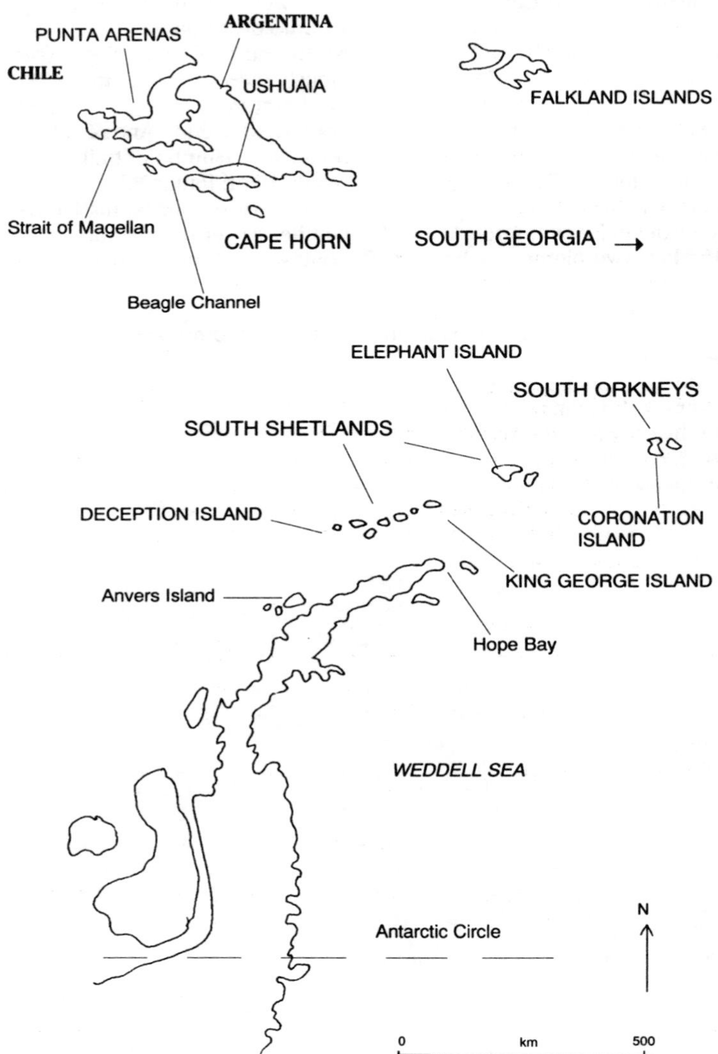

PUNTA ARENAS
ARGENTINA
CHILE
USHUAIA
FALKLAND ISLANDS
Strait of Magellan
CAPE HORN
SOUTH GEORGIA →
Beagle Channel
ELEPHANT ISLAND
SOUTH ORKNEYS
SOUTH SHETLANDS
DECEPTION ISLAND
CORONATION ISLAND
KING GEORGE ISLAND
Anvers Island
Hope Bay
WEDDELL SEA
N
Antarctic Circle
0
km
500

Owing to Antarctica's proximity to South America, and the fact that cruises to this icy wilderness often begin with a brief visit to one or two sites in Argentina and Chile, a taste of what to expect on a cruise has been included in this book.

A few ships visit the edge of Antarctica via the Falkland Islands, South Georgia and various other islands, from ports in South America, usually Punta Arenas in Chile and Buenos Aires in Argentina. Such cruises are very expensive and need to be booked well in advance.

Antarctica is surrounded by the richest oceans in the world, where seabirds, notably penguins, albatrosses (at the nest), and Snow Petrel, are the major attractions. Although the number of species to be expected is very low, and the real Antarctic prize, Emperor Penguin, is rarely recorded, the quality of the birds and the spectacular environment they inhabit may justify the expense. The world's last great wilderness is accessible, if you can afford it.

Ships sail between November and February. The earlier cruises, preferably in November, are recommended because they offer the opportunity to see Light-mantled Albatross at their nests on South Georgia. This eliminates the risk of missing this gorgeous bird, since they may have departed by January, and are tricky to find out on the open ocean.

SITES VISITED ON AN ANTARCTIC CRUISE

For the purpose of this section 'Endemics' refers to those species which usually occur only on Antarctica itself or amongst the surrounding pack-ice and adjacent seas to 63° S.

For birding the **Falkland Islands** see that section of the book, p. 256 to p. 262, for Punta Arenas, Chile, see p. 173, and for Ushuaia, Argentina, another stop-off point, see p. 73.

FALKLAND–SOUTH GEORGIA CROSSING

This two-day trip is one of the highlights of the Antarctica cruise, for the ship passes from the warm sub-antarctic waters to the cold antarctic waters through the Antarctic convergence. This clash of waters creates food-rich upwellings which attract large numbers of sea mammals and seabirds, including the hyper-elegant Light-mantled Albatross.

Specialities

Wandering, Royal, Sooty and Light-mantled Albatrosses, Kerguelen, Great-winged, Soft-plumaged, Atlantic and Blue Petrels, Slender-billed and Fairy Prions, Grey-backed and Black-bellied Storm-Petrels.

Others

Black-browed and Grey-headed Albatrosses, Antarctic, Hall's Giant and Cape Petrels, Antarctic Prion, White-chinned Petrel, Great and Little Shearwaters, Wilson's Storm-Petrel, Southern Skua, Pomarine, Parasitic and Long-tailed Jaegers.

One of Antarctica's star birds is the lovely Snow Petrel

Other Wildlife

Gray's Beaked, Killer, Black Right, Minke, Fin and Southern Bottlenose Whales, Hourglass Dolphin.

SOUTH GEORGIA

This remote island, east of The Falkland Islands and two days by ship, rises spectacularly to 2,934 m (9,626 ft) at Mt Paget. It supports some very special birds including breeding colonies of King Penguin, and Wandering and Light-mantled Albatrosses, as well as the endemic South Georgia Pipit, the only passerine present. Birders with a special interest in seeing Light-mantled Albatross at the nest should book on a November cruise because these birds may have left by January.

Endemics

South Georgia Pipit.

Specialities

King and Macaroni Penguins, Wandering and Light-mantled Albatrosses, Antarctic Prion, South Georgia and Common Diving-Petrels.

Others

Black-browed and Grey-headed Albatrosses, Antarctic Giant Petrel, Yellow-billed Pintail.

Other Wildlife

Southern Elephant Seal (200,000), South American Fur Seal.

The 75,000 strong King Penguin colony at Royal Bay is one of a few on South Georgia.

SOUTH ORKNEY ISLANDS

This small group of islands, southwest of South Georgia and a day by ship, are virtually icebound in the austral winter. They support two penguins endemic to Antarctica and its surrounding oceans, as well as the smashing Snow Petrel.

Endemics

Adelie and Chinstrap Penguins, Snow Petrel.

Others

Cape Petrel, Antarctic Prion, Wilson's Storm-Petrel, Southern Skua, Antarctic Tern.

Adelie Penguins and Lesser Snow Petrels breed on Coronation Island.

Emperor Penguin has been seen on the crossing from the South Orkneys to the South Shetlands.

SOUTH SHETLAND ISLANDS

These islands, west of the South Orkneys, another day by ship, and 966 km from Cape Horn, across the Drake Passage, support large numbers of the endemic Chinstrap Penguin.

Endemics

Chinstrap Penguin.

Specialities

Gentoo and Macaroni Penguins.

Others

Snowy Sheathbill, Antarctic Tern.

Other Wildlife

Weddell, Crab-eater and Leopard Seals.

The Chinstrap Penguin colony near Cape Lookout on Elephant Island usually contains a few Macaroni Penguins as well as Gentoo Penguin.

SOUTH SHETLAND ISLANDS–ANTARCTICA CROSSING

From the South Shetlands the ship may cross the Branfield Strait, dodging gigantic icebergs, on the approach to the Antarctic peninsula, where it is possible to land at Paradise Bay and actually set foot on the continent. There is a Gentoo Penguin colony here. The crossing provides the best chance to see the rare Emperor Penguin. However, they find it a bit hot here, since it is north of their normal range!

Endemics

Emperor, Adelie and Chinstrap Penguins, Antarctic Petrel.

Specialities

Gentoo Penguin.

Others

Antarctic Giant Petrel, Southern Fulmar, Cape Petrel, Imperial Shag, Snowy Sheathbill, South Polar Skua.

Other Wildlife

Humpback and Killer Whales, Crabeater, Weddell and Leopard Seals, Southern Fur Seal.

Paulet Island supports a 800,000-strong Adelie Penguin colony, and Baily Head on Deception Island supports a 650,000-strong Chinstrap Penguin colony, the two most impressive penguin colonies accessible on most cruises. Antarctic Giant Petrel and Cape Petrel also breed here.

ANTARCTICA–CAPE HORN CROSSING

The 1,000-km wide Drake Passage separates Antarctica from South America, and provides an opportunity to see more seabirds. It may be possible to land near Cape Horn, where the rare Striated Caracara occurs, and visit a Magellanic Penguin colony with its attendant Dark-bellied Cinclodes.

Specialities

Wandering and Royal Albatrosses, Blue Petrel, Slender-billed Prion, Black-bellied Storm-Petrel, Magellanic Diving-Petrel, Dark-bellied Cinclodes.

Others

Magellanic Penguin, Black-browed and Grey-headed Albatrosses, Sooty Shearwater, Wilson's Storm-Petrel, Red Phalarope, Chilean Skua.

ADDITIONAL INFORMATION

Ships making the journey to Antarctica from mainland South America include:

(i) the *World Discoverer*, which travels from Punta Arenas, south Chile, to the Falkland Islands, South Georgia, South Orkneys, South Shetlands, the Antarctic peninsula and back through the Drake Passage, past Cape Horn to Puerto Williams;

(ii) the *Society Adventurer*, which travels from Punta Arenas, south Chile, through the Beagle Channel, past Cape Horn, through the Drake Passage to the South Shetlands, South Orkneys, South Georgia, the Falkland Islands and back through the Straits of Magellan to Punta Arenas;

(iii) the *M/S Explorer*, owned and operated by Abercrombie and Kent, which offers a number of cruises. Contact: Abercrombie and Kent International, Inc., 1520 Kensington Road, Oak Brook, Illinois 60521, USA;

(iv) the *Ocean Princess*, which sails from Buenos Aires to the Falklands, then across the Drake Passage to the South Shetlands and the Antarctic peninsula, before returning via Ushuaia, south Argentina, and through the Beagle Channel to Punta Arenas, south Chile. This cruise does not visit South Georgia or the South Orkneys, but it is the cheapest. Contact: Wildwings, International House, Bank Road, Bristol BS15 2LX, UK (tel: 0272-613000).

These cruises can be booked through the contacts given, birding tour companies, other specialist tour companies and travel agents.

CALENDAR

The following is a brief summary of the best countries and regions to visit according to the time of the year. For example, if March is the only time you can get away, then the best birding will be in Bolivia and Venezuela. Furthermore, if you have a year to bird South America then following this schedule will produce the best birding and the most birds. (If anyone tries it please let me know how they get on.)

January: Antarctica, Brazil (Amazonia), Ecuador (north, west and east), Peru (Iquitos), Trinidad and Tobago, Venezuela.
February: Brazil (Amazonia), Ecuador (south and east), Guyana, Venezuela.
March: Bolivia, Venezuela (north and the Llanos).
April: Do not go. Write up your notes.
May: Do not go. Have a rest.
June: Bolivia (Noel Kempff Mercado NP), Galápagos, Venezuela.
July: Ecuador (north and west), Venezuela (Bolivar).
August: Bolivia (Noel Kempff Mercado NP), Brazil (the Pantanal), Ecuador (north), Venezuela.
September: Bolivia, Brazil (the Pantanal), Guyana.
October: Brazil (southeast and east).
November: Antarctica, Argentina, Brazil (Amazonia), Chile, Ecuador (La Selva), Venezuela (Amazonas).
December: Chile, Trinidad and Tobago, Venezuela.

USEFUL ADDRESSES

Clubs

The Neotropical Bird Club: c/o of The Lodge, Sandy, Beds. SG19 2DL, England, U.K.(This club was formed in 1993 with the aim of conserving Neotropical birds and their habitats). Membership costs £25/$50 (founder) and £10/$20 (ordinary), and benefits include receipt of the bi-annual 'Cotinga' bulletin.

Trip Reports

Dutch Birding Travel Reports Service:(Organised and owned by Dirk de Moes): P.O. Box 94, 3956 ZS Leersum, the Netherlands (tel: 31-3434-57501).

Foreign Birdwatching Reports and Information Service: (Organised and owned by Steve Whitehouse): 5 Stanway Close, Blackpole, Worcester WR4 9XL, England U.K. (tel: 0905-454541). To obtain a copy of the catalogue, listing nearly 400 reports from around the world, send £1.

Major Tour Companies

Birdquest: Birdquest Ltd., Two Jays, Kemple End, Birdy Brow, Stonyhurst, Lancashire BB6 9QY, England, U.K. (tel: 0254-826317; fax: 0254-826780; telex: 635159 BIRDQ).

Cygnus: Cygnus Wildlife Holidays, 57 Fore Street, Kingsbridge, Devon TQ7 1PG, England, U.K. (tel: 0548-856178; fax: 0548-857537; telex: 45795 WSTTLX G).

Field Guides: Field Guides, Inc., P.O. Box 160723, Austin, Texas 78716-0723, U.S.A. (tel: 512-327-4953; fax: 512-327-9231).

Naturetrek: Naturetrek, Chautara, Bighton, Nr. Alresford, Hampshire SO24 9RB, England, U.K. (tel: 0962-733051; fax: 0962-733368).

Neotropic Bird Tours: Neotropic Bird Tours, 38 Brookside Avenue, Livingston, New Jersey 07039, U.S.A. (tel: 800-662-4852 and 201-716-0828; fax: 201-740-0256).

Ornitholidays: Ornitholidays, 1 Victoria Drive, Bognor Regis, West Sussex PO21 2PW, England, U.K. (tel: 0243-821230; fax: 0243-829574).

Sunbird: Sunbird, P.O. Box 76, Sandy, Beds. SG19 1DF, England, U.K. (tel: 0767-682969).

Victor Emanuel Nature Tours: Victor Emanuel Nature Tours, P.O. Box 33008, Austin, Texas 78764, U.S.A. (tel: 512-328-5221; fax: 512-328-2919).

Wings: Wings, Inc., P.O. Box 31930, Tucson, Arizona 85751, U.S.A. (tel: 602-749-1967; fax: 602-749-3175).

Travel Agents

Wildwings: International House, Bank Road, Bristol, BS15 2LX, U.K. (tel: 0272-613000).

USEFUL GENERAL BOOKS

Birds

A Parrot Without A Name, Stap, D. 1990. Texas UP.
Birds of the High Andes, Fjeldsä, J. and Krabbe, N. 1990. Apollo.
The Birds of South America: Volume 1: The Oscine Passerines, Ridgely, R. and Tudor, G. 1989. OUP.
The Birds of South America: Volume 2: The Suboscine Passerines, Ridgely, R and Tudor, G. Due 1994. OUP.
Birds of the World - A Check List, Clements, J. 1991. Ibis.
Birds of the World - A Check List: Supplement No. 1, Clements, J. 1992. Ibis.
Birds of the World - A Check List: Supplement No. 2, Clements, J. Due 1993. Ibis.
Birds to Watch 2: The World List of Threatened Birds,Collar, N. *et al.*, 1994. BirdLife International.
Checklist of the Birds of South America, Altman, A. and Swift, B. 1993. Personally published.
A Handbook to the Swallows and Martins of the World, Turner, A. and Rose, C. 1989. Helm.
The Herons Handbook, Hancock, J. and Kushlan, J. 1984. Helm.
Kingfishers, Bee-eaters and Rollers, Fry, C., Fry, K. and Harris, A. 1992. Helm.
New World Warblers, Curson, J., Quinn, D. and Beadle, D., 1994. Helm
Priority Areas for Threatened Birds in the Neotropics, Wege, D. and Long, A., 1994. BirdLife International
Putting Biodiversity on the Map, BirdLife International. 1992.
Rare Birds of the World, Mountfort, G. 1988. Collins.
Seabirds: an identification guide, Harrison, P. 1985. Helm
Shorebirds: an identificatication guide to the waders of the world, Hayman, P., Marchant, J. and Prater, T. 1986. Helm.
South American Birds - A Photographic Aid to Identification, Dunning, J. 1989. Harrowood.
Threatened Birds of the Americas, Collar, N. *et al.* 1992. Smithsonian.
Wildfowl: an identification guide to the ducks, geese and swans of the world, Madge, S. and Burn, H. 1988. Helm.

General

South American Handbook, Box B. (Ed). Annual. Trade and Travel.
South America on a Shoestring, Crowther, G. *et al.* Annual. Lonely Planet. (Lonely Planet also publish guides to individual countries).

REQUEST

If you would like to contribute to the second edition of this guide, please send details of any errors or suggested changes to the site details and species lists included in this edition, and information on any new sites you feel deserve inclusion, to:

Nigel Wheatley, c/o Christopher Helm (Publishers) Limited, 35 Bedford Row, London WC1R 4JH, England U.K.

It would be helpful if information could be submitted in the following format, although this is not essential:

1 A summary of the site's position (in relation to the nearest city, town or village), altitude, access arrangements, habitats, number of species recorded (if known), best birds, best time to visit, and its richness compared with other sites.

2 A species list, preferably using the names and taxonomic order in Clements' *Check List* and supplements.

3 Details of how to get to the site and where to bird once there, with information on trails etc.

4 A map complete with scale and compass point.

5 Any addresses to write to for permits etc.

6 Any details of accommodation.

Any information on the following species would also be very useful:

Black Tinamou*, Chaco Nothura, Buckley's Forest-Falcon, Black-fronted Wood-Quail, Great-billed Hermit, Pink-throated Brilliant, Black-thighed Puffleg*, Perija Metaltail, Black-breasted Puffbird, Citron-throated Toucan, Perija Thistletail, Dusky Spinetail, Cabanis' Spinetail, Fulvous Treerunner, Flammulated Treehunter, Buff-throated Treehunter, Yellow-rumped Antwren*, Yapacana Antbird, Yellow-breasted Antpitta, White-lored Antpitta, Hooded Antpitta*, Black-chested Fruiteater, Buff-throated Tody-Tyrant, Cinnamon-breasted Tody-Tyrant, Reiser's Tyrannulet, Red-billed Tyrannulet, White-bellied Pygmy-Tyrant, Long-crested Pygmy-Tyrant, Yellow-throated Spadebill, Golden-browed Chat-Tyrant, Buff-bellied Tanager.

I would be extremely grateful if you could also include a statement outlining your permission to use your information in the next edition and, finally, your name and address so that you can be acknowledged appropriately.

I would like to take this opportunity to thank you in anticipation of your help. The usefulness of the next edition depends on your efforts.

INDEX OF SPECIES